Astronomy and Society in Britain during the period 4000–1500 B.C.

Edited by

C. L. N. Ruggles & A. W. R. Whittle

BAR British Series 88

1981

B.A.R.

B.A.R., 122 Banbury Road, Oxford OX2 7BP, England

B.A.R. 88, 1981: "Astronomy and Society in Britain."

ISBN 9780860541301 paperback
ISBN 9781407323619 e-book
DOI https://doi.org/10.30861/9780860541301
A catalogue record for this book is available from the British Library

This book is available at www.barpublishing.com

CONTENTS

Please note that additional material is available to download from www.barpublishing.com/additional-downloads.html
The original foldout has been reduced in size to match the A4 format of this book, the image is therefore not as clear as the original foldout. Please refer to the original foldout via the download for the original content.

PREFACE

This volume was originally conceived as the proceedings of a conference organised in March 1980 by Dr Gordon Moir, under the auspices of the Department of Adult Education, University of Newcastle-upon-Tyne, and entitled " Megalithic astronomy and society" . The aims of the conference were twofold: on the one hand to review the theories of Professor Alexander Thom on " megalithic science" in the British Later Neolithic and Early Bronze Age, concentrating on megalithic astronomy (the hypothesis that many megalithic sites were built primarily to serve as astronomical indicators or " observatories" of high precision); and on the other to present an archaeological examination of the period, setting out the social, political and economic context which might have governed any megalithic science.

In the event, the volume has developed into a collection of papers, for the most part expanded versions of presentations at Newcastle, concentrating rather on the following two themes: first the nature and extent of astronomical practice in prehistoric Britain, as evidenced by orientation and related studies; and second how conclusions from such studies bear upon current thought about social organisation throughout the period spanning the fourth, third and mid-second millennia BC in terms of absolute (or corrected radiocarbon) chronology.

This shift of emphasis has led the editors to discourage the inclusion here of two major contributions at the conference which related entirely to the separate issues of megalithic mensuration (the possible existence and widespread use of a precisely-defined " Megalithic Yard") and geometry (the idea that many megalithic rings were laid out according to precise geometrical designs). It was felt that these are hypotheses essentially independent of megalithic astronomy, and which deserve separate and thorough discussion in their own right. We invited one of the contributors concerned, Dr Patrick, to submit an alternative paper for this volume. At the same time we did not try to prevent Dr MacKie discussing these other topics in the course of his article, as for him they represent crucial strands in his general argument.

We sincerely regret that another major contributor to the conference, Colin Burgess, declined (owing to the pressure of other commitments) to provide a paper here. We have attempted to redress the balance by inviting contributions from Nick Thorpe and from Bill Startin and Richard Bradley. Thorpe's paper covers a topic—ethnographic evidence—which was largely unrepresented at the conference. A further contributor, Aubrey Burl, has not included a paper here relating to a second talk he gave at Newcastle, as this formed the basis of a recent article in Antiquity (54, 191-200).

Topics that arose in general discussion at Newcastle have not been carried through to this volume as such. However a number of authors refer to comments and reply to criticisms which were aired at the conference. Many of these points have also been raised elsewhere in print.

Seven of the contributors are archaeologists by training, whereas four (Moir, Patrick, Ruggles and A. S. Thom) have backgrounds in mathematics and/or the numerate sciences such as astronomy or engineering. The Pontings are amateurs; schoolteachers who moved in 1974 to the Callanish area, since when they have devoted a good deal of spare time to the study of the megalithic sites in the vicinity. Such diversity in the backgrounds of the speakers at the conference, and hence of the contributors here, reflects the range of people involved in investigating astronomical practices in prehistoric Britain—an important part of the wider interdisciplinary study which has become known as archaeoastronomy. In contrast, the studies of society presented here are merely a small part of a much wider debate on the nature of prehistoric society, which is now being investigated through a very wide range of archaeological evidence.

A sample of both approaches, therefore, is given together here in the belief that on the one hand archaeoastronomy must not be considered in a social vacuum and that on the other any reconstruction of prehistoric society must attempt to integrate <u>all</u> the available evidence.

C. R.
A. W.

LIST OF CONTRIBUTORS

Prof. R. J. C. Atkinson, Department of Archaeology, University College, P. O. Box 78, Cardiff CF1 1XL.

R. J. Bradley, Department of Archaeology, University of Reading, Whiteknights, Reading RG6 2AA.

A. Burl, 40 St. James Road, Edgbaston, Birmingham B15 1JR.

Dr. E. W. MacKie, Hunterian Museum, The University, Glasgow G12 8QQ.

Dr. G. Moir, Department of Adult Education, The University, Newcastle-upon-Tyne NE1 7RU.

Dr. J. Patrick, Prahran College of Advanced Education, 142 High Street, Prahran, Victoria, Australia 3181.

M. R. & G. H. Ponting, 'Olcote', New Park, Callanish, Isle of Lewis PA86 9DZ.

Dr. C. L. N. Ruggles, Department of Archaeology, University College, P.O. Box 78, Cardiff CF1 1XL.

W. H. Startin, Room 3/41, Fortress House, 23 Savile Row, London W1X 2HE.

Dr. A. S. Thom, Department of Aeronautics and Fluid Mechanics, University of Glasgow, Glasgow G12 8QQ.

I. J. Thorpe, Department of Archaeology, University of Reading, Whiteknights, Reading RG6 2AA.

Dr. A. W. R. Whittle, Department of Archaeology, University College, P.O. Box 78, Cardiff CF1 1XL.

INTRODUCTION: ASTRONOMICAL PRACTICES IN PREHISTORIC BRITAIN

Clive Ruggles and Alasdair Whittle

Thom and after

It has often been suggested that there was an astronomical motivation behind the construction of a range of ceremonial, burial and other sites in Britain, an assertion made on the basis of studies of orientations, alignments and putative horizon indications at such sites. Following recent fashion, we can classify such studies under the label of "archaeoastronomy". Professor Alexander Thom, whose contribution in this field is cardinal, concludes that towards the middle of the second millennium BC "Megalithic Man" (as he labels the inhabitants of Britain at the time) carried out precise and painstaking observations of the sun and moon, whose motions they recorded using standing stones. The sort of precision claimed seems far beyond any conceivable utilitarian benefit. The first six contributions to this volume address themselves, wholly or in part, to the debate on prehistoric astronomy at this level.

Paper 1, written by Alexander Thom's son Dr A. S. Thom, provides a most useful summary of the Thoms' views, methods and conclusions over the years, and presents their most recent evidence supporting the hypothesis of accurate lunar observatories. Dr Thom also takes the opportunity to reply to a number of recent criticisms. The Pontings, in paper 2, present the results of their studies to date at Callanish and suggest a number of ways in which accurate solar and lunar observations could have been made at the sites. Paper 3 is more far-ranging: MacKie discusses archaeological verifications both of Thom's hypothesis of high-precision solstitial indications and of the independent idea of precise geometric designs for sites based upon an accurately-defined "Megalithic Yard"; he then combines these with independent evidence in order to reiterate and develop his hypothesis that there existed in prehistoric Britain a specialist élite of trained astronomer-priests.

The subject has not lacked controversy. It is both scientifically unhealthy and detrimental to reasoned debate that so few first-hand critiques of Thom's work of any substance have been forthcoming until recently. This has prevented worthwhile evaluation and integration of results acceptable to a wide consensus of opinion. Papers 4-6 attempt such a critical evaluation and go on to suggest the direction for future work in this field. Whatever the outcome of the debate presented here, Thom's work will remain the catalyst for archaeoastronomical studies in Britain. For this we owe him a considerable debt of gratitude, which will of course still be due even if general opinion does come to fault a number of his specific conclusions.

The approach of paper 4 reflects the author's view that the value of continued debate on a very few "classic" sites is severely limited, and that progress is only to be made by considering large samples of sites, taken together. The paper thus reconsiders the earlier evidence of the Thoms in favour of precise lunar observatories taken as a whole, and concludes that the case in their favour is both unproven and extremely unlikely on the basis of astronomical considerations. Paper 5, in contrast, addresses itself to a reassessment of two of the most classic Thom sites, the proposed solstitial observatories at Kintraw and Ballochroy. In paper 6 Moir details a range of criticisms of Thom's evidence for astronomical observations of high precision, giving special attention to supposed solar calendrical indicators.

Papers 4-6 all point out the distinction between rough astronomical orientations and much more precise observations and indications. Rougher orientations are a theme central to Aubrey Burl's paper (no. 7) in which evidence for deliberate astronomical orientation in certain classes of monument is traced back to around 4000 BC. Conclusions are drawn on the basis of studies of groups of similar sites rather than a few individual ones; indeed, the dangers of the latter approach are well illustrated by a few poignant examples. Taken as a set, papers 4-7 suggest that the hypothesis of a rougher, secondary, symbolic or "ritual" rather than "scientific" astronomical motivation behind the construction of certain sites may be one which will eventually form the core of statistically provable truth available from archaeoastronomical analyses.

Future work of value in archaeoastronomy lies, in the opinion of several of the authors, in systematic surveys of samples of similar or related sites in certain areas, rather than of a number of archaeologically rather diverse sites as has been the case up to the present. In this manner archaeoastronomy may yield results of lasting value in the study of prehistoric society.

The wider context

The study of astronomical practices in prehistoric communities in Britain raises questions not only of the level of technological and intellectual skills required but also of the social organisation of the communities involved, and five of the papers in this volume (3 and 7-10) are wholly or partially directed to this question. Some perspective on the wider approach by prehistorians to social organisation in this period in Britain—and indeed elsewhere in Europe—is clearly needed.

Linking the many lines of research under way is a common interest in the gradual transformation of society during this period from a less complex into a more ranked or stratified structure. Some would see a simple evolution from egalitarian, as amongst early farmers, to a more ranked society, often labelled chiefdom, from the Later Neolithic period (or the third millennium BC) onwards. Others might prefer a more complicated development through time of degrees of ranking and internal differentiation.

There is probably general agreement as to the complex nature of this social development in terms of the interrelationship of the many factors potentially involved. It is critical therefore in reconstructions of social organisation that as wide an approach as possible be adopted, both theoretically and with regard to the archaeological record. There is probably also general agreement that by the second millennium BC—certainly by the latter part of it—social organisation in many regions of Britain was indeed ranked with an unequal distribution of power, status and resources in many communities, and that this process had been under way since at least the third millennium BC.

What then of the contribution of the study of astronomical practices to the problems of such reconstruction? Useful though it may be, it must be stressed that it is only one line of evidence, to be set alongside those of, for example, prestige indicators in burials and elsewhere, settlement hierarchy, craft specialisation, exchange and trade, land division, control of resources, and so on. Its contribution depends of course on the conclusions reached as to the nature of astronomical practices. Even the level of skill suggested by Professor Thom and others is not a unique indicator of firmly ranked society, since that is evident by the end of the period under discussion without such a hypothesis. Scientific skill of this nature might however be more compatible with more intense ranking than would otherwise be supported by the archaeological record, even though not necessarily supporting any specific model of social organisation (such as that in paper 3). In fact, papers 4-7 all directly challenge the notion of scientific astronomy in Britain, and it may seem to many that the hypothesis of symbolic or "ritual" astronomy proposed in its place offers much greater compatibility with the rest of the archaeological evidence.

Papers 9 and 10 explore some aspects of this evidence. Paper 9 discusses the work organisation behind several classes of monument, and makes the important suggestion that their labour requirements have been substantially overestimated, with the further implication that their construction need not automatically be seen as the product of a highly ranked society. Paper 10 concentrates on the Later Neolithic period and brings in various aspects such as monuments, burials and prestige artefacts which have been individually discussed but not with sufficient reference to the wider debate. This necessitates at one stage detailed examination of the Beaker phenomenon.

Paper 8 reviews some of the ethnographic evidence for a number of astronomical practices which are not of a precise "scientific" nature and are not part of highly regimented societies.

The range of conclusions that emerge from papers 4-10 should not lead to the end of our interest in astronomy from the social point of view. In evolutionary terms, for example, it may be suggested that there was in some sense a social need in the Later Neolithic period for more sophisticated—and perhaps arcane—devices to focus attention on a community's identity in a world in which there was an increasing competition for resources, as set out in paper 10. Paper 7 may help to document the earlier part of such a development. There is also a potential contribution to be made to the study of systems of beliefs, as paper 7 hints.

A final methodological point deserves to be made, and is relevant to the study of both astronomy and society. It is that, apart from the obvious care needed in their initial formulation, hypotheses need to be subjected to the most rigorous testing possible, and of as many different kinds as possible. This is hardly profoundly original, but it is worth repeating.

1. MEGALITHIC LUNAR OBSERVATORIES: AN ASSESSMENT OF 12 LUNAR ALIGNMENTS

A. S. Thom

A presentation of recent developments in archaeoastronomy with the results of survey work and analysis done by Alexander Thom, who has read and approved of the manuscript.

"They bide their time off serpentine
Green lanes, in fields, with railings
Round them and black cows; tall, pocked
And pitted stones, grey, ochre-patched
With moss, lodgings for lost spirits.

Sometimes you have to ask their
Whereabouts. A bent figure, in a hamlet
Of three houses and a barn, will point
Towards the moor. You find them there,
Aloof lean markers, erect in mud.

Dolerite, porphyry, gabbro fired
At the earth's young heart: how those men
Handled them. Set on back-breaking
Geometry, the symmetries of solstice,
What they awaited we, too, still await."

from "Ancient Monuments" by John Ormond.

(Published in his book Definition of a Waterfall and reproduced by permission of Oxford University Press.)

LIST OF SYMBOLS

β = $(\delta - \delta_e)$

c = curvature correction near crest of lunar declination curve

c_s = sum of curvature corrections

δ = declination of Moon or point in space (after fig. 1.6 δ_o is used in place of δ)

δ_e = nominal or expected declination (= $\pm(\varepsilon \pm i \pm \Delta \pm s)$)

δ_o = declination of notch or foresight after observing its altitude above horizontal and its azimuth and correcting for parallax, refraction, etc. (Before fig. 1.6, δ is used for δ_o)

Δ = perturbation of the orbit of the Moon, about $\pm$ 9'

ε = obliquity of the ecliptic

h = observed altitude of horizon

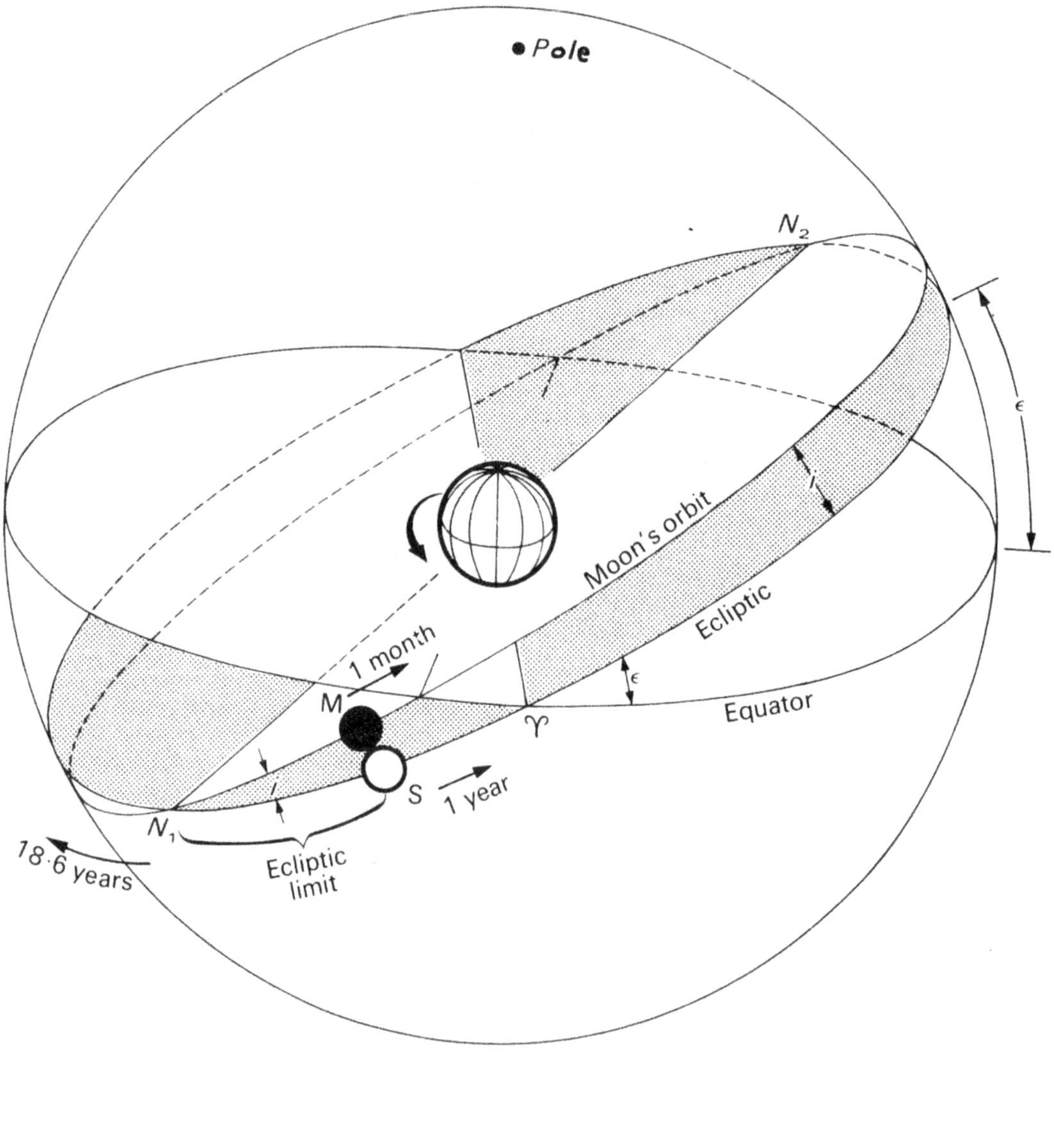

Fig. 1.1 Celestial sphere

i = mean inclination of Moon's orbit to the ecliptic = 5°08'.7

p = parallax of the Moon

Q = expected value of β; $Q = \delta_e - (\varepsilon \pm i \pm s)$ for each site. Q is the angle by which δ_e is above the top of the chain-dotted curve in fig. 1.2

R = $\beta - Q$ Residual

s = semi-diameter of the Moon

r = r.m.s. of Residuals R.

1.1 Introduction

With passage of time, each generation has shown interest in the mysterious single menhirs, rows, groups and rings of standing stones to be found in N.W. Europe. Why were they erected, sometimes at great expense? Our thesis is that at one stage in prehistory many of them were used in conjunction with the horizon to observe the rising and setting of the Sun and Moon. No written evidence exists; the evidence is only that obtainable from the positioning of the stones themselves. Erected within the last 10,000 years, many of the stones seem to act as focal points in the landscape involving the horizon, the sky and the land. Each lichen-covered stone or group of stones seems timeless, with character formed by the weathering effects of wind, rain, frost and sun. We attempt to show it is highly probable that about 2000 BC erectors made and used a Solar calendar and later were observing and recording the Moon's movements so carefully that they could have predicted eclipses. A sensitive method of observing the Moon on the horizon at standstill was developed and we now know that it showed up the effect of the small wobble Δ of the lunar orbit due to the perturbing effect of the Sun.

1.2 Astronomical concepts

Megalithic Man could only make and record accurate observations of the azimuth (clockwise bearing from true north) of the Sun and Moon on the horizon at turning points, namely the solstices and lunar standstills. Explanations are of necessity complicated and modern man's knowledge of astronomy has to be invoked. The following is a brief attempt to explain the relative movements of the Earth, Moon and Sun.

Astronomical declination δ of a heavenly body is the angular distance of the body from the plane of the equator, taken at the centre of the earth. As the declination of a body varies, its azimuth on rising or setting will also vary. Thus an observation of sunset (for example) is at that time an observation of the Sun's azimuth from which its declination can be ascertained. Because the obliquity of the ecliptic ε the declination of the Sun varies from $+\varepsilon$ to $-\varepsilon$ and back to $+\varepsilon$ during a year (fig. 1.1). The Moon's orbit is inclined at an angle i to the plane of the ecliptic and the places where it crosses that plane are referred to as the nodes. The Moon's declination at a period of major lunar standstill varies from $(\varepsilon+i)$ to $-(\varepsilon+i)$ and back in one lunation (fig. 1.2). Because of the rotation of the nodes of the lunar orbit, 9.3 years later the limits of the Moon's declination in one lunation are $+(\varepsilon-i)$ and $-(\varepsilon-i)$, and this is referred to as a period of minor lunar standstill. After 18.6 years (the period of rotation of the line of nodes) the declination limits are again $+(\varepsilon+i)$ and $-(\varepsilon+i)$. Superimposed on the combined motion described, the varying direction of the Sun's attraction adds to the Moon's declination a cyclical perturbation or wobble having amplitude Δ of about 9 arc minutes and

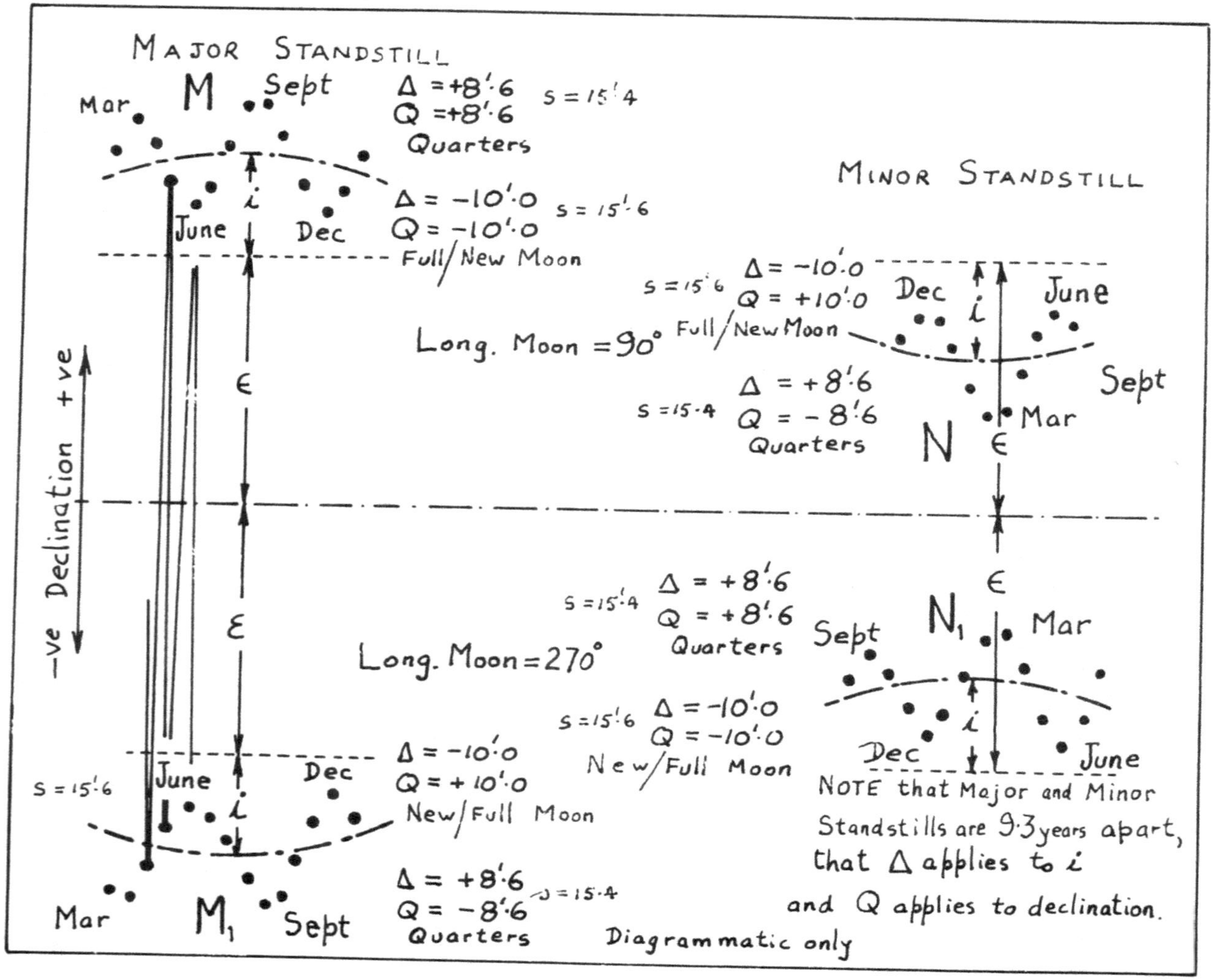

Fig. 1.2 Conditions at the standstills

period 173.3 days (fig. 1.3). Δ is applicable to i; the large dots in fig. 1.2 show the actual peaks at the monthly declination maxima and the curve joining these dots comes to a maximum every 173.3 days. At the major and minor standstills, when the rate of change of ($\varepsilon \pm i$) is zero, the perturbation is most apparent. This change in the Moon's declination is visible upon successive horizon observations taken one lunation apart at rising or setting. As solar and lunar eclipses at major standstills can occur only at the new or full Moon nearest the date when the perturbation is near its maximum (twice a year of 346.6 days, the period in which the Sun revolves relative to a node), a trained observer would be able to predict eclipse occurrences. Fig. 1.3 is used to show declination maxima (north and south) at the 1969 major standstill, with the eclipses at that time occurring within the ecliptic limits, and these happen at the crests of the perturbation wave. A similar type of figure can be drawn to show conditions at the 1978 minor standstill when the ecliptic limits occur at the troughs of the perturbation wave.

With the approach of the major standstill, the observers would be watching the gradual increase of the declination peaks (the black dots on fig. 1.2) and so it would be natural for them to record the maximum in the vicinity of M. Similarly for the negative declination they would tend to use the minimum at M_1. At the minor standstill on the other hand they would have been watching the gradual decrease in the declination hollows, and so we assume that they tended to mark the lowest at N. For the negative declination here we assume they tended to use N_1.

Marking, as they do, the minimum declination range at first or third lunar quarters at the equinoxes at minor standstills, the troughs/crests at N/N_1 are not unique, like the crests/troughs at M/M_1, at the major standstills. The Moon must pass through these former positive/negative declinations N/N_1 on every cycle for the next 18.6 years.

1.3 Method of observing

Megalithic Man constructed his observations empirically, probably with little awareness of any of the above concepts except that of azimuth. Undoubtedly he could count. He had no instruments with accurately graduated circles, but made use of the great circle of the horizon. Instead of graduations he used foresights, far enough away on the horizon to permit of great accuracy. As the foresight could not be moved about, he moved his observing position, the backsight. The observer stands, for example, in such a position that the Sun or the Moon will, as it goes down, graze a distant foresight, in fig. 1.4 assumed to be a hill top. He marks the final position by a rod. Repeating this procedure for several nights about the time of maximum azimuth, if the hill top is in the north-west he ends up with with several rods on a line AA lying across the line of sight. If we assume that he steps back a short constant distance each night he sets out a curve on the ground which is, to some scale, a graph of the azimuth, and so of the declination, plotted on a time base. The maximum point on this curve was then his final backsight. According to Roslund (1979) this procedure was recorded for solstitial sunrise at Ale in Sweden, where the day-to-day observing positions are laid out in parabolic form with a stone for several sunrises before and after the solstice. For the Moon the turning point would not always occur at the only time available for observing, namely moonset (or moonrise), and a method of interpolating between two evening (or morning) recordings had to be developed (see later). Two separate solar solstitial sites have been discovered where the Sun had actually set behind a mountain before re-appearing momentarily in the notch between that mountain and the next. Several sites were carefully chosen in

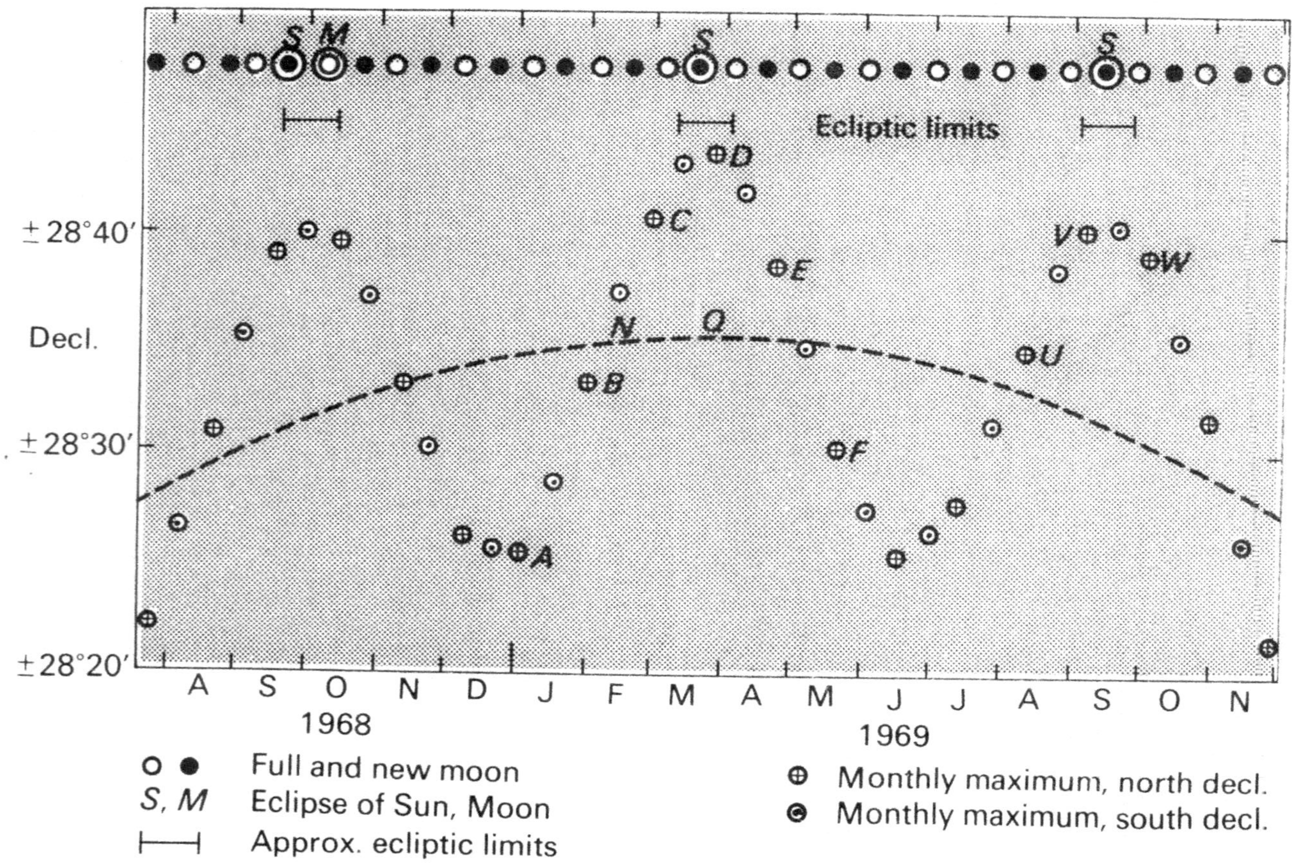

Fig. 1.3 1969 standstill

order to make simultaneous use of two solar lines, each at a different calendar date; of both a lunar and a solar line; or of several lunar lines.

1.4 Two types of observing line used

Megalithic Man has left remains of two totally different types of line. The first type (a) was certainly the earlier and consists of two marks which may be of two circles, or a circle and an outlying stone. The line joining these, extended to the horizon, shows the rising or setting points of the Sun at an important date in the calendar, such as the solstice or the equinox. These are little more than symbolic however, and can not possibly give us an accurate measurement of the declination involved.

The second type of line (b) has a backsight which may be a standing stone, the line from which, joined to a notch on the distant horizon, gives a really accurate declination of one or other of the limbs of the Sun or Moon. Apparently Megalithic Man did not use (b) until about 2000 BC and here we shall interest ourselves only in this type of line. By using a suitably distant natural foresight he obtained an accuracy of better than one arc minute in declination.

It may be that at some places Megalithic Man made use of artificial foresights of some perishable material. Possibly this happened at Stonehenge but we do not know for certain, nor do we know for certain what they looked like. We can guess that they might have consisted of huge mounds of turf or brushwood. To be clearly visible to the naked eye when silhouetted against the Moon's disc they would need to have subtended an angle angle of over one arc minute which is about a foot and a half per mile, or 0.3 m per km. If we assume that the foresight was 10 miles (15 km) away this would have required a target of about 15 ft (5 m) square, quite a possible construction.

If we stand at a marked backsight and make careful measurements of the profile of part of the horizon which turns out to contain a significant declination we can assume that we are at a real observing point. An observing point or backsight may consist of a large stone, sometimes particularly large as at Ballinaby in Islay, and if this is slab-shaped, the long side will point to the foresight. (The Comet stone at Brogar (Thom & Thom 1978a, hereinafter "MRBB":ch. 10) unequivocably indicates the Mid Hill notch in this way.) In other places there are two or more stones ranged up so that the line points to the foresight, e.g. at Dunskeig, Kintyre; at Brogar in Orkney there are earth mounds which serve the same purpose. (See table 1.1 for references.) A foresight can be used for either the upper or lower limb, where upper and lower refer to declination and not to altitude.

When Megalithic Man wanted to mark the setting point of the Moon's centre the only method available was to get one man to range himself in for the upper and a second man to range himself in for the lower limb. Midway between the two positions (on relatively even ground) was the sighting position for the centre of the orb and in a few places this is marked by a stone.

For all cases, as the Moon slipped past the foresight the observer moved about until his position was exact. He then marked the position by a stake. He did this for consecutive nights, month after month at each turning point and then deduced the final position from the stakes on the ground. Thus if he had missed an observation due to bad weather he could interpolate the position. (See section 1.25 below.)

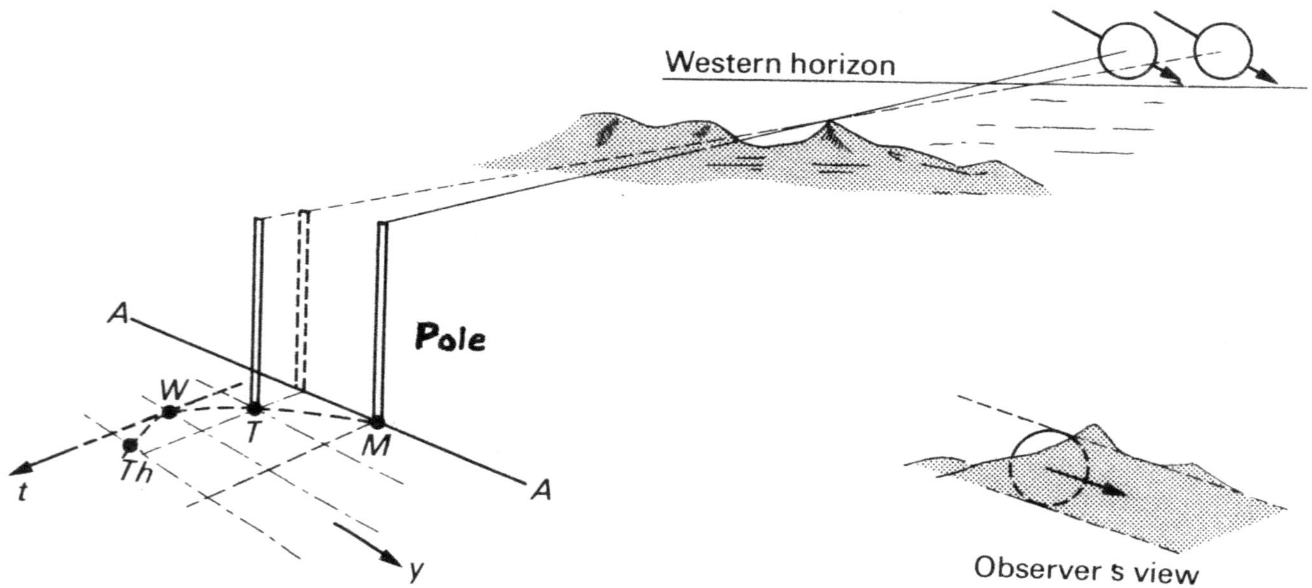

Fig. 1.4 The method of observing

1.5 Ideal foresights: Kintraw

The ideal foresight is perhaps that which we find for the Sun near Kintraw (Thom 1967, hereinafter "MSB":154-156; 1971, hereinafter "MLO": 37-40), where the setting orb had been out of sight for some time before it reappeared momentarily. Before finally vanishing the upper limb twinkled in the Vee between Ben Shiantaidh and Ben a Chaolais. This must have been quite dramatic. As the winter solstice approached the observer would find it necessary to move to the right by a small amount each sunset until the actual evening of the solstice. The site is described fully by MacKie (1974), who has reported that only to the left of the backsight which is sited on a track or pathway, was the platform levelled with small stones.

1.6 Other types of foresight

We have classified four other types of foresight that we find Megalithic Man used (Thom & Thom 1980a). All types are capable of defining the position of the Moon's limb with an accuracy of better than one minute of arc. The backsight must be on a reasonably flat area of ground and it can not have been particularly easy to find a position for a foresight and a backsight. The task was made more difficult by the evident desire to have two or more foresights using the same backsight or at least having their backsights grouped together. Examples of this are the lunar sites at Brogar (MRBB:ch. 10) and Temple Wood (MLO:45-50) or the solar site at Ballochroy (MSB:151-154; MLO:36-37).

1.7 Accurate surveying necessary

It soon became apparent to us that the lunar observatories had been set up with much greater accuracy than at first had been thought possible. It was nonsense to talk about a foresight being within half a degree of its ideal position. We had to think of half minutes of arc and to survey accordingly.

To survey a megalithic observatory we require to record, as accurately as possible, the azimuth and altitude of each foresight from the observer's position at the backsight. After correcting the observed altitude for refraction and parallax, the declination of the line is evaluated. The azimuth of the foresight is usually obtained by taking the mean of three time/azimuth observations of the Sun using a good theodolite and chronometer. Readings are taken which enable a profile of the horizon to be plotted. The chronometer is checked daily by radio, to within half a second, and the position of the instrument is carefully recorded along with the referring object, some easily recognised feature in the distant landscape. Darkened eye-pieces are used while observing firstly the leading and secondly the trailing edge of the Sun. It is of course necessary to maintain the theodolite in good condition, and frequently to check the zero of its vertical circle.

1.8 Presentation of results of surveys

Over the years, as our surveying of sites has proceeded, the subject ha has grown and expanded and consequently the technique of presentation has developed. Refinements have continually been developed as more and more knowledge is forthcoming. The reader is referred to the reference list and to the further bibliography which follows it.

When my father began to survey megalithic sites he surveyed alignments of at least three stones or two suitably oriented slabs, and outliers from

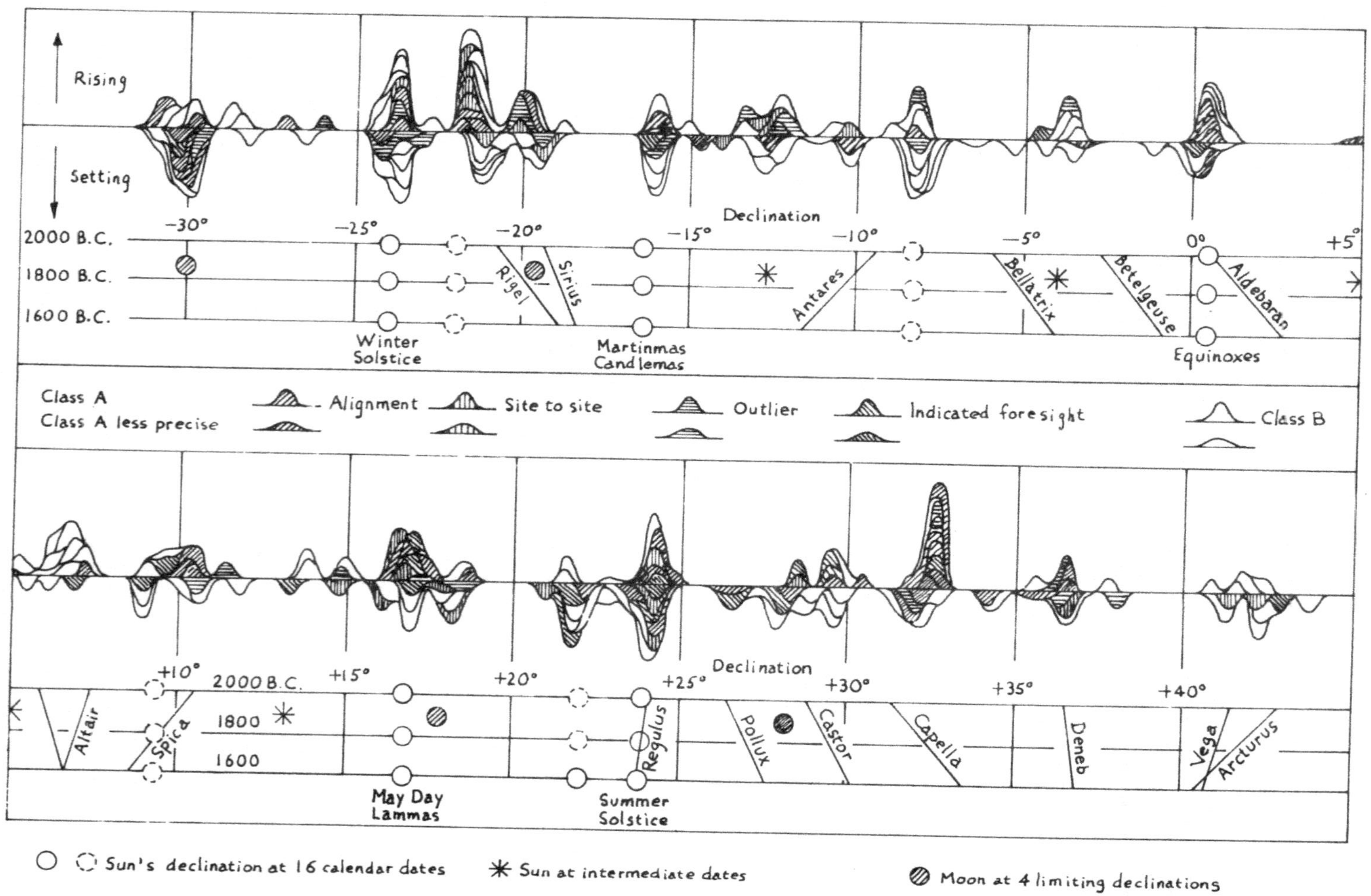

Fig. 1.5 Histogram of observed declinations δ

circles. From the observations obtained he plotted the histogram of declinations shown in fig. 1.5 (MSB:fig. 8.1) using a small gaussian curve at each observed declination and showing risings and settings.

1.9 Solar calendar

Among other things shown by this histogram are the Mayday/Lammas and the Candlemas/Martinmas dates in the year. These, with the equinoxes and solstices, divide the year into eight parts. The histogram also shows that there are concentrations midway between the above, thus dividing the year into 16 parts. Had the lines been set up at random, the histogram would be completely different. The manner in which the shaded areas tend to form definite groups cannot be explained on the assumption that the observed lines got there by accident. From this we infer that Megalithic Man was using a Solar calendar dividing the year into 16 parts. Because the Earth's orbit round the Sun is elliptical, Megalithic Man's recorded declinations of the Sun at the equinoxes are off by a certain amount, and in the correct direction. This is because Megalithic Man was placing his equinoctial backsights by choosing and checking that the Sun was again on the same line exactly half a year later, whereas today we define the equinox as that time when the declination of the Sun is zero. This proof of our thesis is clearly seen near the zero declination line on fig. 1.5.

Space does not permit presentation of the solar date here, and the reader is referred to other publications (MSB:chs. 8, 9, 11 & 12; MLO:ch. 4).

Twenty-nine miles to the south of Kintraw is the site at Ballochroy where, from eye level on the line of stones and kist, the midwinter sun sets on the fall of Cara (fig. 1.9). Equally important is the orientation of the widest stone, which unambiguously indicates Ben Corra on Jura. Here it is seen that the outline of the horizon is slightly steeper than the slope of the Sun's path, and conditions are such that the Sun would take about 3 minutes of time after setting and passing behind Ben Corra before it reappeared momentarily. This site has been fully described (MSB:151-154), and is important because of its double purpose, for observing winter <u>and</u> summer solstice.

An observing position could be obtained on the day of the solstice, but to give the date to within a day or so observations must be carefully made, because change in declination of the Sun is very small near the solstices. There are no other points throughout the year so obvious as these turning points and so Ballachroy and Kintraw are extremely important.

It has to be pointed out here that one observation on the exact day of the solstice, if missed, could be easily replaced by halving the number of days between observations made at a fixed point near to the final backsight for recording the equinox. The first observations could be made, say, 5 days <u>before</u> and the second 5 days <u>after</u> solstice. Relatively flat ground is <u>necessary</u>; or at least there has <u>to</u> be freedom to move along a straight line passing through the final backsight.

The remaining 14 calendar dates throughout the year do not require such accuracy because the Sun's declination is then changing rapidly— as much as 21' per day at equinoxes.

1.10 Distribution of suspected lunar lines in fig. 1.5

Groupings of observed declinations are also seen at salient dates for the Moon, and so after producing fig. 1.5 in 1967 an analysis was made of the lines near to the Moon's position at the four limiting declinations shown on the histogram by shaded circles. All the declinations δ which lay within 0°.8 to 1°.0 of the expected values δ_e were examined and $\beta = (\delta - \delta_e)$ was plotted for the 37 relevant lines on fig. 1.6 (MSB:fig. 10.1). The remarkable way in which the gaussians form a peak for each limb of the Moon is notable. This diagram came as it were out of the blue and showed quite clearly the two peaks corresponding to the two limbs of the Moon. It has to be noted that we are discussing upper and lower limb here in the sense of declination. No attempt was made to remove the perturbation before the histogram was plotted. As no other explanation could be found for the declinations involved it had to be accepted that they possessed lunar significance. This was the first clear indication that the Moon had been observed. (For the sake of comparison later, see figs. 1.14 & 1.15.)

Now that my father knew that Megalithic Man had had a calendar he was ready to think of them predicting future events, such as eclipses. The histogram, fig. 1.5, had shown the turning points of the Moon at $\varepsilon \pm i$. Did the megalithic observers know about the small wobble or perturbation $\pm\Delta$ in the Moon's orbit? For this, accurate surveys of alignments were necessary, and we now survey to within 1 minute of arc in azimuth and 20 seconds in altitude. Although the wobble is going on continuously it can only show up at the standstills, and so we are only interested in lunar declinations within bands such as $(\varepsilon + i + \Delta + s)$ to $(\varepsilon + i - \Delta - s)$ (fig. 1.7). Thus when we go to a site we have to look at the places on the horizon which give these declinations. There are eight of them; whenever one is indicated we have to measure the altitude and azimuth of points on the horizon profile so as to be able to plot it. If there is a notch or mark or a clearly defined foresight of any kind on the profile then its declination must be calculated, and if it turns out to be a pertinent declination then we know that we are at a megalithic backsight.

1.11 More lunar lines

Since producing figs. 1.5 & 1.6 in 1967, 58 more lunar observing sites have been surveyed, bringing the total number under discussion to 95. The results of a careful analysis of 42 of these, where we consider each line to be clearly defined and well measured, are given in table 1.1. The plot on fig. 1.8 shows a much more refined histogram of β. For continuity, however, the 42 lines have also been used to produce fig. 1.15 for comparison with fig. 1.6. As before, β for each observed line is shown by a gaussian. The "expected" or theoretical values, called Q, are shown by arrows as indicated in the key in fig. 1.8 (see section 1.18 below). Ideally, we expect each β to have a value equal to Q, and it will be seen how the observed β values cluster round the expected Q values. This shows unequivocally that Megalithic Man was observing the top and bottom of the wobble shown in fig. 1.2. We are perfectly sure about this because we know how the analysis has proceeded since 1967, the original date of publication of fig. 1.6. Others were not convinced; they pointed out that probably our choice of foresight was not objective. Accordingly we set up strict terms of reference for the selection of certain sites from our collection, namely (a) that the backsight mark must give some indication of the direction in which to look to find the foresight, and (b) that the foresight must be within one of the lunar bands described above (fig. 1.7). There must only be one point which could have been used to indicate the declination. The use of these strict criteria cuts down the number of sites to 20, but from these alone we found a probability level of about 0.2% (Thom & Thom 1978b: 177).

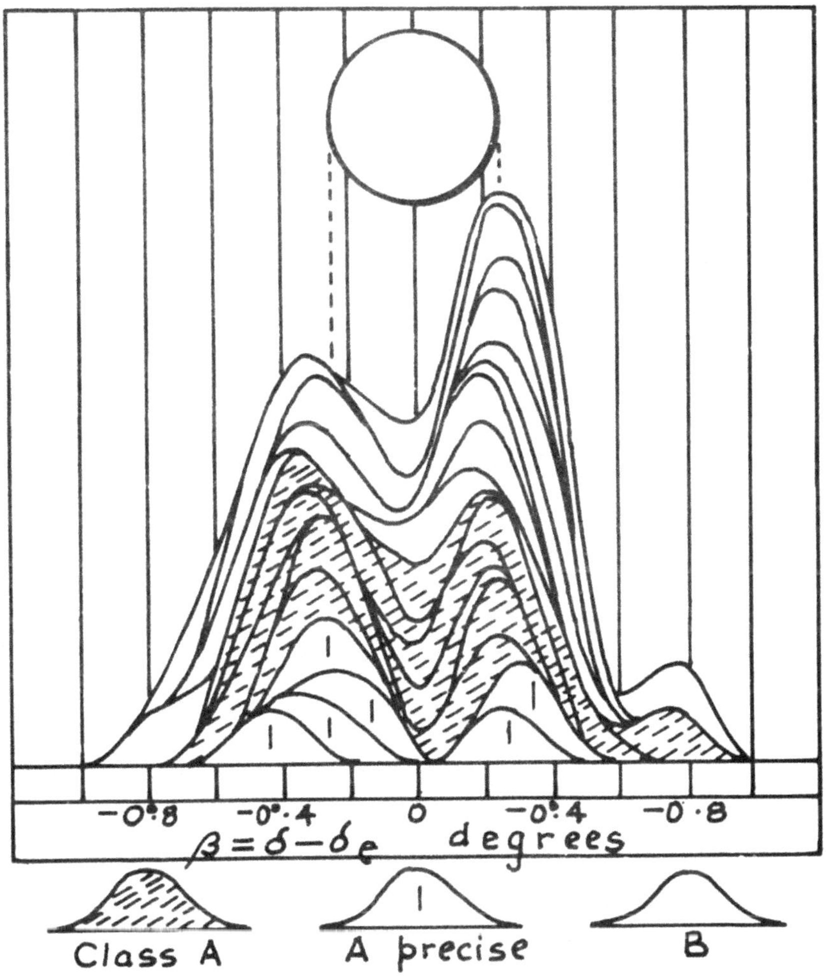

Fig. 1.6 Distribution of suspected lunar lines plotted as $(\delta - \delta_e)$

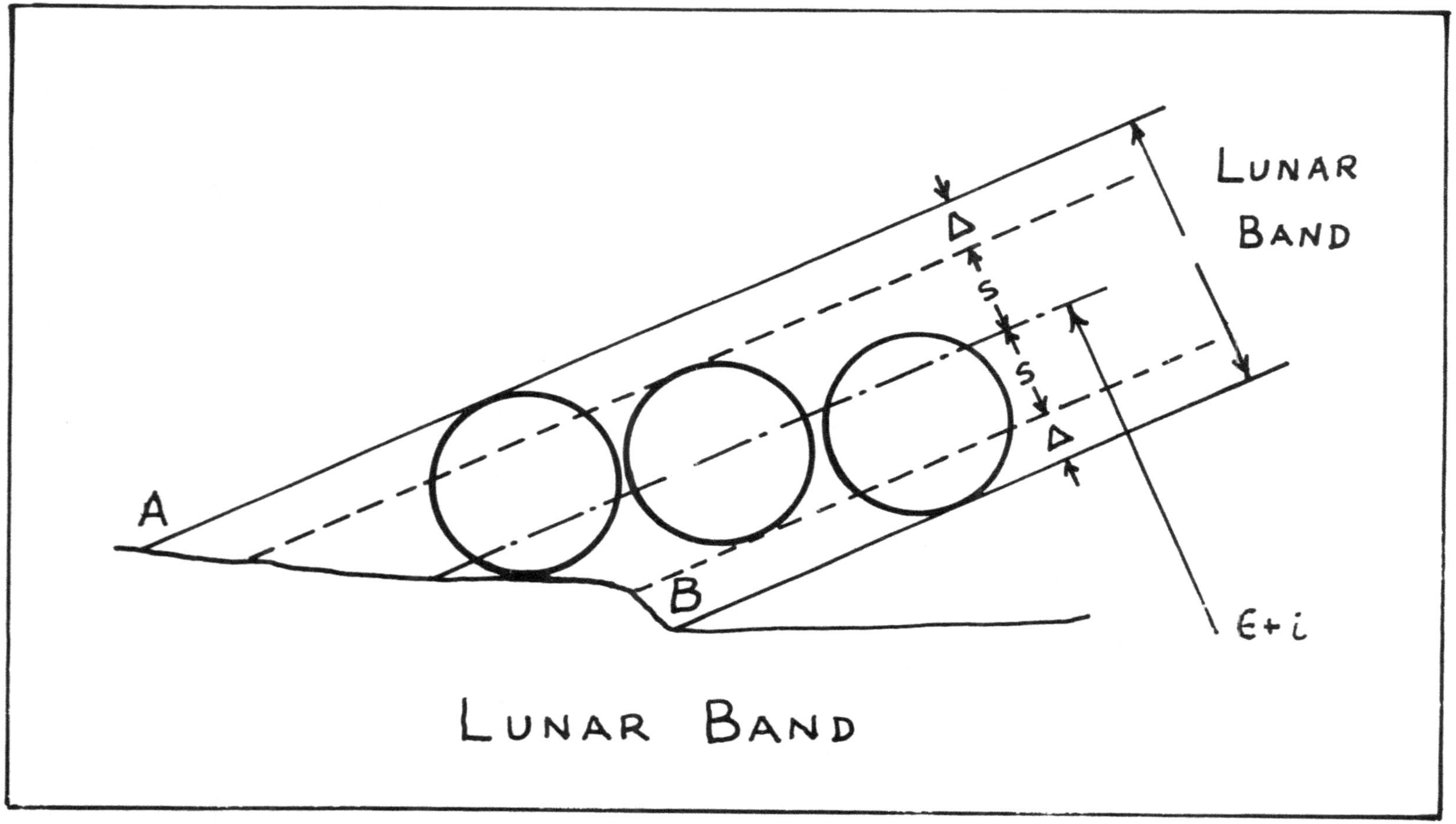

Fig. 1.7 Lunar band

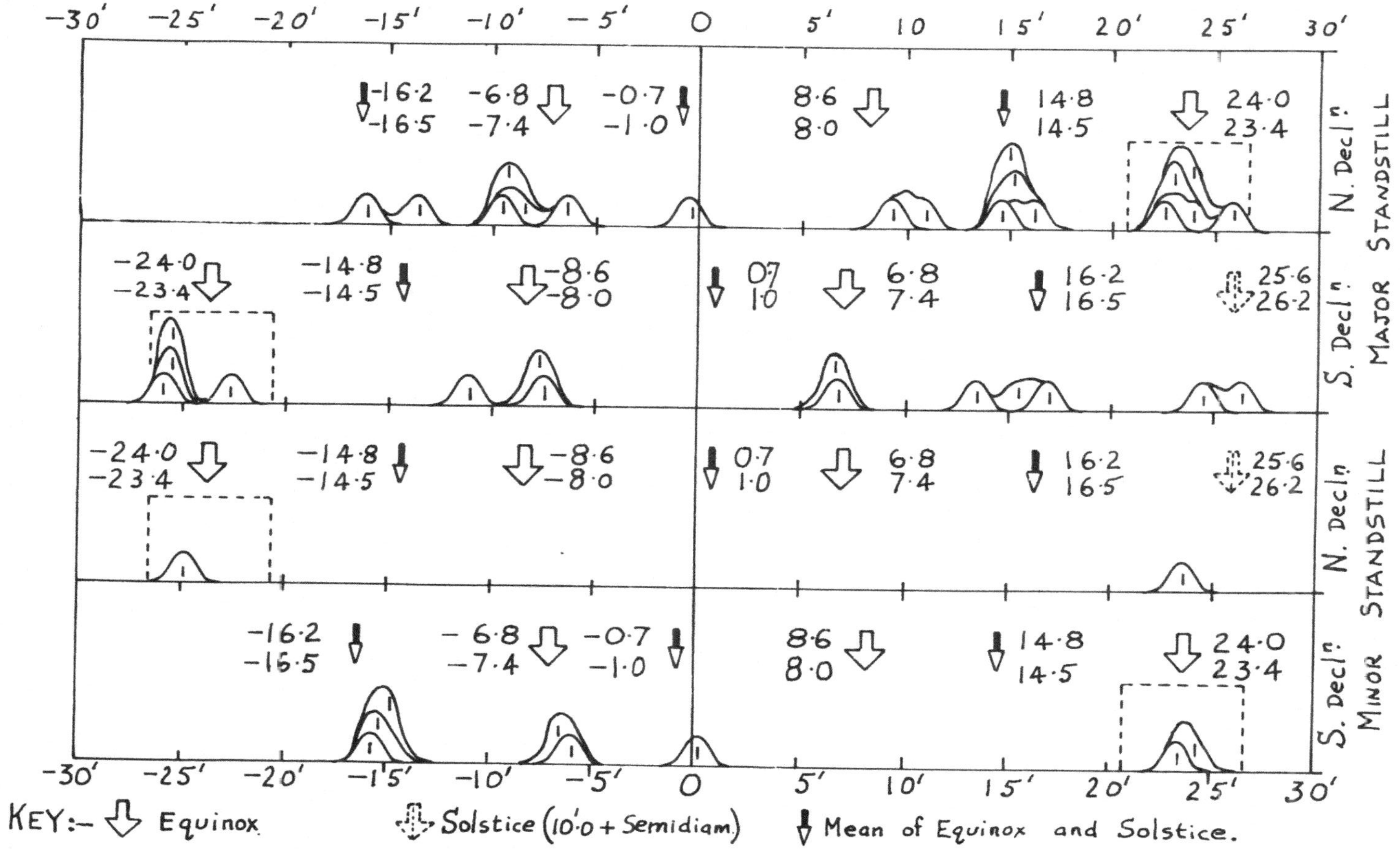

Fig. 1.8 Four histograms of β values where $\beta = \delta_o - (\varepsilon \pm i)$; the expected values of β, namely Q, are shown by arrows as in key. Upper numbers of pairs of numbers beside each arrow show Q for $c_S = 0$. Lower numbers show Q for c_S= 0'.6 (equinoctial or solstitial values) or for c_S= 0'.3 (mean of equinoctial and solstitial values) for $(\varepsilon \pm i)$ cases.

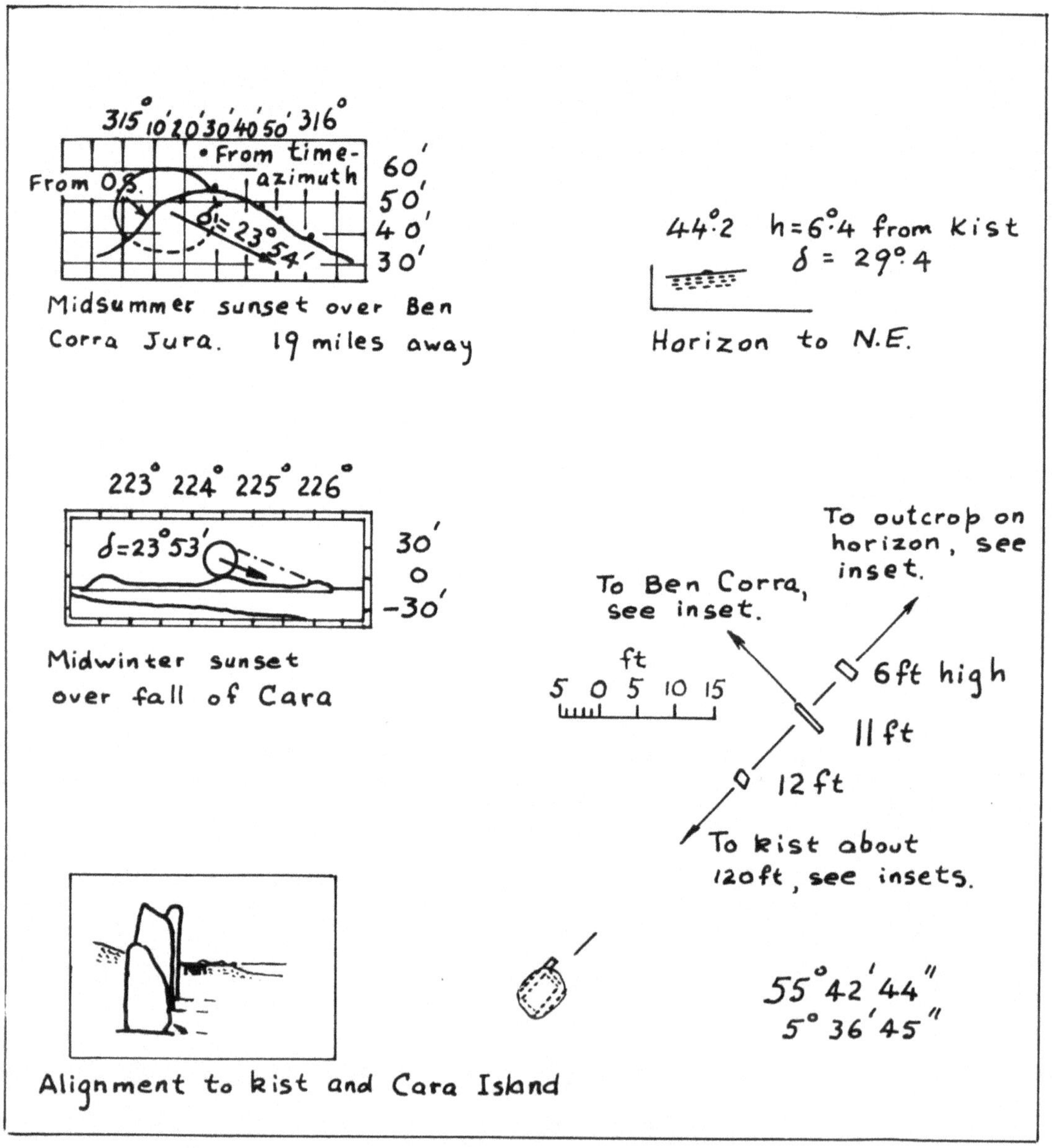

Fig. 1.9 Ballochroy, Kintyre

1.12 Forty-two measured foresights

Having established that the above 20 lines were real lunar foresights, we now feel that we are entitled to relax the strict terms of reference (a) and (b) mentioned above and to use our measured declinations in order to find out just how Megalithic Man worked. As a result we are able to plot the histogram for 42 lines shown in fig. 1.8. For comparison with the earlier way of presentation we also plot fig. 1.15 showing the 42 lines. So far, in MSB and MLO, we had published our measurements on about 80 British lunar lines. By 1971, 28 of these 80 had been discarded. In addition to the 80 we now include a further 15 lines, namely those at Wormadale Hill, Brogar, Stenness, Ballinaby, Skipness and Dunskeig.

1.13 Twenty-five sites not used

For the record, we give in appendix 1.1 a list of the remaining 25 lines unused in the present analysis of 42 lines. In fig. 1.14 we show these discarded data in the same way as in figs. 1.6 & 1.15. In each case we state the reason for rejection: see tables 1.3 & 1.4 in appendix 1.1. The subject is continually growing. Our knowledge develops with passage of time as more data become available, and so it is necessary to concentrate on the latest analysis.

The information used to plot the histogram involving 42 lines is listed in table 1.1, with, for each line, a column for δ_o the observed declination; the nominal declination; Q (for $c_s = 0$); ε; β and R. Table 1.1 also has a column with Q (for $c_s = 0'.6$ or $0'.3$) and the relevant R (see section 1.15 below). Q is discussed in section 1.18.

1.14 Re-calculation of observed declinations

The "observed" declination δ_o in the above analysis for each of the 42 lines was again calculated starting with the azimuths and altitudes as previously published, taking the appropriate times of day when finding refraction and using mean parallax values. Details of an example calculation are given by Thom & Thom (1978b). Mean parallax values applied were $56'.4$ at equinoctial and $57'.4$ at solstitial seasons (see section 1.24). A value is needed for ε, the obliquity of the ecliptic, but this varies slowly with time and so since archaeologists cannot give us exact dates for the sites we have to determine ε from the observations themselves. From each line a value of ε can be calculated. Any differences in δ_o from previously published values are due to these re-calculations; e.g. for Ballinaby (table 1.1) we now have $29^o16'.8$, whereas previously (MRBB: 170) we had $29^o17'.7$.

1.15 Curvature correction c_s

In tables 1.1 & 1.2 we refer to c_s, a curvature correction involving the sum of two corrections $c_1 = 0'.3$ and $c_2 = 0'.3$ as described in appendix 1.2: column four in table 1.1 gives the Q values to be expected from Megalithic Man's having taken a number of observations of the maximum over three or four 18.6 year cycles and having made only one recording, so we have put $c_s = c_1 + c_2 = 0$. This is on the assumption that he had enough information to record the crest of the perturbation wave near the crest of the (ε + i) wave. (For a full explanation of the c correction see Thom & Thom (1980b).)

For the maximum positive declination at the major standstill, for example, we must subtract c_s from the Q values already mentioned, before we can compare the results with the observed β values. This is because if the observers erected each backsight as a result of an observation made at a single standstill we could expect the resulting declination to be, on the average, low by c_s. However if the determinations of the positions of the backsights were spread over a number of standstills, then correction c_s is not appropriate because at M (for example) the erectors would probably have used the maximum of the values obtained from the various standstills. One of the objects of the present paper is to find out if this method of observing was used, and so a trial is made with and without the above c correction.

Column eight in table 1.1 gives the Q values to be expected for one of two c_s values: either $c_s = c_1 + c_2 = 0'.6$ for equinoctial and solstitial occurrences, or $c_s = c_1 + c_2 = 0'.3$ for the mean of these two occurrences (top and bottom of wobble). In fig. 1.8 the broad-shafted arrows (of width 0'.6) and the barrow-shafted arrows (of width 0'.3) purport to show the c_s band-width involved.

Column five in table 1.1 shows the value of ε calculable for each line (see section 1.16 below). The mean of the 42 values of ε is $23°53'.1$ and this mean value is used each time to calculate the values of

$$\beta = \delta_0 - (\varepsilon \pm i) \qquad .. (1.1)$$

which are given in table 1.1 column six; there are also columns for the residual R $(= |\beta - Q|)$, the difference between actual and expected values irrespective of sign.

1.16 Obliquity ε obtainable from each site

Equation (1.1) gives

$$\varepsilon = \delta_0 \pm i - \beta,$$

and for ideal conditions $\beta = Q$; thus

$$\varepsilon = \delta_0 \pm i - Q. \qquad .. (1.2)$$

Mean i remains constant over the centuries, and so each line yields a value for ε. The mean obliquity found from the observations themselves (column five in table 1.1) can then be used in an attempt to date the erection. A specimen calculation is given by Thom & Thom (1978b: table 2).

1.17 Graze

Graze is the extra bend experienced by a ray of light which passes over a ridge, especially in the early hours of darkness when the ground sometimes cools more quickly than the air. We assume that the graze effect occurs from one hour before sunset to one hour after sunrise, but this may not be exact. In table 1.1 we show calculations involving graze of -0'.6. Graze is applied in addition to refraction. It will be seen from table 1.1 that we can extract means ε_N and ε_S for obliquity obtained from N (+) and S (-) declinations. Half of the difference between ε_N and ε_S will give the correction to the graze, if any is required (Thom & Thom 1980b). For the results of the set of calculations given in table 1.1 the

Key for table 1.1

Values of ε and β deduced from δ_o the observed declinations for mean of Day and Dark

$\beta = \delta_o - (\varepsilon \pm i)$

$Q = \delta_e - (\varepsilon \pm i)$ where δ_e = nominal or expected declination

R = residual = β-Q irrespective of sign

r = r.m.s. of R

ε for example for major standstill and positive declination is
$\varepsilon = \delta_o - i - Q$. Mean i = 5°08'.7.

From the above 42 values of β, r = 1'.34 for assumption 1 and 1'.53 for assumption 2. In assumption 1 the observers had determined their position over a span of about 179 years; for this c_s would be zero (first 7 columns). In assumption 2 the observers had determined the position of the stone from one standstill. For this c_s may be 0'.3 or 0'.6 (columns 8, 9, see also text). The broad-shafted (0'.6) and narrow-shafted (0'.3) arrows in fig. 1.8 purport to show to scale the corrections involved.

Key to reference numbers given in final column

2 Thom & Thom 1978a ("MRBB")
4 Thom 1971 ("MLO")
7 Thom & Thom 1978b
15 Thom & Merritt 1978
17 Thom 1974a
18 Thom & Thom 1979a

TABLE 1.1

Site	δ_0	Nominal δ	Assumption .1. Q for $c_\delta = 0$	Assumption .1. ϵ 23°	Assumption .1. β	Assumption .1. R $= \lvert \beta - Q \rvert$	Assumption .2. Q for $c_\delta = 0'.6$ or $c_\delta = 0'.3$	Assumption .2. R $= \lvert \beta - Q \rvert$	Ref.
Brogar JK	−18 44·1	$-(\epsilon - i)$	−0·7	52·1	0·3	1·0	−1·0	1·3	2
" M	−18 59·2	$-(\epsilon - i + \sigma)$	−16·2	51·7	−14·8	1·4	−16·5	1·7	2
" Comet	−18 50·3	$-(\epsilon - i - \Delta + \sigma)$	−6·8	52·2	−5·9	·9	−7·4	1·5	2
" MLJ	−18 20·1	$-(\epsilon - i - \Delta - \sigma)$	24·0	52·8	24·3	·3	23·4	·9	2
" Kame l	29 24·3	$+(\epsilon + i + \Delta + \sigma)$	24·0	51·6	22·5	1·5	23·4	·9	2
" Kame u	29 24·8	$+(\epsilon + i + \Delta + \sigma)$	24·0	52·1	23·0	1·0	23·4	·4	2
" L	−18 50·9	$-(\epsilon - i - \Delta + \sigma)$	−6·8	52·8	−6·5	·3	−7·4	·9	2
" Ravie Hill, Comet	28 48·2	$+(\epsilon + i - \sigma)$	−16·2	55·7	−13·6	2·6	−16·5	2·9	2
Brogar Ravie Hill, H	28 52·3	$+(\epsilon + i + \Delta - \sigma)$	−6·8	50·4	−9·5	2·7	−7·4	2·1	2
Stenness	19 08·0	$+(\epsilon - i + \Delta + \sigma)$	25·6	51·1	23·6	2·0	26·2	2·6	2
Callanish I	−29 12·8	$-(\epsilon + i + \Delta)$	−8·6	55·5	−11·0	2·4	−8·0	3·0	4 H 1/1
" V	−28 35·4	$-(\epsilon + i - \Delta - \sigma)$	25·6	52·3	26·4	·8	26·2	·2	4 H 1/5
Mid Clyth A_1	−29 09·2	$-(\epsilon + i + \Delta)$	−8·6	51·9	−7·4	1·2	−8·0	·6	4 N 1/1
" " A_2	−29 27·1	$-(\epsilon + i + \Delta + \sigma)$	−24·0	54·4	−25·3	1·3	−23·4	1·9	4 N 1/1
Wormadale Hill	−29 09·5	$-(\epsilon + i - \Delta + \sigma)$	−8·6	52·2	−7·7	·9	−8·0	·3	15
Ballinaby	29 16·8	$+(\epsilon + i + \sigma)$	14·8	53·3	15·0	·2	14·5	·5	17
"	29 25·6	$+(\epsilon + i + \Delta + \sigma)$	24·0	52·9	23·8	·2	23·4	·4	2
Kintraw	−18 20·9	$-(\epsilon - i - \Delta - \sigma)$	24·0	53·6	23·5	·5	23·4	·1	4 A 2/5

TempleWood S_1A_1	28	55·4	$+(\epsilon+i+\Delta-\sigma)$	− 6·8	53·5	− 6·4	·4	− 7·4	1·0	4	A 2/8
" " $S_{54}A_1$	29	18·1	$+(\epsilon+i+\sigma)$	14·8	54·6	16·3	1·5	14·5	1·8	4	A 2/8
" " QA_1	29	01·5	$+(\epsilon+i)$	− 0·7	53·5	− 0·3	·4	− 1·0	·7	4	A 2/8
" " A_2	−28	48·3	$-(\epsilon+i-\sigma)$	16·2	55·8	13·5	2·7	16·5	3·0	4	A 2/8
Ballymeanach A_1	29	12·8	$+(\epsilon+i+\Delta)$	8·6	55·5	11·0	2·4	8·0	3·0	4	A 2/12
" A_2	29	27·7	$+(\epsilon+i+\Delta+\sigma)$	24·0	55·0	25·9	1·9	23·4	2·5	4	A 2/12
Dunadd	29	16·4	$+(\epsilon+i+\sigma)$	14·8	52·9	14·6	·2	14·5	·1	4	A 2/13
High Park A_1	−29	27·7	$-(\epsilon+i+\Delta+\sigma)$	−24·0	55·0	−25·9	1·9	− 23·4	2·5	4	A 4/2
" " A_2	−28	46·3	$-(\epsilon+i-\sigma)$	16·2	53·8	15·5	·7	16·5	1·0	4	A 4/2
Skipness	−29	24·4	$-(\epsilon+i+\Delta+\sigma)$	−24·0	51·7	−22·6	1·4	− 23·4	·8	18	
Beacharr	29	11·1	$+(\epsilon+i-\Delta+\sigma)$	8·6	53·8	9·3	·7	8·0	1·3	4	A 4/5
Crois Mhic-Aoida	−28	37·2	$-(\epsilon+i-\Delta-\sigma)$	25·6	54·1	24·6	1·0	26·2	1·6	4	A 4/9
Campbelltown	−28	44·8	$-(\epsilon+i-\sigma)$	16·2	52·3	17·0	·8	16·5	·5	4	A 4/14
"	−28	54·9	$-(\epsilon+i+\Delta-\sigma)$	6·8	53·0	6·8	0·0	7·4	·6	4	A 4/14
Knockstaple	28	52·5	$+(\epsilon+i+\Delta-\sigma)$	− 6·8	50·6	− 9·3	2·5	− 7·4	1·9	4	A 4/19
Gigha	29	25·7	$+(\epsilon+i+\Delta+\sigma)$	24·0	53·0	23·9	·1	23·4	·5	4	A 4/17
Dunskeig	−19	00·0	$-(\epsilon-i+\sigma)$	−16·2	52·5	−15·6	·6	− 16·5	·9	7	
Knockrome	−29	27·0	$-(\epsilon+i+\Delta+\sigma)$	−24·0	54·3	−25·2	1·2	− 23·4	1·8	4	A 6/4
Stillaig	28	53·4	$+(\epsilon+i+\Delta-\sigma)$	− 6·8	51·5	− 8·4	1·6	− 7·4	1·0	4	A 10/5
Glen Prosen	−28	55·1	$-(\epsilon+i+\Delta-\sigma)$	6·8	53·2	6·7	·1	7·4	·7	4	P 3/1
Lundin Links	−18	59·7	$-(\epsilon-i+\sigma)$	−16·2	52·2	−15·3	·9	− 16·5	1·2	4	P 4/1
Fowlis Wester	29	17·0	$+(\epsilon+i+\sigma)$	14·8	53·5	15·2	·4	14·5	·7	4	P 1/10
Parc y Meirw	18	19·6	$+(\epsilon-i-\Delta-\sigma)$	−24·0	52·3	−24·8	·8	− 23·4	1·4	4	W 9/7
Blakeley Moss	28	45·8	$+(\epsilon+i-\sigma)$	−16·2	53·3	−16·0	·2	− 16·5	·5	4	L 1/16

graze effect was initially taken as -0'.6 and the correction turned out to be 0'.07. (See case 1 in table 1.2 below.) For any further calculations the graze would have to be -0'.5: this small change would have its effect on each of the 42 values of δ_Q and finally on the mean value of ε, but any alteration would be negligible.

In table 1.2 we summarise the results of repeating several times the whole calculation shown in the first seven columns of table 1.1, using in each instance the graze found and the appropriate value of refraction. We see from Thom & Thom (1978b: table 2) that there are always two dates in the year when the observation might have been made, namely in March or September (or June or December). Usually, the Moon was on the foresight in darkness at one of these dates, and six months later was again on the foresight in daylight. We shall assume first of all that both of these could be observed, and so we call this case "day and dark".

In table 1.2 case 1, we make the assumption that c_S is zero and the graze -0'.6. We then find eventually that the half difference between ε from the south declination and that from the north is + 0'.1, so the graze assumed as -0'.6 becomes -0'.5. We assume here that $\partial\delta/\partial h$ is approximately unity. At the same time we find r, the root mean square of R, equal to 1'.34. (The calculation will be found in the first seven columns of table 1.1.)

Now consider case 2. Here we do not make the assumption that c_S is zero but use the value appropriate to a single observation which would have resulted had each backsight been placed following a single observation as opposed to the mean of several observations. The calculation for assumption 2 will be found in columns eight and nine of table 1.1, yielding the final value of r of 1'.53.

In the next three cases, 3, 4 and 5 (table 1.2), we entirely neglect possible observations when the Moon was on the foresight in daylight and use only the dark condition. In case 3 we find that the graze was -0.2. We use this in case 4 and find r = 1'.41. Case 5 shows the same calculation but with the value of c_S appropriate to a single observation. Again we find the graze -0'.2 but r has risen to 1'.55. We thus see that in both comparisons the value of the r.m.s. residual is lowest when c_S is assumed zero.

In both comparisons made above we get a lower deviation (albeit by a small amount) with correction $c_S = 0$. If this means anything it shows that each backsight was not erected following observations made at one standstill, but that the work at each was spread over a number of standstills, perhaps as many as 10. The evidence is obviously very slight, so fortunately there is another method of showing the same thing and this we shall describe below (section 1.29).

1.18 Expected value of β, called Q

Q is the theoretical value of β found by a combination of the relevant perturbation Δ with the relevant s (the Moon's semi-diameter) (see 1.22 & 1.23 below). The Q values shown in fig. 1.2 are for the Moon's centre only; on the histogram (fig. 1.8), however, the upper and lower limbs are included as and where necessary. The Q values therefore occur in groups of three. β is the angle by which the observed declination δ_Q is above the top of the chain dotted curve in fig. 1.2. This curve shows the locus of the monthly declination maxima, had there been no wobble of

Table 1.2 Collected results of calculations under various assumptions regarding conditions, c_s and the graze

Case	Condition assumed	c_s	Graze assumed	ε From N decl.	ε From S decl.	ε Mean	Graze found	r = r.m.s. of R
1	Mean of day and dark	0'	-0'.6	$23^o53'.0$	$23^o53'.2$	$23^o53'.1$	-0'.5	1'.34
2	Mean of day and dark	appropriate c_s	-0'.6	------ See table 1.1 ------				1'.53
3	Dark	0'	-0'.5	$23^o52'.6$	$23^o53'.0$	$23^o53'.0$	-0'.2	
4	Dark	0'	-0'.2					1'.41
5	Dark	appropriate c_s	-0'.2	$23^o53'.09$	$23^o53'.14$	$23^o53'.1$	-0'.2	1'.55

the orbit. Some of the values of Q to be expected are shown by arrows on fig. 1.8. The purpose of the broad and narrow shafts on the arrows is indicated in section 1.15 above.

1.19 Plot of histogram of β

In the plot of the histogram of β on fig. 1.8 where $\beta = \delta_o - (\varepsilon \pm i)$, ε is the mean obtained from the 42 lines and δ_o is the "observed" declination. A small gaussian curve is again used at each observed declination, as in fig. 1.8. Ideally, β should be equal to Q.

1.20 Clustering of β values round expected Q values

The first thing to notice about the histogram in fig. 1.8 is the manner in which the β values cluster round the expected values, i.e. the Q values. To show how closely these follow the Q values let us make a crude estimate of the probability level. Note first that in fig. 1.8 the Q values (the "nodes") are roughly uniformly spaced. Any inequalities of spacing will have but little effect on the following estimate. In MRBB we have described in detail Broadbent's method of dealing with uniformly spaced "nodes". Using Broadbent's notation, the spacing 2δ is equal to about 8 arc minutes. Using R for the residual error in each case (see table 1.1), the "lumped variance" is $(\Sigma R^2)/n$, where n is the number of observations (in this case 42) and $R = |\beta - Q|$. The square root, or what we call here the root mean square (r.m.s.) is 1.34 (for c = 0) and so we find S^2/δ^2 to be $(1.34)^2/4^2 = 0.113$. We find the probability level to be about 0.01% or 1/10,000. This very low value shows how far the β values are from being randomly distributed in the histogram. Looking at the way in which the values are grouped the result is not surprising.

1.21 Lunar declinations at solstice

There are only 3 alignments which show directly the lunar declinations at the solstices. These are Callanish V, Stenness and Crois Mhic-Aoida (MacKay's Cross). The upper limb values of Q for these are shown in fig. 1.8 by dotted broad arrows at $Q = s + \Delta = 15'.6 + 10'.0 = 25'.6$. As we cannot simply ignore these, all three are retained.

No attempt has been made to show the other combinations of Δ and s for the solstitial case. They would complicate fig. 1.8 unnecessarily and even if we attributed the β values in their vicinity to them the probability level would still be low.

1.22 How the observers used the sites

Two examples are given to show how we believe the observers worked, namely the sites at (a) Ballinaby and (b) Stenness. See appendix 1.3.

1.23 The perturbation

In appendix 1.4 we show how to evaluate Δ at the vernal or autumnal equinoxes and at the solstices. At the equinoxes we find the Moon at first or third quarter with $\Delta = 8'.6$ and at the solstices the full or new Moon with $\Delta = -10'.0$. These values are applicable to i.

1.24 Parallax p and semi-diameter s

Unless we know the actual date when each monument was erected and thereby know the parallax, we have to use a mean value for p of 56'.4

at the equinoxes and 57'.4 at the solstices. Mean semi-diameter s equals 0.2725 x p, and so s = 15'.4 and 15'.6 respectively at the equinoxes (quarters) and solstices (full and new). These values have been used in the calculations involved in table 1.1. They were provided in a private communication by L. V. Morrison of the Royal Greenwich Observatory.

1.25 Extrapolation to maximum observed declination

Observations of the Moon on the horizon can be made only at intervals of one lunar day. Between consecutive recordings, one on either side of the maximum, the Moon's declination can change greatly, reaching the maximum and then decreasing rapidly. The observer would in general miss the actual maximum. We believe that Megalithic Man overcame this difficulty by a method of extrapolation to maximum declination (MLO: ch. 8). Wood (1978: ch. 7) gives a full discussion of extrapolation.

1.26 One missed observation at the maximum

A criticism often raised is that if bad weather interfered with an observation of the Moon on the horizon, the whole process would fail. This need not be so. Suppose, for example, that the observation of moonrise was lost at a time similar to W on fig. 1.4; then it would have been possible to fix W on the ground from stakes fixed the day before (T) and the day after (Th). To a trained observer watching the Moon rise or set at its maximum, such interpolation would not have been difficult. (The ground distance from the mean position of T and Th to W is a constant at each site equivalent to a declination change of about 48'.) The value of 48' has of course been obtained from the Nautical Almanac.

1.27 A missed maximum

Should no observations have been possible over three consecutive nights as discussed in section 1.25 above, the two adjacent maxima, one lunation before and one lunation after, could have been used. The method would be similar to that mentioned in section 1.25 except that the sagitta would be the equivalent, on the ground, of 4 arc minutes. This exercise would have been perfectly feasible to workers accustomed to observing the Moon's movement over long periods.

1.28 Dating of sites

From the observations, assuming $\beta = Q$, for each line a value of ε can be obtained and the average, which is $23^{o}53'.1$, yields a date of 1575 BC ± 100 years. The date can not be estimated any closer than this because mean parallax values have to be used and because of the difficulty of estimating refraction at low altitudes. The graze effect will itself vary from the mean. Faulty measurements of altitude and azimuth may have been made by ourselves, the erectors may themselves have made errors in their final choice of backsight positions, and geological movement and erosion or growth of vegetation may have taken place. It should be remembered that the observatories were probably not all erected at the same time and so the actual value of ε may vary, and this introduces another possibility of error. With all these effects it is amazing that the root mean square residual is as low as the results obtained in table 1.1 (namely 1'.34), and this makes it all the more certain that Megalithic Man obtained a mean position over about ten standstills.

1.29 Calculable adjustment to β over 179 years

We have shown (MLO: 81-82; MRBB: 13-16) the peculiar manner in which the parallax at the standstills is subject to a sinusoidal oscillation of period 179 years and amplitude about 3'. This is borne out by calculations made by A. T. Sinclair of the Royal Greenwich Observatory showing the variation in apparent declination (adjusted for parallax). We give an example in fig. 1.13 chosen from one of 16 possibilities. If we assume that the values are evenly scattered along a sinusoidal curve with amplitude b, then the r.m.s. value is about

$$\left\{ \frac{1}{2\pi} \int_0^{2\pi} (b \sin \theta)^2 \, d\theta \right\}^{\frac{1}{2}} .$$

This has a value of $b/\sqrt{2}$ and becomes 2'.12 when b = 3', an amount definitely larger than the values we obtained from the actual observations (table 1.1). The r.m.s. deduced from 20 points on fig. 1.13 between 1800 and 1630 BC, for example, lies between 3'.0 and 3'.5.

We assume that the dates of erection of the various observatories were spread over a number of years and so if the values of R we have found from our field measurements in table 1.1 depended on stones, each of which had been set up individually as a result of observations made at one standstill, then we might expect them to be distributed roughly as in fig. 1.13 and so have a value of r of the order of 3' as above. But r which we actually find is sometims as low as 1'.3 or 1'.4, and this must have been raised already by the various factors mentioned in section 1.28 above. The only explanation of the discrepancy is that the observations made by Megalithic Man before he erected a permanent backsight were spread over a period of the same order as the parallax period which we have seen is about 179 years.

1.30 Geographical division of the sites

We tried the effect of dividing the data geographically into two groups. The NW group contained—arbitrarily—Shetland, Orkney, Caithness and the Hebrides; 17 sites in all. The SE group contained all the remainder. The mean values for the obliquity of the ecliptic found for the NW group was a fraction of a minute smaller than that found for the SE group. The differences, while not statistically significant, might indicate that the NW sites were built about 100 years after the SE sites.

1.31 Argyllshire sites alone, both lunar and solar

At present a large proportion of the lunar lines available for analysis are to be found in Argyllshire. In a report on these 22 lines (Thom & Thom 1979b) it will be seen that the mean of 19 calculated dates tabulated is 1657 BC ± 54 years. The three solar lines in Argyllshire give a date of 1750 BC ± 100 years. (In the first of these dates the ± figures indicate the standard error of the mean. It has to be remembered, however, that the uncertainty due to parallax variation has to be considered in addition.)

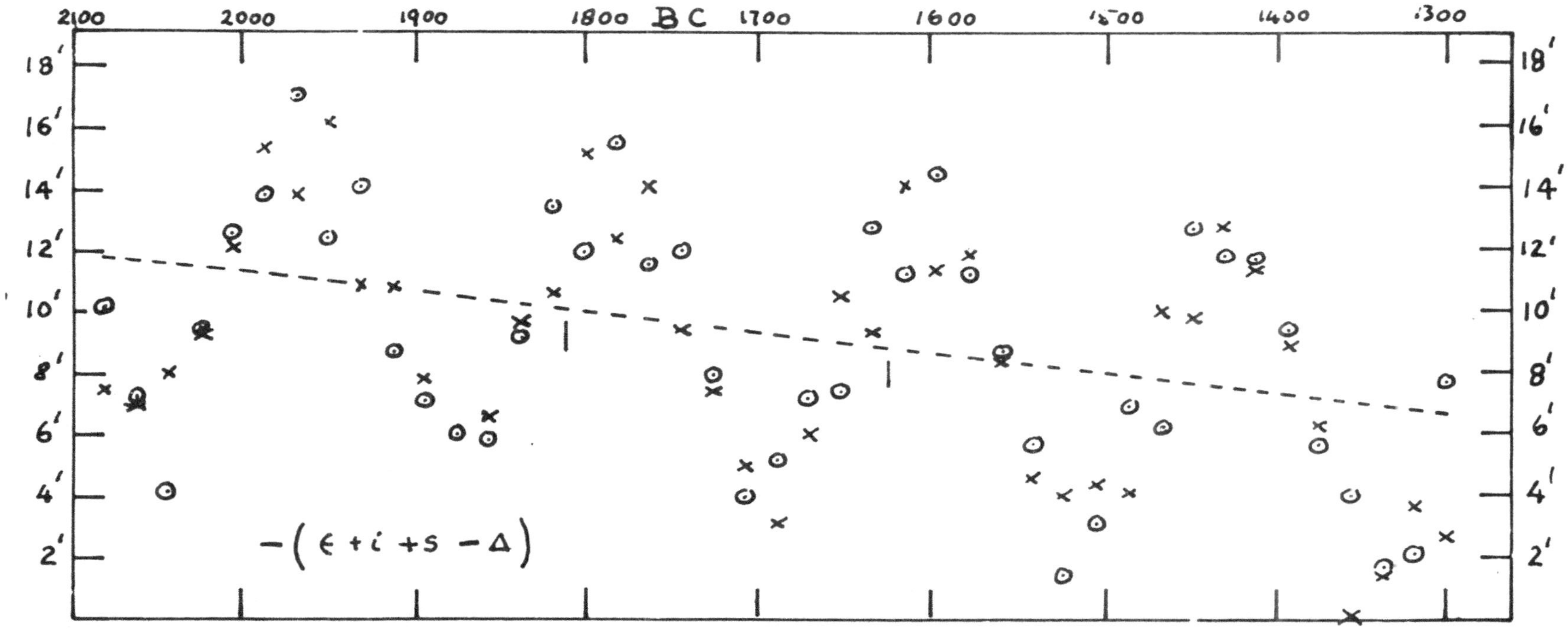

Fig. 1.13 Calculated lunar declinations at standstills over the period 2100-1300 BC. Values deduced from Greenwich calculations of declination and parallax so as to be ready for direct comparison with observed declination obtained by using a parallax of 57'.0 (MRBB: section 2.6). ⊙ = Autumn, x = Spring. The dotted line shows how $-(\varepsilon + i - \Delta + s)$ decreases numerically at about 39" per century. This figure is for use for Wormadale Hill.

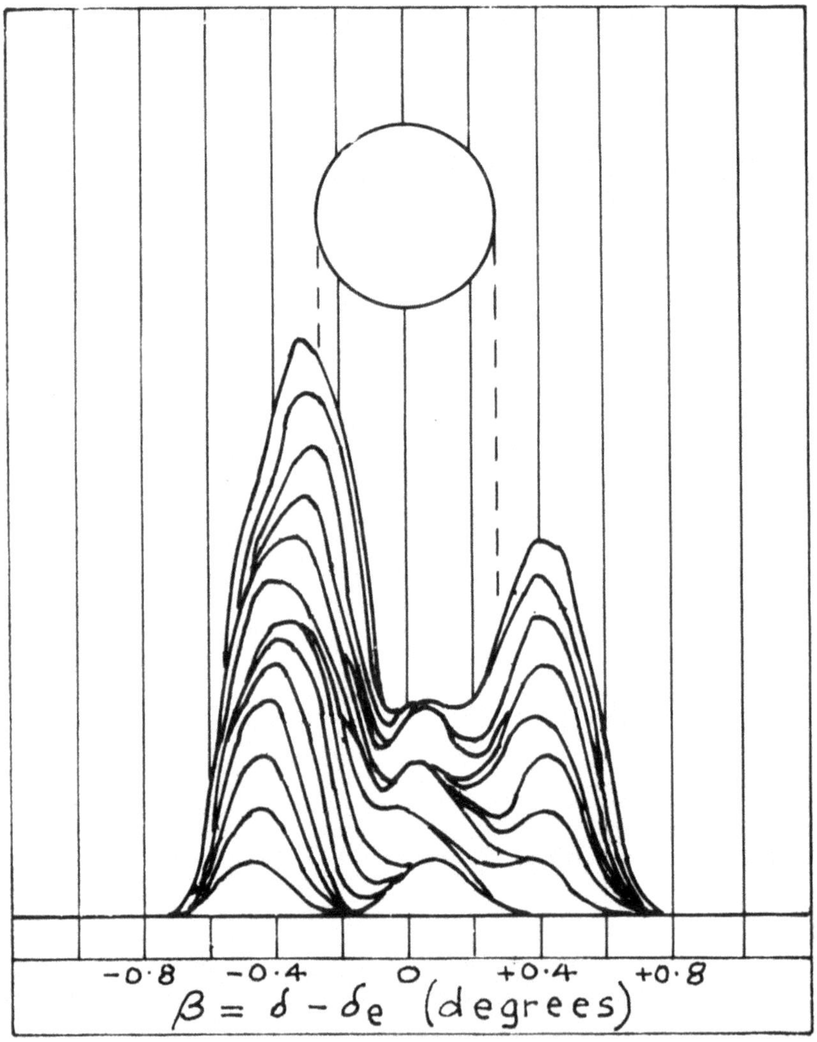

Fig. 1.14 β for 25 lines used in MLO (tables 7.1 & 7.2) and not used with the 42 lines in the 1980 analysis.

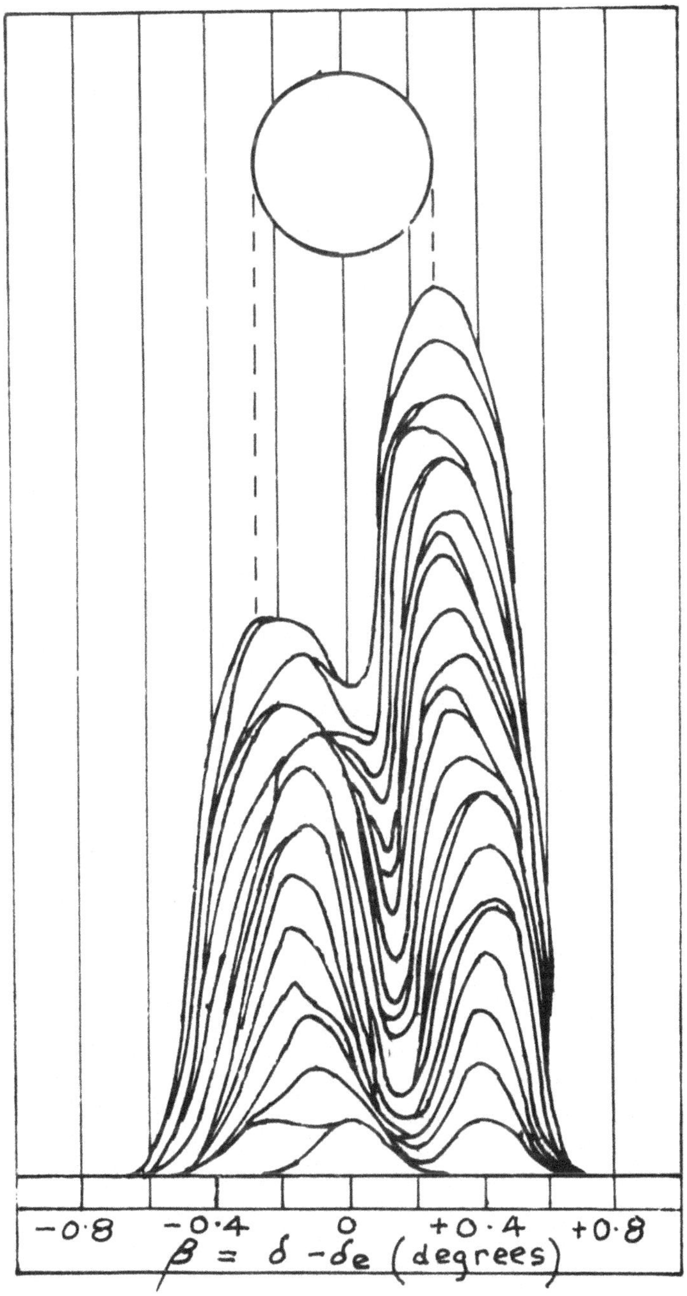

Fig. 1.15 β for 42 lines used in the 1980 analysis.

1.32 Discussion on dating

If we assume that at any one observatory the observations were spread over 3 or 4 standstills, then what cannot have happened is that at one observatory the 3 or 4 were centred (for example) on 1600 BC and at another on 1500 BC; for this would give a difference of about 8 minutes in the declination on fig. 1.13, which on the evidence of our results is unallowable.

If by some other means we knew the dates at which the various centres were operative it would make matters much clearer. In the meantime we can say that the scatter of the observations is so small that they cannot have been made individually at single standstills. The observers probably took the mean position, on the ground, of a number of observations made at standstills throughout a period of perhaps 150 years (see fig. 1.13).

1.33 Rising versus setting lines

There is another point which strengthens the case that we are dealing with intended observing positions. It is obviously much more easy to observe the setting than the rising Moon. The setting Moon can be followed along the horizon until it vanishes, whereas the rising Moon appears suddenly. To overcome this at Brogar, special mounds (Salt Knowe and Mound A) were erected so that a watcher could be placed to give warning to the observers that the Moon was about to emerge (MRBB: 128). Similarly the special row of boulders on the hill behind Mid Clyth served the same purpose (MLO: 94). Nevertheless the setting Moon was preferred. Witness that we find in the list used here that there are 28 setting and only 14 rising examples. We can either accept the possibility that this peculiar distribution is the result of chance (probability about 1 in 45) or we can accept that Megalithic Man really was observing the Moon.

1.34 Warning stances

A warning of the imminent appearance of the rising orb behind a sloping horizon, for example, can be given on flat ground by another observer standing nearby. If the ground rises behind the observer, the helper need only be above the backsight position to give the warning. Was this the purpose of the stone above the observing platform at Kintraw (MLO: 40)?

The above idea of using the help of a warner would indicate the ability of the observers to pin-point the upper limb of the Moon without having to wait until a fairly large segment had risen. At sunset in the clearer skies of Megalithic times the observers would almost certainly use the green flash as a criterion that they were in the correct position.

1.35 Date of Brogar ring

So far no reliable solar line has been found near to Brogar to enable us to date the ring using the Sun's declination. People in the locality have pointed out to us two or three places where they had either seen a standing stone years ago or had been told of its previous existence (Odin Stone, neighbour to Watch Stone, etc.). There are other stones in the district.

The difficulties of accurate dating by the Moon have been discussed above (sections 1.28-1.32), and the best date for the use of Brogar as a lunar observatory is 1485 BC ± 100 years.

The ditch round the Ring of Brogar has been excavated from the rock (Renfrew 1979: 42) but no conclusive evidence has yet been discovered to help in dating the monument. Determinations of dating, obtained by Dr. Graham Ritchie (1975: 50) during his work in the ditch cut in the rock round the Stones of Stenness – the nearby sister monument, yield dates from the organic level at ditch bottom of 2356±65 bc and from charcoal from the central feature, 2238±70 bc. Archaeological evidence indicates therefore that Stenness—the Brogar by its proximity, were constructed late in the third millennium bc, i.e. early in the third millennium BC.

From the above two dates it would appear possible that in the beginning the Brogar site was chosen so that from within the ring the three or possibly four foresights showed the Moon rising and setting at standstills, but without the fine accuracy which was developed later.

1.36 Astronomical discussion

Over the years, as we have surveyed more and more sites, the subject was opened up and our technique of presentation has expanded and developed. This is why the latest histogram of lunar data, for example, differs from previous ones published (see section 1.40).

We must avoid laying down the law as to the kind of foresight (or backsight) which Megalithic Man may have used. For this we are 4,000 years too late and we shall never obtain a fraction of his experience of choosing and using the lines he established. We must learn from him by examining every example we can find and we now have information about a sufficient number to allow us to judge how the sites were used.

There is no definite connection between the actual date of the standstill and the date of the observed maximum declination. The date of the standstill may be taken as the maximum of the chain-dotted curve on fig. 1.2, occurring when the node is at the First Point of Aries.

A method of extrapolating (sections 1.25 & 1.26) was absolutely essential.

The small r.m.s. residual (1'.34) calculated for (β - Q), with $c_s = 0$, leads us to the conclusion that their observations were very carefully made and recorded. This has emerged from calculations based on our precise surveying. Megalithic Man's careful work must have been spread over a long period.

1.37 Sites with more than one line

There emerges a convincing argument that solar and lunar observatories existed, when we find at a given observatory that there were several foresights in use, all seen from closely adjacent backsights. Included in this category are (for the Moon) Stonehenge, Temple Wood and Brogar; and (for the Sun) Ballochroy. At Kerlescan (Carnac in Brittany) stone S is a backsight for both a lunar and a solar line (Thom & Thom 1971: 152).

1.38 Foresight with several backsights

The summit of Boreray, the most northerly island of the St. Kilda archipelago in the North Atlantic, is used as a foresight for the setting Sun from four backsights; namely one each on South Uist, Benbecula, North Uist and Harris (MSB: 130). All of these lines give important calendar declinations.

In the Carnac area, both the Manio Stone and Grand Menhir Brisé were used as lunar foresights with many backsights (MRBB: ch. 9).

1.39 Warning positions

Outside the Brogar ring, the Comet Stone clearly indicates the notch on Mid Hill. It gradually became clear that the small mounds were indicators of other lunar lines but it was not at first evident why mound A and Salt Knowe had been constructed. The rising Moon appears suddenly and we believe that both mounds were erected so that a watcher could be placed to give warning of the Moon's imminent appearance.

At the time when the Moon was expected to rise it was necessary to have some warning so that several observers placed along the line of their movement could be prepared for the emergence of the upper limb. Since the horizon at Mid Hill rises towards the south, this meant that a man watching should have been stationed to the north-east. But here the site is limited by Loch Harray and to compensate it was necessary to raise the watcher's eye level. This explains the large mound at A placed as far as possible to the north-east and as high as necessary. In partial darkness it was desirable to have an indication of the direction in which the small notch lay on Mid Hill. There could have been poles on both mounds A and B on the line of sight from A to the notch.

For Kame, Salt Knowe served the same purpose, placed as it was well to the left; but not too far to the left or it would have been on lower ground.

The presence of these mounds greatly strengthens the case that we are dealing with intended observing positions.

1.40 A revised way of presenting β

Although initially, in 1967, the histogram (fig. 1.5) of observed declinations had been plotted as plus or minus $\beta = \delta - \delta_e$, the $\pm\ \beta$ values later came to be plotted numerically irrespective of sign: for example in 1971 (MLO: fig. 7.1), when the 40 lines then relevant were collected together. Since 1980 however, the initial format has again been employed as our knowledge of the variations of Q with Δ and s developed (Thom 1980a; Thom & Thom 1980b). The need to explain the small values of r resulting from the analysis demanded separation of the plus and minus β angles.

1.41 General discussion

Until recently it was unthinkable that our rude forefathers, living 120 generations ago in small isolated communities, illiterate peasant farmers and hunters, could possess the brain power and skills necessary to support the observatorial and geometrical techniques attributed to them. The evidence we put forward here supports our thesis, evidence collected from Brittany to Shetland and from Stonehenge to the Hebrides. No one who has not studied our work is in a fit position to pronounce at all upon the astronomical, geometrical and civil engineering achievements of Megalithic Man.

Our thesis is that north-west Europe was inhabited by people some 4,000 years ago who had a basic knowledge of applied mathematics and astronomy. The intelligence required is of the same order as that needed to develop a first class language. This survey of megalithic astronomy is written in English and to do so we must have a working knowledge of its vocabulary, grammar and syntax. Classicists maintain that the most perfect language of all is probably ancient Greek, which had a complex set of rules. Greek existed 3,000 years ago. From where did it come? How long did it take to develop? We cannot answer this question. We infer however that Megalithic Man was able to develop in the same way a working knowledge of elementary astronomy. Why should his intellectual abilities be any less than those of the early Greeks or their predecessors who left the Greek language?

The description of Megalithic Man's lunar observations is of necessity extremely complicated because the motion of the Moon is so intricate. We now know the precise movements of the Moon; Megalithic Man could observe its movements on the horizon and only at each maximum or minimum turning point of azimuth were his observations usable. It is extremely unlikely that he used the concept of declinations as we now know it. He could only record and observe horizon azimuths accurately at rising or setting.

It is highly probable that Megalithic Man knew how to predict eclipses and that the purpose of his observations was to correct his lunar count, that is the number of days between possible eclipses. He must have been able to count. His observations certainly pinpointed the major standstills which included the perturbation at the very maximum. There is no reason to believe other than that Megalithic Man knew of the perturbation Δ. By observations made 87 days apart he would have been able to make a fair assessment of its value, at least in its effect on the azimuth of the Moon on the horizon; at 14 sites (see table 1.1) no perturbation appears in the nominal declination and so we reason that he was using the means of observations made 86 days apart at crest and trough of the wave (see for example fig. 1.2).

Men in prehistoric times lived mostly in the open air, and seeing and remembering natural phenomena was part and parcel of their very existence. Obviously they were extremely conscious of their environment. The Moon itself was by night an important lamp which enabled night-time journeys to be made much more quickly. Knowledge of the lunar cycle is obviously common to all mankind. Why should prehistoric man in 2000 BC not have been able to devise some means of making a precise calendar to predict and pin-point future movements of the Moon, including eclipses?

It is almost a certainty that in the minds of these people, the tidal cycle would be connected with the lunar cycle. Safe movement by sea along the shores of continental Europe depended on navigators having an accurate knowledge of tidal ebb and flow (MLO: 9-11). Did this supply a secondary reason for spending the time and effort needed to study and record the Moon's movements? The observatories were not however erected to predict tides.

Much time and energy was obviously expended on the initial observing necessary for the choice of place and then for the building of a megalithic site. Was the incentive purely scientific curiosity and were these people recording their knowledge, by the stone positions, simply for interest? Scientific curiosity is a human trait and the human intellect can not have altered greatly in the 120 generations since these prehistoric observatories were erected.

Living standards must have been sufficiently high to enable the expenditure of such a large amount of energy on making the observations and on erecting the sites. Were there "astronomer priests" with sufficient power to have these hundreds of stones erected (MacKie 1977)?

A calendar for prediction of seasonal variation is a necessity for any efficiently run society, either nomadic, savage or agrarian; and there is strong evidence that a solar calendar was used with 8 and probably 16 "months" per year. The activities of the "astronomer priests" certainly allowed them to pin-point what we now call the lunar perturbation maxima at major and minor standstills, and it is almost certain that they had noticed the occurrence of eclipses some 7 days earlier or later.

No decipherable records remain. The only evidence is contained in the actual positioning of the stones, markers or backsights. Are the cup and ring markings often found on standing stones or on flat rock surfaces nearby (Morris 1977; 1979) a method of writing or recording?

Our prehistoric ancestors were certainly our equals in intellectual ability. Men of great capacity would come to the fore in every generation. The fact that they did not have the use of pen and paper would not have prevented them from thinking, speculating and expounding.

How much did Megalithic Man know about the solar/lunar system? We firmly believe that he knew that the Moon went round the Earth. From high ground on the island of Unst, the most northerly of the Shetland Islands (Thom & Merritt 1978), the Moon's visible path at the time of major standstill is such that it almost does not set. It is then almost circumpolar. Some navigators would have sailed further north and would have reported that the Moon sometimes never sets. In the same way, knowledge of the "land of midnight Sun" would come south to the north-west Europeans and we believe that Megalithic Man was capable of reasoning that the Sun went round the Earth and that the Earth was spherical. He could have deduced this from two points of view: firstly from reasoning by observing the shadow of the Earth on the Moon, and secondly from observing the curvature of the sea's surface, a phenomenon obvious to thinking men observing the horizon during a voyage or while climbing a hillside and observing more and more islands (and so on) coming into view.

Only half of all lunar eclipses are visible from one position on Earth, but on this count alone, about 22 eclipses occur during each Saros period of 18.6 years. Did Megalithic Man's recording system include solar as well as lunar eclipses? This we can not be so sure of; fewer solar eclipses occur at any one position on Earth but there is no reason to believe other than that they could have predicted the possibility of solar eclipses, because these, like lunar eclipses, all occur about 7 days before or after the perturbation maximum or minimum changing points at major (or minor) standstills.

It would be evident that the Sun at its eclipse was being hidden by the Moon; the curved limb of the Moon would be seen against the Sun.

We have searched for the scientific, mathematical or practical aspect rather than for the ritual astronomical evidence, for the accurate alignments rather than for the orientations in the standing stone sites; as engineers what else are we able to do? In the absence of written records, the only evidence is in the positioning of the stones themselves, and we consequently measure them up as carefully as possible. We do not excavate, and only very rarely are we invited to see the "bones" of an excavated sight. Our work has consisted in making accurate plans of visible stones which appear to form a geometrical shape or to have some kind of astronomical significance. Our findings are that many sites were scientifically laid out, indicating the existence, at the end of the epoch, of a precise scientific astronomy.

The requirements of biological survival have little in common with those which produce ethical and mathematical development. Man must go beyond biological evolution if he is to realise his full potential, and progress depends on improving the evolutionary process with an ethical system, and with the science of magnitude and number.

No problem existed in the handing down of information from generation to generation without the aide-memoire of modern writing. For the continuance of this part of the culture over more than several centuries there must have been in power a number of persons possessing enquiring minds and interested in their environment from the purely scientific viewpoint.

The archaeologist can deal with these elements of culture only with great difficulty. Likewise the archaeoastronomer initially faces only the material remains of the past. At first sight not much science and mathematics would appear to exist in a few standing stones yet overall there are many hundreds of signs of consciousness of astronomical phenomena. The question of exactly what such a knowledge of astronomy actually meant to its practitioners remains unanswered.

There has been much comment upon and criticism of our work. Many points that have been raised have been dealt with in this paper. A few more miscellaneous points are covered in appendix 1.5.

1.42 Conclusions

Using the 42 lunar lines we have shown that the backsights were not erected as a result of measurements made at a single standstill. It seems likely that the work leading to the erection of a backsight was spread over about 150 years. Does this explain why many of the lunar backsights consist of huge stones erected to last for generations?

The mean obliquity of the ecliptic appears to be about $23^{o}53'.1$ which would indicate a date of 1590 BC ± 100. The three good solar solstitial sites discussed in 1971 (MLO: ch. 4) give 1750 BC ± 100.

It now appears certain that by the beginning of the second millennium BC Megalithic Man recorded not only the larger lunar movements but also the 9 arc minute perturbation of the Moon's orbit. Continuous observing would be taking place and eclipses could have been forecast. We are convinced that apart from any practical use, Megalithic Man was scientifically interested in the relative movements of Sun, Moon and Earth.

We place strong emphasis on the smallness of the residuals in table 1.1. Other investigators have been glad to get an agreement (of β) of about one degree in declination, but we find the residuals are so small that their root mean square does not approach 2 arc minutes: in table 1.1 it is about 1.4 arc minutes, the biggest residual being 3.0 arc minutes.

Attempts to analyse the results statistically indicate the extremely low probability of the 42 lunar lines having occurred by chance.

We shall in all probability never know why the megalith builders expended so much energy in building so many dolmens, cairns, mounds, rings, alignments etc. Few other prehistoric cultures in Europe have left so much massive material behind. We believe that, over the period 3000 BC to 1500 BC, they left considerable evidence of their interest in the Sun and Moon; evidence remaining in the accurate positioning and orientation of many of their monuments, which occurs especially towards the end of the period, 1500 BC.

ACKNOWLEDGEMENT

Since 1939 the author has helped Professor Thom periodically with the field work involved. Often the author has acted as a sounding board for discussions on the theories put forward, and he has helped considerably in the compiling, writing and checking of books, reports, etc. He thanks Professor Thom for helping to compile this paper and for reading over the manuscript.

REFERENCES

MacKie, E. W., 1974. Archaeological tests on supposed prehistoric astronomical sites in Scotland. Phil. Trans. R. Soc. London, A276, 169-194.

MacKie, E. W., 1977. Science and society in prehistoric Britain. London: Paul Elek.

Moir, G., 1980a. Megalithic science and some Scottish site plans: Part I. Antiquity, 54, 37-40.

Moir, G., 1980b. A review of megalithic lunar lines. Northern Arch., 1, 14-22.

Morris, R. W. B., 1977. Prehistoric rock art of Argyll. Poole: Dolphin Press.

Morris, R. W. B., 1979. Prehistoric rock art of Galloway and the Isle of Man. Poole: Blandford Press.

Renfrew, C., 1979. Investigations in Orkney. London: Society of Antiquaries.

Ritchie, J. N. G., 1975. The stones of Stenness, Orkney. Proc. Soc. Ant. Scotl., 107, 1-60.

Roslund, C., 1979. Ale—forntidsimatematiker och astronom? Forskning och Framsteg, 5, 6-11.

Ruggles, C. L. N. & Norris, R. P., 1980. Megalithic science and some Scottish site plans: Part II. Antiquity, 54, 40-43.

Thom, A., 1954a. Some refraction measurements at low altitude. J. Inst. Navigation, 7(3), 301-304.

Thom, A., 1958. An empirical investigation of atmospheric refraction. Empire Survey Review, 14, 248-262.

Thom, A., 1967 ("MSB"). Megalithic sites in Britain. Oxford: Oxford University Press.

Thom, A., 1971 ("MLO"). Megalithic lunar observatories. Oxford: Oxford University Press.

Thom, A., 1974a. A megalithic lunar observatory in Argyll. J. Hist. Ast., 5, 50-51.

Thom, A. S., 1980a. Solar and lunar observatories of Megalithic Man. In Archaeoastronomy and the roots of science, 146th A.A.A.S. national meeting on "Science, our common heritage". San Francisco, in press.

Thom, A. & Merritt, R. L., 1978. Some megalithic sites in Shetland. J. Hist. Ast., 9, 54-60.

Thom, A. & Thom, A. S., 1971. The astronomical significance of the large Carnac menhirs. J. Hist. Ast., 2, 147-160.

Thom, A. & Thom, A. S., 1978a ("MRBB"). Megalithic remains in Britain and Brittany. Oxford: Oxford University Press.

Thom, A. & Thom, A. S., 1978b. A reconsideration of the lunar sites in Britain. J. Hist. Ast., 9, 170-179.

Thom, A. & Thom, A. S., 1979a. Another lunar site in Kintyre. Archaeoastronomy (supplement to J. Hist. Ast.), 1, 96-97.

Thom, A. & Thom, A. S., 1979b. The standing stones in Argyllshire. Glasgow Arch. J., 6, 5-10.

Thom, A. & Thom, A. S., 1980a. Astronomical foresights used by Megalithic Man. Archaeoastronomy (supplement to J. Hist. Ast.), 2, 90-94.

Thom, A. & Thom, A. S., 1980b. A new study of all megalithic lunar lines. Archaeoastronomy (supplement to J. Hist. Ast.), 2, 78-89.

Wood, J. E., 1978. Sun, moon and standing stones. Oxford: Oxford University Press.

FURTHER BIBLIOGRAPHY

Thom, A., 1954b. The solar observatories of Megalithic Man. J. Brit. Ast. Assoc., 64, 396-404.

Thom, A., 1966. Megalithic astronomy: indications in standing stones. Vistas in Ast., 7, 1-57.

Thom, A., 1969. The lunar observatories of Megalithic Man. Vistas in Ast., 11, 1-29.

Thom, A. & Thom, A. S., 1972. The uses of thealignments at Le Ménec, Carnac. J. Hist. Ast., 3, 151-164.

Thom, A. & Thom, A. S., 1973. A megalithic lunar observatory in Orkney: the ring of Brogar and its cairns. J. Hist. Ast., 4, 111-123.

Thom, A., Thom, A. S., Merritt, R. L. & Merritt, A. L., 1973. The astronomical significance of the Crucuno stone rectangle. Curr. Anthrop., 14, 450-454.

Thom, A., 1974b. Astronomical significance of prehistoric monuments in Western Europe. Phil. Trans. R. Soc. Lond., A276, 149-156.

Thom, A., Thom, A. S. & Thom, A. Strang, 1974. Stonehenge. J. Hist. Ast., 5, 71-90.

Thom, A., Thom, A. S. & Thom, A. Strang, 1975. Stonehenge as a possible lunar observatory. J. Hist. Ast., 6, 19-30.

Thom, A. & Thom, A. S., 1975. Further work on the Brogar lunar observatory. J. Hist. Ast., 6, 100-114.

Thom, A., Thom, A. S. & Gorrie, J. M., 1976. The two megalithic lunar observatories at Carnac. J. Hist. Ast., 7, 11-26.

Thom, A. S. & Foord, T. R., 1977. The island of Eday. J. Hist. Ast., 8, 198-199.

Thom, A. & Thom, A. S., 1977. "Megalithic astronomy" (the Duke of Edinburgh lecture). J. Navig., 30, 1-14.

Thom, A. & Thom, A. S., 1979c. Rings and menhirs: geometry and astronomy in the neolithic age. In E. C. Krupp (ed.), In Search of Ancient Astronomies. London: Chatto & Windus.

Thom, A. S., 1980b. A solstitial site near Peterborough? Archaeoastronomy (supplement to J. Hist. Ast.), 2, 95.

Thom, A., Thom, A. S. & Burl, A., 1980. Megalithic rings: plans and data for 229 monuments in Britain. Oxford: B.A.R. British Series 81.

APPENDICES

APPENDIX 1.1

List of 25 sites not used in this analysis

In 1967 (MSB: table 8.1) there were 40 lunar lines which seemed worthy of consideration. Three of these were unsuitable for inclusion in the table (MSB: table 10.1) of observed declinations assumed to be lunar—see table 1.3 here. The remaining 37 declinations were given to within 0°.1 and were used to plot the histogram shown in fig. 1.6 (MSB: fig. 10.1). In that histogram any values of β numerically greater than 1°.1 were excluded. Declinations within the band of ± 1°.0 were looked for and selected. Many of the sites were thereafter re-surveyed and the analysis followed in 1971 as detailed in MLO (tables 7.1 & 7.2). A total of 52 declinations are listed there, 40 of which are given to within 0'.1. During this stage in the analysis, 28 lines were discarded from the 1967 group. The total number of lines which had been examined in the lunar bracket up to this stage was 80.

Since 1971, a further 15 lines have been added (see table 1.3), namely Brogar (9), Stenness (1), Dunskeig (1) (Thom & Thom 1978b: 173), Skipness (1) (Thom & Thom 1979a), Wormadale Hill in Shetland (1) (MRBB: 166), and Ballinaby, Islay (2) (Thom 1974a). The total number of lines examined to date is now 95, and in this paper we use 42 of them, giving our reasons for inclusion or rejection throughout. Generally (see table 1.4) we omit all sites where the position of the backsight is not sufficiently accurately known, where map work is unacceptable, where woods between backsight and notch make notch azimuths extremely difficult to measure and so on. We have been criticised for including Blakeley Moss (MLO: 71-72) as a backsight. This ring was reconstructed in 1925, when 8 of the original 13 stones were missing and the remaining 5 were re-erected. We cannot be absolutely certain that there was a prehistoric ring there; with 5 stones lying in a curve however, the re-erector would have a good idea of the required diameter and presumably found the base end of each stone and its stone hole before putting it upright. The distance to Screel Hill, the foresight, is 30.7 miles (49.1 km). A movement of half of the circle diameter alters the azimuth of Screel Hill by 0'.57 and therefore the observed declination by about 0'.17, which is sufficiently accurate. Screel Hill was seen from the ring but the Sun did not appear and so the azimuth and altitude were taken from the one-inch Ordnance Survey map.

Table 1.3 Lunar declinations discussed in MSB & MLO with 15 more lines added

By	New lines	Total lines to date	Unaccept-able lines	Total of lines unacceptable to date	Lines used in MSB, MLO or here	Notes
1967	37	37	None	None	37	MSB: fig. 10.1
1971	43	80	28	28	52	MLO: tables 7.1 & 7.2
1980	15	95	25	53	42	Here: figs. 1.14, 1.8 & 1.15.

Table 1.4 List of 25 of the total of 52 lunar lines discussed in MLO, with reasons why each of the 25 lines is unacceptable for the present analysis. Figure numbers refer to MLO

Site	Name	Reason
H3/11	Leacach an Tigh Cloiche (fig. 6.15), A_2	Map work only
H1/1	Callanish I	Map work only; unsure of hill behind the hollow
M1/3	Quinish, A	Survey too hurried; no Sun
H1/5	Callanish V, to N.W.	Position of backsight unknown
H1/5	Callanish V (fig. 6.14), A_2	Only foresight A_1 admissible because stones indicate Mor Monach
H1/5	Callanish V, A_4	Only foresight A_1 admissible because stones indicate Mor Monach
N1/17	Dirlot (fig. 9.5), A_1	Unsure of backsight
N1/17	Dirlot, A_2	Unsure of backsight
H3/11	Leacach an Tigh Cloiche, A_1	See above
A4/1	Escart (fig. 6.1), A	Survey uncertain; trees
A2/5	Kintraw (fig. 4.4), A_2	Ambiguous notch
G3/2	Haggstone Moor (fig. 6.17)	Uncertainty of backsight
A4/17	Gigha (fig. 6.5), A_1	Foresight uncertain
P4/1	Lundin Links (fig. 5.5)	Map work only for Cormie Hill
A2/13	Dunadd (fig. 6.8), B	No indication
A6/1	Camus an Stacca (fig. 6.9), B	Too many notches; uncertain
A4/9	Crois Mhic-Aoida (fig. 5.6), A_1	Line uncertain
G3/2	Haggstone Moor (fig. 6.17), A_1	See above
A6/1	Camus an Stacca, B	Too many notches; uncertain
M1/3	Quinish, B	See above
G3/2	Haggstone Moor (fig. 6.17)	Uncertainty of backsight
G3/2	Haggstone Moor (fig. 6.17)	Uncertainty of backsight
P1/10	Fowlis Wester (fig. 5.4), B	No foresight on near horizon
A2/8	Temple Wood (fig. 5.1), B	Azimuth uncertain; surveying accuracy marred by two plantations
G9/13	Kell Burn (MLO:70)	Short line; no foresight

APPENDIX 1.2

Effective perturbation: adjustment c_s applicable to Δ

Certain adjustments c_s to Δ at M (fig. 1.2) are necessary, for example, at the time of major standstill in order to allow evaluations of (β-Q) for analysis of the results.

Unless we know the date we cannot tell how far the observation was from the top of the chain-dotted curve, and so we cannot calculate the amount by which the declination was low.

Since the top of the perturbation wobble (of period 173.3 days) shown near M does not in general happen at the top of the long-period oscillation of $\pm(\varepsilon \pm i)$, the crest of the wave can be before (or after) the top by 86.6 days. It has been shown (MRBB: 10) that the discrepancy due to curvature might be as large as c_1= 1'.0. To take the time mean over the 86.6-day interval appears to be the only thing to do and this gives c_1= 0'.3. Thus the mean value of c_1, i.e. 0'.3, has for example to be subtracted from the Q values already mentioned in order to obtain the maximum positive declination at the major standstill.

Similarly, looking at the black dots on fig. 1.2 (the declinations able to be seen and recorded by the observers), the highest one (at M for example) may be as much as half a lunation before (or after) the top of the wobble. The mean correction for this is also about c_2= 0'.3. When we are taking the mean of the top and bottom of the wobble (March to June or June to December etc.), c_2 cancels itself, leaving the total correction c_s equal to c_1. At other times c_2= 0'.3.

Corrections c_1 and c_2 have to be applied to Q in the correct sense at major and minor standstills and for N and S declinations. Q applies to the declination and ideally, Q = β. Relevant semi-diameters s are applied. Q is the theoretical value of β found by the correct combination of Δ the perturbation and s the semi-diameter. See fig. 1.2 for Q- values.

APPENDIX 1.3

How the sites were used: two examples

(a) Ballinaby, major standstill. Here we find a tall slender stone perhaps 17ft (5 m) high. At the base it is roughly 3ft by 9 inches (1 m x 0.2 m) and the long sides are oriented on the col at B, 1 1/4 miles (2 km) away (fig. 1.10). The col B gives a declination of about (ε + i + s), and since there is no Δ in this the stone must show the middle position between the equinoctial and the solstitial stances. In table 1.1 there are 14 sites where no Δ appears. We believe that Megalithic Man finally knew about the Δ, and that the backsights at these stations averaged two observing stations found half of 173 days apart and embracing two of the three turning points on the wobble curve nearest the (ε + i) maximum. Thus they could have averaged a June and a September pair, some 87 days apart: see fig. 1.2, top left.

We estimate that the Ballinaby site was used at the spring equinox about 4 a.m. and at the summer solstice about 10 p.m. when the temperatures are about 40° and 50° F (5-10°C). A sample calculation for these times is given by Thom & Thom (1978b). Accordingly we used a refraction of 32'.4 and a parallax of 57'.2. These give, with the values of azimuth and altitude shown on the figure, a declination of 29°17'.7. Deducting a

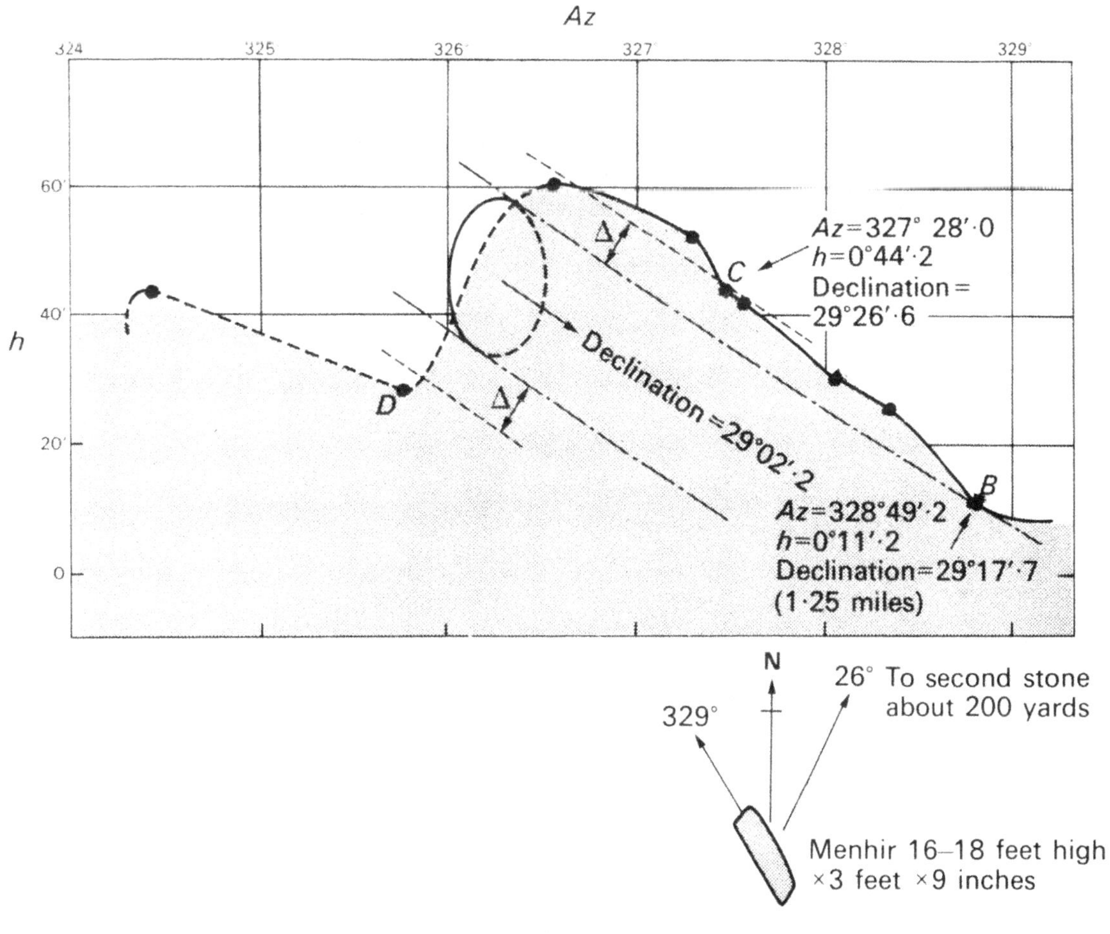

Fig. 1.10 The menhir at Ballinaby, Moon setting with extreme $\delta = (\varepsilon + i) = 29°02'.2 \pm 0'.5$

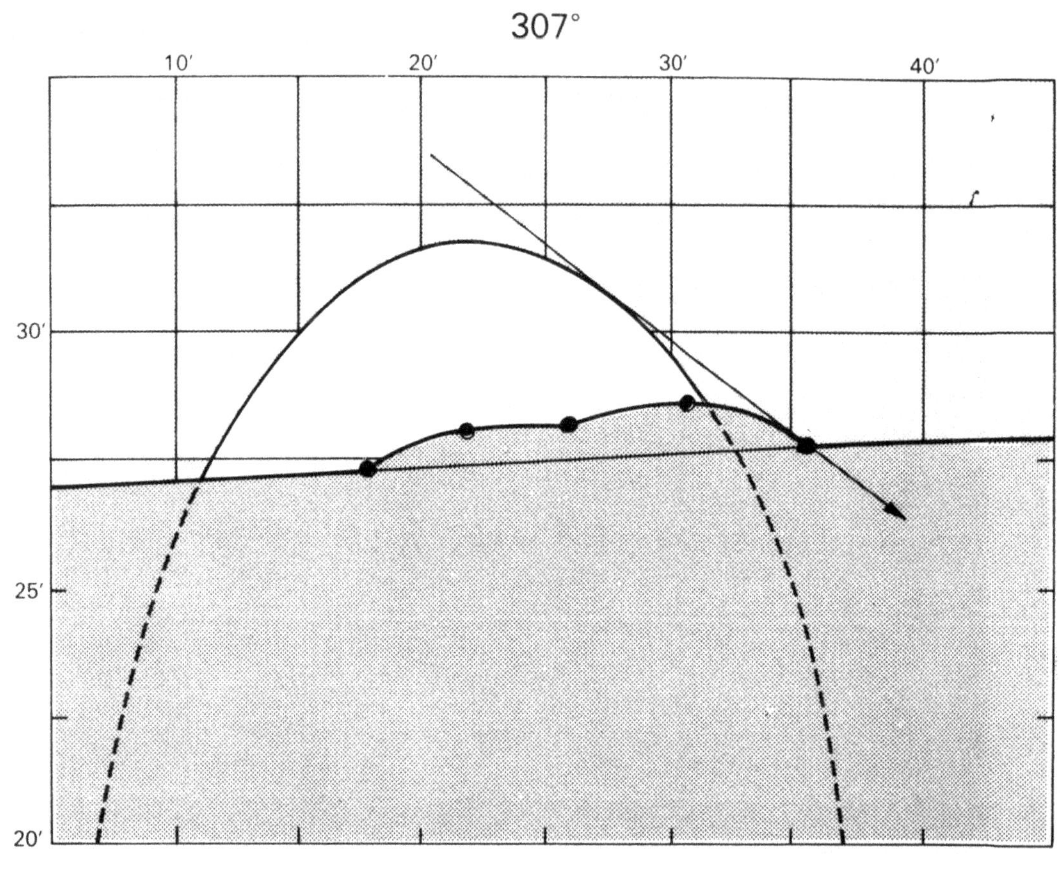

Fig. 1.11 Moon setting at minor standstill on cairn as seen from Stenness circle with declination ($\varepsilon - i + \Delta$)

semidiameter of 15'.5 and i, gave $\varepsilon = 23°53'.5$. The above declination value (MRBB: 169) has been revised slightly since 1976, as described in section 1.14. The same remark applies for the declination of upper notch C.

There is ample room for side movement when observing, and altogether this must be looked on as one of the most convincing single stone observing sites in Britain. The midpoint of the wobble was recorded by the stone using notch B. Observed from a stance some distance to the left of the stone the notch would show $\delta = +(\varepsilon + i + \Delta + s)$; from a stance some distance to the right of the stone the notch would show $\delta = +(\varepsilon + i - \Delta + s)$. The mean of the two stances, the stone's position, shows $(\varepsilon + i + s)$. The recorders would know about the wobble and would be able to determine the requisite June maximum occurring in the trough.

The observers here, as in many other sites, were interested in the mid point of the wobble using notch B and in the crest of the wobble using notch C (fig. 1.10). This is better seen at several other sites where there are foresights for the crest and the trough or the mean of the wobble; see table 1.1 for Campbeltown, Callanish V, Mid Clyth, etc.

From the stone, the notch at C was indicating the crest of the wobble $(\varepsilon + i + \Delta + s)$. For each notch, B and C, the Moon had set and reappeared, making Ballinaby an excellent observing stance. If they missed the crest due to bad weather their observation of the trough would still be usable.

(b) <u>Stenness, minor standstill</u>. Near the Ring of Brogar is the Ring of Stenness, from the centre of which the declination of the right hand side of a prehistoric mound or cairn is found to be 19°08'.4 (MRBB: 130). This $(\varepsilon - i + \Delta + s)$ value gives for ε a figure almost identical with that obtained from the north declinations at Brogar. The reason for having this line at Stenness instead of at Brogar was that there was no natural foresight for the $(\varepsilon - i)$ minor standstill line at Brogar and no suitable position where an artificial foresight could have been erected, whereas from Stenness the cairn is on the horizon, as seen in fig. 1.11. The nominal declination indicated by the line from the Ring centre to the cairn is $+(\varepsilon - i + \Delta + s)$.

In fig. 1.12 we show how the upper limb of the Moon would set near minor standstill. As shown in fig. 1.2, the ecliptic limits occur in March and September at quarters at the troughs of the perturbation wave. At this Stenness site the measured declination indicates a nominal value of $+(\varepsilon - i + \Delta + s)$, which occurs near Full Moon in June or December at a wave crest. About a dozen maxima, spaced about 28 days apart, are shown in fig. 1.12.

From the straightforward requirement of fixing the lowest maximum (as at N in fig. 1.2) for checking the lunar eclipse count, it would appear superfluous to have the above Stenness line. We can only infer that the observers were putting on record their observations involving the perturbation wobble. How many more sites have yet to be discovered?

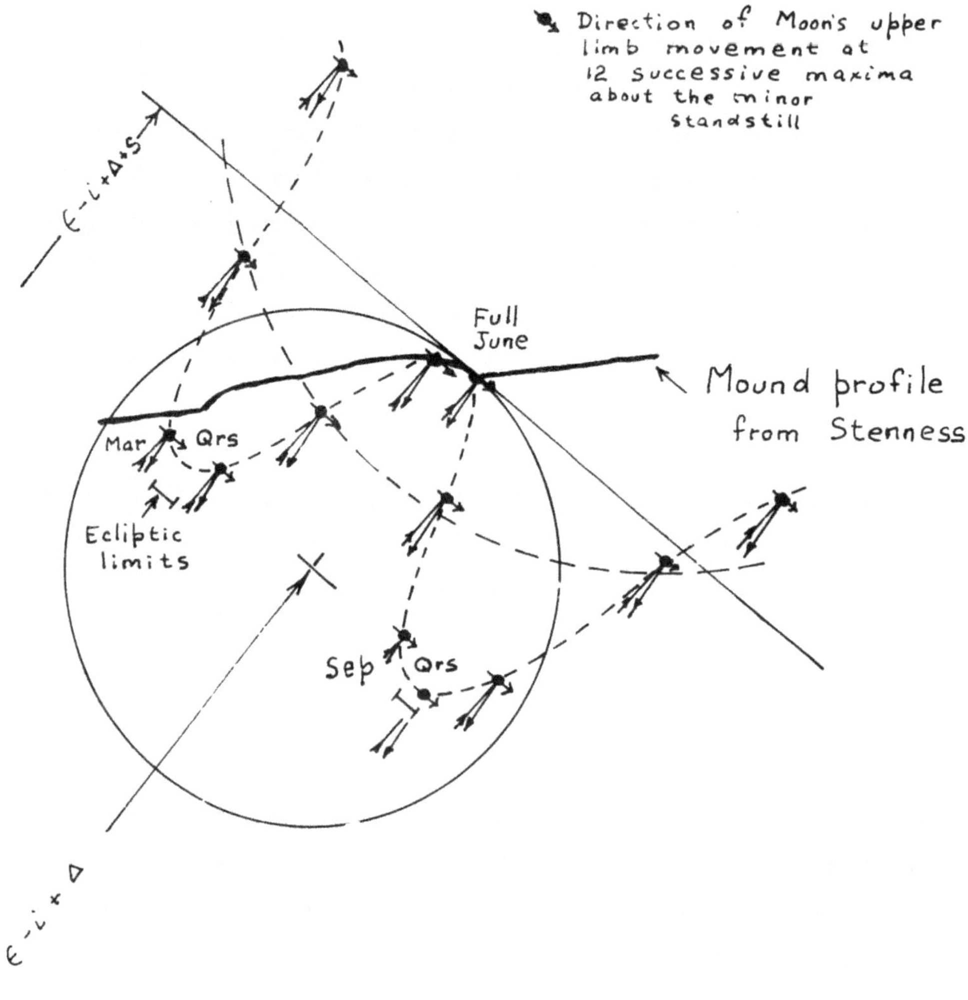

Fig. 1.12 Setting full Moon at Stenness, minor standstill; Moon's declination at crest of perturbation wave is $\delta = (\varepsilon - i + \Delta)$.

APPENDIX 1.4

The perturbation Δ, applicable to i

Values of Δ have been calculated from

$$\Delta = 519'' \cos 2(\lambda' - \Omega) - 42'' \cos 2(\lambda - \lambda') + 39'' \cos 2(\lambda - \Omega)$$

where

λ = longitude of the Moon

λ' = longitude of the Sun,

and Ω = longitude of the node.

(The above formula and values differ from the values given in MRBB and were provided in a private communication by L. V. Morrison of The Royal Greenwich Observatory.) Longitude (celestial) is measured along the ecliptic anti-clockwise from the First Point of Aries. At a major standstill, with the Moon at its extreme declination, longitude of the node is 0° or 180° and longitude of the Moon is 90° for N declination (+) or 270° for S declination (-) (MRBB: 9).

Only when the node is in conjunction with or in opposition to the Sun can the Moon's monthly maximum declination attain its greatest values; these are separated by 173.3 days, i.e. half of an eclipse year. When the node is near the First Point of Aries the Sun's longitude is therefore near to 0° or 180°. Thus for a maximum declination we must have

(1) Moon's longitude near 90° or 270° and
(2) Sun's longitude near 0° or 180°.

As a result of (1) the Moon is at its first or third quarter when the maximum occurs, and as a result of (2) the maximum declination occurs near either the vernal or the autumnal equinox. We should probably choose the Sun's longitude which gives the phenomenon in darkness. The turning points are clearly shown in the values for 1969 plotted in fig. 1.3, where the three consecutive maxima occurred at the equinoxes at the end of September or the end of March, whereas the minima happened at the end of December or the end of June, i.e. at the solstices. Thus the relevant turning points, maxima or minima, never occur at new or full Moon but always when there is a reasonable amount of the disc illuminated.

At the equinoxes we thus find the Moon at first or third quarter with $\Delta = 8'.6$, and at the solstices it is full or new with $\Delta = -10'.0$.

When talking about the hollows in the declination wave, $(\varepsilon + i - \Delta)$, the longitude of the Moon is, as before, + 90° or 270°; but the longitude of the Sun is now also 90° or 270°.

Every standstill is different in detail but each has the wobble coming up to practically the same maximum declination. In general, Megalithic Man would miss the absolute maximum, as this probably occurred in between the two observing times 173 days apart. We have to consider this in our analysis.

APPENDIX 1.5

Response to some miscellaneous criticisms

(i). Refraction. Different weather conditions from hour to hour and from day to day give different terrestrial and astronomical refraction. Present day phenomena would also be seen by Megalithic Man and he would know of the different apparent altitudes, to him inexplicable, seen from day to day when a distant hill top (for example) might sometimes be visible, sometimes not, over a nearer ridge. Conditions similar to those which cause what is called "Arctic mirage" can occur in these latitudes. For us, in calculating dates of erection of the monuments from declinations obtained using observed apparent altitudes, it is not in general possible to estimate with any accuracy the value of the coefficient of refraction for altitudes near to, or lower than, zero apparent altitude. The reader is referred to A. Thom's recordings of empirical measurements of refraction (Thom 1954a; 1958), the results of some 600 observations of the angular altitudes of a number of hill tops, etc., from one station. At Kintraw (MLO: 42), the efficacy of which site has been queried, the observed altitude is well above the horizon, being +26.1 arc minutes. Tables of refraction give the value of 28.3 arc minutes and at this observed altitude the rate of change of refraction is 0.18 arc minutes per minute of apparent altitude. We claim to be able to read vertical angles to within ± 2 seconds of arc and so any uncertainty in dating cannot therefore be attributed entirely to the uncertainty of refraction at the altitude of the Kintraw foresight. Whether or not the platform at Kintraw was man-made, the midwinter Sun set in the way described about 1750 BC ± 100 years.

(ii). Star rising and setting; theodolite work. It will be noticed that we do not now include the star rising and setting points. We simply do not know if these are real, and certainly several lines which were attributed to Capella in 1967 (MSB:ch. 8) could be lunar, looking in the reverse direction. To quote A. Thom here:

> "Those who have been following my more recent writings will know that I tend today to play down some of the material in MSB. The modern radiocarbon dating, and especially re-calibration, have put a different complexion on the matter. The subject is growing and I must be allowed to change my ideas. This I have done several times. Much of my earlier work was done using an old theodolite which I lightened by machining and filing until it could be easily carried long distances, but of course it does not have the accuracy found to be necessary for measuring the profiles at the foresights. Later I bought a good (Tavistock) theodolite. When I began this work I generally worked alone, then later I often had an assistant. As the work expanded I found myself with larger and larger numbers of willing helpers. At one time in Carnac the party consisted of 14 workers."

It will be noted in passing that no material from our Carnac surveys is included in the 42 lines.

(iii). Leacach an Tigh Cloiche: "The slab(s) at the house of stones". In South Uist, the line which we did not use here, from slab A at Leacach an Tigh Cloiche (H3/11) (see table 1.4) was not included for the reason that the profile of Wiay Island was obtained entirely from map contours (MLO: fig. 6.15). A. Thom was at the site for only about an hour, visibility was poor, and he used only his prismatic compass. To quote him:

> "When our ship was lying in anchorage off the west coast we were always anxious to get back into sheltered waters. Anyone who knows the coast knows that it is one of the most exposed in the world. Hence some of the Hebridean surveys were done hurriedly."

The possible error in position of one of the four larger stones in A. Thom's small scale sketch plan (MSB: fig. 11.7a) referred to by Moir (1980a) and Ruggles & Norris (1980) in no way invalidates his discovery that the Moon rose at major standstill over Wiay. The plan is used to show how the lone slab A is separate from the other structural remains and is oriented on Wiay. The July 1979 theodolite survey by Ruggles & Norris gives the position of the dozen or so stones. At the time of recording his work at this site, A. Thom chose to draw on his plan an ellipse through four major stones. This ellipse was never used in any analysis, nor is it referred to anywhere in the text; the plan was never claimed to be anything more than a sketch. From the archaeoastronomical point of view it is the position of the backsight which is important.

(iv). Choice of lines to be included in the 42 assessed here. We have stressed the idea of the lunar band (fig. 1.7) and inevitably decisions had to be made as to whether or not to include any particular site. Such a place is Knockstaple, A4/19 (MLO: fig. 6.6). Here we find a slab, 11ft. ($3\frac{1}{2}$ m) high by 5ft. ($1\frac{1}{2}$ m) wide, with flat faces indicating a bearing of 326°, which is within about 10 arc minutes of the notch B. Moir (1980b) does not agree withour choice of the notch position. Has he remeasured this? We decided to include notch B in the analysis, and we consider this is the only possible choice, since the Moon would show it clearly as it set. This site was included in the present analysis only because of the size and orientation of the slab.

(v). Azimuths of the Moon at major standstills. At southern declinations, major standstill, rising or setting, the rate of change of declination with respect to azimuth is small for azimuths between 165° and 195°. Moir (1980b) notes that at Glen Prosen (MLO: table 7.1 & section 6.21) "two of the four stones are fallen" and there is "no distant foresight". Nonetheless we use the site in this analysis. At the azimuth of the line of stones, the amount by which the declination changes owing to a change in azimuth of one minute is only 0.194 minutes. Consequently, for a band width of ±2 minutes, the azimuth need be known only to ±10.3 minutes, which can be obtained with no difficulty from the plan on the drawing board which shows the upright and fallen stones. When observing over a row of stones such as these, with no horizon marker visible today, we consider that observation within the band width of ±2 minutes of arc is feasible. Perhaps in spite of the above reasoning there may have been an artificial marker on the horizon which now no longer exists.

2. DECODING THE CALLANISH COMPLEX—SOME INITIAL RESULTS

Margaret and Gerald Ponting

2.1 INTRODUCTION: THE MEGALITHIC COMPLEX AT CALLANISH

The Standing Stones of Callanish are widely recognised as one of the major megalithic sites of the British Isles. However their remote position on the Isle of Lewis inevitably has meant that they have not been studied as intensively as, say, Stonehenge or Avebury. In a book to be published shortly (Ponting & Ponting 1981) we present a detailed study of all the old maps, plans and illustrations of the site. More recent work is described in the following two sections.

The Callanish area is rich in megalithic monuments. Six "minor sites" in addition to the main site have been described by various authors, notably Worsaae (1846), Mackenzie (1904), Somerville (1912a), the Royal Commission on Ancient and Historical Monuments (Scotland) (1928), Thom (1967, hereinafter "MSB"), Burl (1976) and Ponting, Ponting & Curtis (1976). Names given to the sites differ, often confusingly. Accordingly we have collected them all together in table 2.1 for direct comparison in each case. We consider the local Gaelic names the best to use, but for convenience in this and our other publications we have adopted the site numbering system (Callanish I, II, etc.) invented by Thom. The main site is designated Callanish I.

Professor Thom (MSB: ch. 11) described seven sites in the Callanish complex which he designated Callanish I to Callanish VII. A Glasgow University team subsequently extended the nomenclature to include additional sites VIII to XII. We have studied further possible sites and designated them Callanish XIII to XIX. Notes on those sites not covered in table 2.1 are given in table 2.2, where appropriate references are cited. It should be noted, however, that the megalithic status of sites VII, IX, XIII, XIV, XVIII and XIX has not been confirmed. Fig.2.9d shows the relative positions of the sites around the head of East Loch Roag.

Aubrey Burl (1979) has objected to the use of the "Thom nomenclature" of the Callanish sites (i.e. I, II, etc.) on three grounds:

(i) Each site already has an accepted and traditional name;
(ii) The "Thom names" are misleading, a number of the sites being some km from Callanish village;
(iii) The use of the numbers implies an inter-relationship which may be non-existent.

We accept the third of these most readily, although we believe that evidence is accumulating for a relationship between the sites (e.g. Curtis 1979). Burl's own names for sites II and IV are less precise than the Gaelic names

Table 2.1 The Callanish "Minor Sites" — 1

Site No.	Callanish II	Callanish III	Callanish IV	Callanish V	Callanish VI	Callanish VIII
Gaelic name	Cnoc Ceann a' Gharaidh	Cnoc Fillibhir Bheag	Ceann Hulavig	Airigh nam Bidearan	Cùl a' Chleit	Cleiter
Approximate pronunciation	kroc kyain-a-gaaree	kroc fill-i-ver veg	kyain oo-la-vig	aa-ree nam bidge-ya-ran	cool-a-cletch	cletch-er
Grid Ref: NB	222326	225327	230304	234299	247303	164342
Worsaae 1846	Knock Ceann a Gharie	Knock ...elivir (illegible in ms.)	Ceann Hulivig	-	-	-
Mackenzie 1904	p 190 (no name)	p 190 (no name)	p 191; Tursachan Ceann Thulabhig	p 193; Tursachan Airidh nam Bidearan	-	p 189 (no name)
Somerville 1912a	Circle 2	Circle 1	Circle 3	-	-	Circle 4
RCAHMS 1928	90; Tursachan nr. Callanish	91; Tursachan, Cnoc Fillibhir Bheag	93; Tursachan, Garynahine	-	95	86; Tursachan, Barraglom
Thom 1967 (MSB)	H 1/2; Cnoc Ceann	H 1/3; Cnoc Fillibhir Bheag	H 1/4; Ceann Thulabeg	H 1/5; Airidh nam Bidearn	H 1/6	H 1/7; Great Berneray [sic]
Burl 1976	Lewis 8 Loch Roag	Lewis 4 Cnoc Fillibhir	Lewis 7 Garynahine	Lewis 2; Airidh nam Bidearan	Lewis 5 Cul a Chleit	-
Ponting, Ponting & Curtis 1976	p 57	p 57	p 57	-	p 57	p 57

Table 2.2 The Callanish " Minor Sites" — 2

Site No.	Gaelic Name	Grid ref.	Origin of Designation	Other References
VII	Cnoc Dubh	232302	MSB: 128	-
VIIIA	Aird a' Chaolais	165340	PPC76: 58	MSB: 128
IX	Druim nam Bidearan	233297	UG78	PPC76: 58
X	Na Drommanan (= Druim nan Eun = Druim nan Eum)	230336	UG78	RCAHMS 1928: 92; Burl 1976: Lewis 6; PPC76: 58
XI	Airigh na Beinne Bige	222356	UG78	PPC76: 58; PPC77: 32
XII	Stonefield	215350	UG78	RCAHMS 1928: no. 88 (" Blair")
XIII	Sgeir nan Each	215341	PPC76: 58	-
XIV(E)	Cnoc Sgeir na h-Uidhe	228329	PPC76: 59	-
XV	Airigh Mhaoldònuich	117346	PPC76: 59	-
XVI	Cliacabhaidh	213338	PPC76: 59	PPC77: 32
XVII	Druim na h-Aon Chloich	237320	PPC76: 59	MacGregor 1933
XVIII	Loch Crogach	244292	PPC76: 59	-
XIX	Buaile Chruaidh	218331	PPC77: 32	-
-	Cnoc a Phrionnsa	211356	-	Henshall 1972: 461

PPC76 denotes Ponting, Ponting & Curtis 1976; PPC77 denotes Ponting, Ponting & Curtis 1977; and UG 78 denotes University of Glasgow 1978.

we have used, and his name for site II ("Loch Roag") does not have the authority of use in the literature. In addition, he made the unfortunate slip in his letter to Current Archaeology (1979) of calling site IV "Loch Roag" rather than site II, thereby adding to the confusion.

2.2 ARCHAEOLOGICAL BACKGROUND

There is, as yet, no evidence of habitation sites in the Callanish area contemporary with the megalithic monuments. In an attempt to discover such sites, we investigated stone walls which were becoming exposed during peat cutting. Our work acted as the impetus for more extensive peat excavation in the summer of 1979 by the Central Excavation Unit. Trevor Cowie's results (1979) confirm our conclusion that the walls are later than the megalithic remains, probably of Iron Age date. The application of radio-carbon dating and pollen analysis to peat columns, taken in association with the excavations and within 1 km of sites I, II, III and XIX, will provide a vegetational history of the area and hence an environmental context for the stone rings. Early results from the palaeobotanical study (Bohncke 1979) show that, at the 15 cm level in sample CN-1, there is a decline in birch and hazel pollen, with an increase in grasses—this is most likely to have been due to deliberate scrub clearance. The presence of cereals and ribwort pollen also indicates that this level corresponds to the first marked effects of human influence on the local environment. This level has been dated (Cowie 1979) to 1270 bc ± 65 (GU-1171). A confirmatory test on the 16 cm level from the same sample produced a result with a satisfactory correlation—1110 bc ± 95. Calibrated, 1270 bc gives a value of approximately 1550 BC. Provisionally, we chose this date for our astronomical calculations. We accept that new archaeological evidence may make it necessary to adjust our declination lines (see section 2.5). Fig. 2.1c shows the amount of adjustment needed to a given declination line for a given change of supposed date of construction of the indication.

The recent excavation at Callanish I (Ashmore 1980) and the second phase due to take place in 1981 may produce such evidence. This is the first modern scientific excavation at any of the Callanish sites and as such should produce many new data. Beaker pottery and ard marks representing prehistoric ploughing have already been found. Further analysis of the relationship between these finds and the relative stratigraphy of the cairn and the circle stones (see section 2.7) together with radio-carbon dating of peat layers, may produce more reliable dates for the monument itself. It may also be possible to discover whether Callanish I was built in several phases, or whether the idea that it was a single conception is still tenable. This would also have a bearing on any astronomical work, as it might be necessary to use different dates for different parts of the site.

A conclusion of our documentary studies (Ponting & Ponting 1981) which has a bearing on the astronomical work is that Callanish I has not suffered major restoration, despite the claims to this effect by Thom (e.g. Thom 1966: 50; Thom & Thom 1978a, hereinafter "MRBB": 3). Our studies suggest that few stones have been lost from the site in the last 400 years, and that only one (or possibly two) stones were re-erected last century.

2.3 A REVIEW OF ASTRONOMICAL THEORIES AT CALLANISH

Published speculation on the origins and purpose of the Callanish sites, especially the large and unique site at Callanish I, has been rife since John Morisone, c. 1680, wrote that the stones were "men turned to stone by ane Inchanter" (Macfarlane 1907). Indeed, there is nothing new about the idea that Callanish was designed as a "rude astronomical observatory", as Headrick (1808) wrote over 150 years before Alexander Thom's first book was published. His basis for this theory seems to have been his opinion that stones in a circle often marked the four cardinal points and sometimes the finer divisions of direction.

The southern row of stones at Callanish, having its orientation lying within 2° of true N/S, and the more approximate N/S orientation of the whole monument have attracted many theories.[1] Headrick thought that the worshippers at Callanish "probably offered sacrifices when the sun was in the meridian". James Kerr (1873) presumably had a similar idea when he wrote:

> "Walking up the avenue at twelve o'clock noon, and looking towards the great centre stone while the meridian sun throws his rays right athwart it, one can hardly fail to see the great object for which this rude memorial was erected."

Other authors who published quasi-astronomical theories about Callanish included Toland (1726), MacCulloch (1824) and Callender (1854).

More scientific archaeoastronomical speculation began with Sir Norman Lockyer's investigations (1909). These were based, not on a visit, but on a plan prepared by James (1867). Lockyer considered that the avenue was orientated to the rise of Capella in 1720 BC and that the "sight-line from the chamber" was orientated to the rise of the Pleiades in 1330 BC.

Rear Admiral Boyle Somerville, ashore from a hydrographic survey in 1909, prepared and published (1912a; 1912b) a detailed but somewhat inaccurate plan, which introduced the stone numbering system for the site (fig. 2.3b) that is in general use today. He identified various alignments at the main site, including one to a lunar extreme—the first suggestion that prehistoric Man anywhere had been concerned with accurate observation of the moon. We return to this suggestion later in this paper. Somerville also surveyed four other sites in the area (1912a).

Following his controversial paper "Stonehenge Decoded" (1963), Gerald Hawkins published another paper titled "Callanish, a Scottish Stonehenge" (1965), in which he subjected the main site to a computer analysis similar to that which he had used at Stonehenge. The stone positions fed to the computer were those on Somerville's plan, now shown (University of Glasgow 1978; Ponting & Ponting 1981) to be inaccurate in several respects; Hawkins had not visited Callanish at that time. Checking every possible line between two stones, the analysis produced twelve possible alignments with solar and lunar significance, including three to extreme south positions of the midsummer moon.

Of Hawkins' suggested alignments, the one which has since caused most controversy is that southwards along the avenue to Clisham, the highest mountain on the island of Lewis-with-Harris. Inspection of the site on a clear day shows that Clisham cannot be seen along the line of the avenue, due to an outcrop of rock (Cnoc a Tursa) immediately to the south of the site. Hawkins (1973:240) later admitted this when he described a visit to Callanish, and the outcrop was also clearly shown in his photogrammetric survey (Hawkins 1971) of Callanish—still the only published plan of Callanish with contours. However, Dr. Jean-René Roy (1980) claimed that Hawkins had not known of the outcrop which obscured the avenue/Clisham alignment; Hawkins (1980) justifiably criticised Roy for publishing a paper without being aware of all the relevant literature.

Like Hawkins, Thom (1954; MSB: ch. 11) used Somerville's plan; his type A flattened circle geometry for the central ring at Callanish I has required re-orientation since the Glasgow Survey (see below) revealed Somerville's inaccuracies. As he did not have clear weather on his visit to Callanish, Thom produced horizon profiles by calculation from the O.S. maps. His profile of the Clisham range, as it would be seen along the avenue but for the rocky outcrop, has unfortunately found its way into secondary literature. We clarify the position in section 2.7.2. Thom studied the minor sites in the area and proposed eleven inter-site lines with possible astronomical significance. He also stressed the need for much more detailed studies of the Callanish sites and their relationships.

In 1974, a team from the Geography Department of Glasgow University performed such a survey (University of Glasgow 1978). Their plan of the main site (used as the basis of figs. 2.3a and 2.3b) agrees closely with Hawkins' photogrammetric survey.

Cooke, Few, Morgan and Ruggles (1977) carried out their own horizon surveys at several of the Callanish sites in 1975. Their analysis of these surveys was intended as a test of a new approach to archaeoastronomy. They defined rigid selection criteria for the sections of horizon which they considered to be indicated; and then limited themselves to the study of the declinations derived from these sections of horizon. Their intention was to accumulate a corpus of data from many sites suitable for statistical analysis—a corpus which would be comparable to that of Thom's, but gathered according to <u>their</u> selection criteria. At Callanish, their method eliminated many of the astronomical indications found by earlier authors—but we believe this to be an inevitable consequence of their method. They admit (Cooke et al. 1978) that their selection criteria are to some extent arbitrary—"Provided the criteria are laid down and strictly adhered to throughout, the results will be valid. Any set of criteria might be chosen. The trick is to choose ones which are most likely to produce a positive result". Freeman and Elmore (1979) using a different statistical method, confirmed that the results in the paper by Cooke et al. gave no support for the existence of astronomical alignments within the Callanish complex.

Gordon Moir (1978) is critical of much of Thom's work in the Outer Hebrides. His comments concerning the main site are essentially a detailed critique of Thom's remarks in MSB. This was published back in 1967, and Thom himself would be the first to admit that there is much material to retract in the light of more recent studies. In any case, several of Moir's points are very minor ones.

2.4 METHODS

Our own archaeoastronomical objective is more ambitious than that of previous workers, whose researches inevitably have been limited by the short duration of their visits to the island and, frequently, by adverse weather conditions. We hope, ultimately, to survey the whole of the horizon from each site; to plot the rising and setting sun and moon at their extreme positions (for dates consistent with the construction of the monuments); to check for indications at each site; and thus to provide a corpus of data for statistical analysis. The present paper includes only initial results.

Thanks to the survey carried out by the Geography Department of Glasgow University and to triangulation work by Ron Curtis, we have an accurate National Grid reference for each site in the Callanish area, and a number of permanent triangulation points for use as Referring Objects.[2] This means that we do not need to use the midday sun to achieve accurate bearings, as it is a simple matter to use Glasgow Survey data to calculate the true bearing of any Referring Object from any position at a site. Thus, when we carry out the on-site survey work with our 20" Watts theodolite, we can always record true azimuth readings direct from the instrument. We have occasionally needed to survey from satellite stations to avoid modern obstructions such as houses.

We use a Minolta camera with 165 millimetre telephoto lens to record the same segment of horizon on colour slides from the theodolite position. Each photograph covers about 12 degrees of azimuth. Our horizon profiles are drawn up on large sheets of graph paper at a scale of 3 inches to one degree so that points can be plotted to the same degree of accuracy as that read from the theodolite (see, e.g., fig. 2.1b). Nineteen sheets are required for a complete 360 degree profile. After the theodolite readings of apparent altitude and true azimuth are plotted on as points, the slides are projected to fit, and the intermediate sections of horizon carefully drawn in.

The next stage is the calculation of the paths of the sun and moon. Points representing the moon's centre are plotted onto the graph paper, and the limbs and perturbations are added by direct measurement. Each survey and calculation is performed at least twice before the figures are plotted. On our horizon profiles, when the indication is rough, we have drawn the moon in its average position and shown maximum and minimum perturbations where the limbs cross the horizon. Continuous lines are used above the horizon. Dashed lines are used below the horizon; short for minimum perturbation, long for maximum perturbation, and alternating long and short for the moon in its average position (see fig. 2.1a). Where we consider that one or other perturbation is clearly indicated, we have drawn the moon at that perturbation and indicated the opposite one. We check which moon phase is indicated to avoid suggesting observations of an unlit limb (see table 2.3).

Using our method, we produce horizon profiles with declination lines for any lunar or solar event, whether or not there is any on-site or inter-site indication of the piece of horizon concerned. When selecting indications, we have given at this stage equal validity to rows of stones, inter-site lines and the attitudes of individual flat slabs or of the flattened faces of stones.

Table 2.3 The relationship between the moon's phases and the perturbations at the major standstill

We have used this table mainly to determine the moon's phase and possible month and time of day at which indicated horizon markers could have been used. It is intended more as a pragmatic "observers' guide" than as a strictly accurate astronomical chart. Megalithic Man would not have been concerned with the strict definition of a lunar limb, but with, for instance, the moment of the first gleam of the moon.

We have used the terms "upper limb" and "lower limb" in this table to refer respectively to the first and last parts of the moon's disc to emerge above the horizon (i.e. maximum and minimum altitude of the lunar disc, rather than maximum and minimum declination (± s) of the moon). At the high latitude of Callanish, at southern extreme of major standstill, the moon rises and sets at such a shallow angle that the effect of this distinction is negligible.

At full moon, both of these "limbs" could have been observed on the horizon at moonrise and at moonset, giving four positions that could have been indicated consecutively - gleam, rising moon tangential on horizon, setting moon tangential on horizon, and last gleam. When only part of the moon's disc was illuminated, we have assumed that an observer would have noted only the appearance of a "limb" on the horizon, not the first or last appearance of a cusp, horn or terminator. Two of the four positions cannot be observed on the horizon. For example, the "upper limb" of a rising third quarter moon would be unlit and so unlikely to have been indicated. Therefore, if a megalithic alignment appears to have indicated this direction, it must have been the September first quarter moon which was observed, not the March third quarter moon. See Moir (1979:126).

Times of day and directions given in the chart are approximate; in particular, the "before" and "after" sunrise or sunset may not be correct in every case, as the June full moon may sometimes set a few minutes <u>after</u> sunrise.

Major Standstill	Perturbation	Month	Moon Phase	Rising Moon: Direction	Rising Moon: Time	Rising Moon: Illumination of upper and lower limbs	Setting Moon: Direction	Setting Moon: Time	Setting Moon: Illumination of upper and lower limbs
Southern Extreme	Maximum	March	Third quarter	S. E.	Night	Lower limb (tangential)	S. W.	Day	Upper limb (last gleam)
	Minimum	June	Full moon	S. E.	After sunset	Both fully visible	S. W.	Before sunrise	Both fully visible
	Maximum	Sept.	First quarter	S. E.	Day	Upper limb (first gleam)	S. W.	Night	Lower limb (tangential)
	Minimum	Dec.	New moon	S. E.	After sunrise	Neither visible	S. W.	Before sunset	Neither visible
Northern Extreme	Maximum	March	First quarter	N. E.	Day	Upper limb (first gleam)	N. W.	Night	Lower limb (tangential)
	Minimum	June	New Moon	N. E.	Before sunrise	Neither visible	N. W.	After sunset	Neither visible
	Maximum	Sept.	Third quarter	N. E.	Night	Lower limb (tangential)	N. W.	Day	Upper limb (last gleam)
	Minimum	Dec.	Full moon	N. E.	Before sunset	Both fully visible	N. W.	After sunrise	Both fully visible

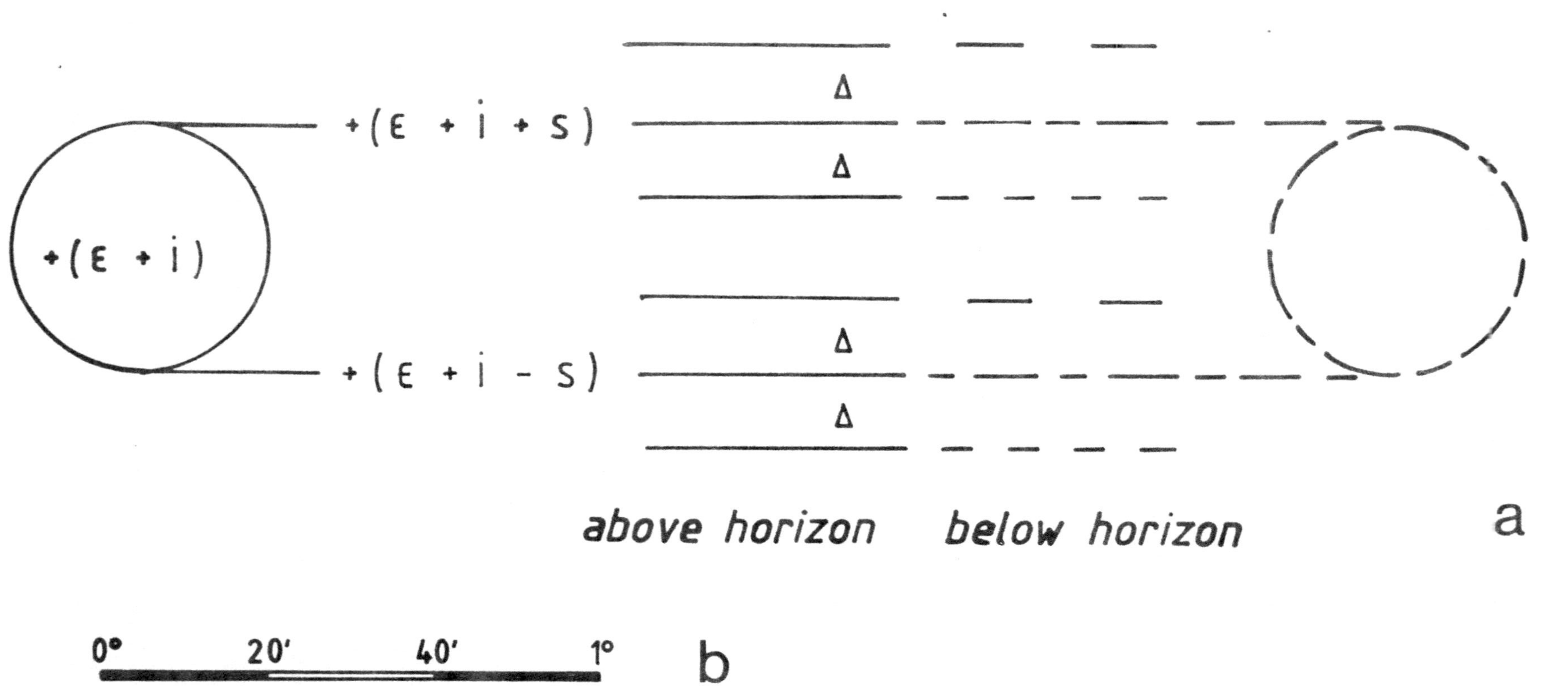

Fig. 2.1 (a) The moon's path, with perturbation lines, drawn to the size and style used on the horizon profiles. This is drawn for northern extremes. At southern extremes, the maximum perturbation lines occur below the average lines, and the minimum above. (b) The scale of degrees used for both azimuth and apparent altitude on the horizon profiles. The profiles in this paper are reduced to about 15% of the size of the originals, which unfortunately renders some clear horizon features invisible. (c) While all our declination lines are drawn for 1550 BC, an overlay can be used to see the effect of altering the supposed date of construction. This is to the same scale as 2.1a and 2.1b.

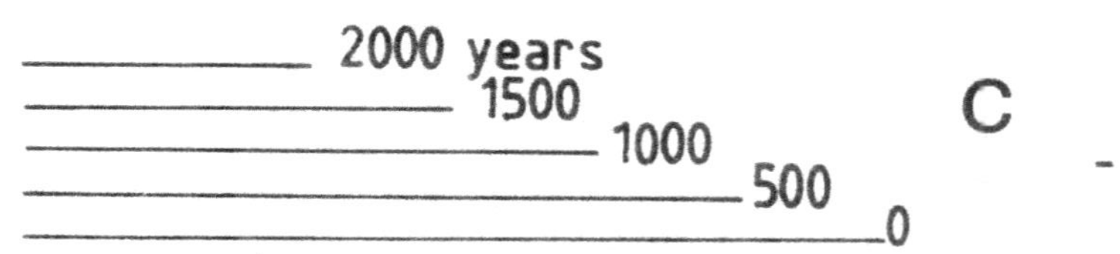

We have also considered the geometric axes of stone rings which appear to be ellipses or flattened circles, although any horizon indication from these must be considered rough or symbolic.

In the present state of knowledge, we prefer not to use rigid criteria like those of Cooke et al. (1977), nor to make assumptions about the types of foresight and backsight which might have been used by Megalithic Man.

2.5 BACKGROUND DATA

We do not have access to computer facilities, and all our calculations are performed on a TI57 programmable pocket calculator. We list here the data which we use, in order to demonstrate our method.

In order to calculate rising and setting paths we use the following standard formula to determine the azimuth corresponding to the apparent altitude at suitable intervals:

$$\text{Cos } A = \frac{\sin\delta - \sin\phi \sin h}{\cos\phi \cos h}$$

where A = azimuth
δ = declination
ϕ = latitude
and h = apparent altitude + parallax - refraction.

Declination. All our calculations, at present, are based on the date 1550 BC (see section 2.2). The obliquity value for that date was obtained by the de Sitter formula (Thom 1971, hereinafter "MLO":15). Other values are from standard texts.

ϵ (obliquity of the ecliptic, 1550 BC)	=	23°	52'	49"	.15
i (inclination of the moon's orbit)	=	5°	08'	43"	
Δ (perturbation)	=	0°	08'	42"	
s (semi-diameter of the moon)	=	0°	15'	30"	
s (semi-diameter of the sun)	=	0°	16'	00"	

Latitude. This ranges from 58° 10' 16" for Callanish V to 58° 13' 20" for Callanish XI. While a difference of a few seconds of latitude makes little difference to azimuth values, we work to the nearest second on the principle of greatest possible accuracy for each value used.

Parallax. On the same principle we include here the value for the sun's parallax, although the effect is negligible.

Moon (major standstill)	0°.950
Moon (minor standstill)	0°.858
Sun	0°.002

Refraction. Refraction values, including temperature corrections, were obtained by interpolation from Thom's table (MLO:34). In our surveys during the period October 1979-March 1980 we were fortunate to have long spells of clear weather, and quite a number of the surveys were carried out at sub-zero temperatures.

2.6 INTERVISIBILITY OF SITES

At an early stage in our research, we made an attempt to establish the inter-visibility of the Callanish sites in an objective manner. Each of the sites in the Callanish Complex was visited numerous times, at all seasons of the year, and under different weather conditions, for the purpose of determining the visibility of the other sites from it. Site X and some others with fallen stones were marked by erect poles of approximately the same heights as the fallen megaliths.

A grading system was used to classify the lines according to the degree of visibility—this was, admittedly, a subjective process. The seven grades which were used in the field have been combined into four grades in the following outline of results. The first grade includes all easily intervisible sight lines. The second grade includes those lines which required binoculars, or especially good weather conditions, to see the distant site. These were included as it is not known whether or not megalithic man increased the visibility of distant sites by fire markers, by the use of brushwood foresights, or even by the clearance of avenues through surrounding tree cover. If it is possible that lines, which are not clearly intervisible with the naked eye today, were usable in the past by such methods, then the lines should be included in the study until the use of such methods is disproved. A third grade includes lines whose status is doubtful, sometimes due to the intervention of modern buildings. The fourth grade includes all those lines which are definitely not intervisible.

Of 342 possible sight lines examined, 89 (26%) fell in Grade 1, 123 (36%) in Grade 2 and 8 (2%) in Grade 3, giving a total of 220 inter-site lines which need to be studied for astronomical indications. When surveying at any site, we always include and plot the positions of any other visible sites, and consider them as possible deliberate foresights for solar or lunar events on the horizon beyond.

The authors cannot agree with Cooke et al. when they state (1977:117) that "sites visible from each other are a comparative rarity"; although they agree that the boundary between what was "visible" and what was "not visible" to the megalith builders has not been defined.

2.7 SITE I—THE STANDING STONES OF CALLANISH

2.7.1 Horizon indications

Fig. 2.3 shows the layout of the stones at the main site. A ring of tall stones, about 12 metres in diameter, is surrounded by radiating rows of stones. Two rows of 4 stones lie to the east[3] and west; five stones form a row which runs southwards; and two rows form an avenue which runs approximately northwards. The largest monolith of all, $4\frac{1}{2}$ metres tall, stands within the circle. A chambered cairn also lies inside the circle.

Stones 9 and 34 both occupy somewhat anomalous positions in the layout of the site. In addition, the northwestern face of stone 34 has a regular flat surface and is aligned with the north edge of stone 9. Somerville claimed that the alignment from 9 to 34 (long dashed line on fig. 2.3a) indicated the full moon rise at its greatest northern declination, and this was the first suggestion that prehistoric man anywhere had observed and recorded lunar extremes. Standing at stone 9 today and looking along the flat face of stone 34 gives a view of a fairly featureless horizon (fig. 2.6a). Even allowing for the parallax problems caused by a slight movement of the observer's head, our profile shows that the indication was clearly for the moon tangential on the horizon at major standstill, north extreme. With no distant foresight, this cannot be considered an accurate indication—the perturbations could not have been distinguished.

Stone 34 may form part of the east side of the avenue. At the other end of this row, stone 8 also has a flattened face which, used as an attitude to indicate a foresight, clearly points to a conspicuous horizon notch (fig. 2.3b, short dashed line). This notch was surveyed from stone 8 and from a number of other positions, but it was only from stone 8 that it gave a definite result for moonset at the north extreme of the major standstill (fig. 2.6b)—the same event for which the moonrise is indicated by the 9 to 34 alignment. The setting moon at maximum perturbation around 1550 BC would have disappeared and then gleamed again in the notch; that is assuming the notch to have been the same shape at that time as it is today. We have visited the base of the notch, probed, and discovered the peat to be nearly 2 metres deep. The base of the notch, about 1 kilometre from stone 8, must be lowered by 5 minutes of altitude, as shown by the short thick line, in order to "remove" the layer of peat. Thus, the moonset would have been seen briefly to gleam again at the minimum perturbation also. This horizon distance gives a 4G value (see section 2.9 below) of about 70 metres, very close to the distance from stone 34 to stone 8.

Two solar events also seem more likely to have been viewed from the north end of the avenue than from the circle. When viewed from the circle, the upper limb of the sun at midwinter setting just missed a conspicuous notch. However when viewed from the north end of the avenue the sun would have set and reflashed in the notch (fig. 2.6d). The path of the upper limb is such that the disappearance and reappearance of the sun would have occurred for only 3 days on either side of the solstice around 1550 BC. Furthermore this phenomenon would only have occurred at the solstice for a few centuries before and after our chosen date—from 2000 BC to 1325 BC approximately. (Fig. 2.1c shows the rate of change of a declination line over the centuries.) There is no obvious stone attitude or alignment to be seen at the site today to indicate this notch. However, our documentary researches suggested that there may have been one or two more stones in the west side of the avenue, between stones 18 and 19. A socket for one of these stones, provisionally designated 18A, was found in the 1980 excavation (Ashmore 1980). (Recent building work within 100m of this position revealed a broken stone over 1 m tall which may be part of a missing megalith from the north end of the site.) It is conceivable that 18A may have acted as an indicator for this notch for an observer standing by stone 8.

Even people with little knowledge of, or interest in, astronomy seem to expect us to have found some indication of midsummer sunrise; but there is no convincing alignment. There is a slight notch in an otherwise featureless horizon, where the upper limb of the midsummer sun first appeared in 1550 BC, viewed from the north end of the avenue but not when viewed from the circle. The horizon, at 5 kilometres, is close for a solar event, and the possible foresight is an upright slab of stone on an intervening hillock.[4]
So far in this section we have discussed horizons indicated only by two small parts of the site, the alignment of stone 9 with stone 34, and the region of stone 8 at the north end of the avenue. Other workers have attempted to find significance in the more obvious stone rows, sometimes including star alignments.

2.7.2 The Callanish avenue and the visibility of Clisham

When Hawkins (1965) and Thom (MSB:ch.11) claimed that the two long rows of the avenue were aligned to the moonset on Clisham at the southern extreme of the major standstill, both had worked from maps. Although Thom had surveyed on site, he had not obtained theodolite readings of the Clisham range. After his photogrammetric survey and site visit, Hawkins stated (1971) "It is now evident ... that (Clisham) is not visible when one stands at the north end of the avenue" and (1973) "The moon, of course, could still have been seen in 1500-2000 BC if one walked up the avenue to the outcrop. Perhaps the site was chosen deliberately to make the moon appear to enter into the rock."

Although Cooke et al. (1977) and the authors (Ponting & Ponting 1977a; 1977b) have also pointed out that Clisham cannot be seen from the avenue, the alignment continues to be mentioned in popular accounts of megalithic astronomy. However, the Clisham range <u>can</u> be seen if an observer moves 7 metres westward from the north end of the avenue. Our surveyed and photographed horizon profile (Fig. 2.6c), taken from a position just west of the circle, can be compared directly with those of Thom (MSB:123; MLO: 69) and shows the omission of one hill from Thom's calculated profile. The two arrows show the two azimuths indicated by the two sides of the avenue.

According to our 'best-fit flattened circle', the long axis of the central ring also indicates this direction. (Wood (1979) has proposed an egg shape for this ring with a different axis.) If the avenue lines and circle axis (see continuous lines on fig. 2.3a) were intended to indicate the southern extreme of the major standstill, they could have done so only symbolically, as the Callanish avenue and circle are situated in one of the few places in the vicinity from which much of the distant southern horizon is <u>not</u> visible.

We surveyed virtually the complete horizon from the vicinity of the circle and also from the north end of the avenue. We have already described profiles taken from the north end where 2 solar and 1 lunar events were associated with obvious horizon features; but this was not the case when surveyed from near the circle. So we decided to draw up the profile of the southern horizon including the stones and rocky outcrop as seen from the north end of the avenue. This is shown, considerably reduced, in fig. 2.10a. It will be seen that, at the southern extreme of the major standstill at this latitude, the path of the moon traced a spectacular course through the southern sky. On a horizon with an altitude of one degree, the moon rose at an azimuth of approximately 163°, transitted with the moon's centre at just over 2° apparent altitude, and set at an azimuth of about 196°. The moon skimmed over the southern horizon for only about 2 hours between rise and set. (At more northerly latitudes, such as the north of Shetland, the moon did not rise at all at the same event; while further south, the moon rose and set further from due south and transitted higher.) As seen from the north end of the Callanish site, looking southwards towards the circle, with most of the Clisham range hidden by the site and the outcrop, the moon rose out of the "Sleeping Beauty" (see below), skimmed the stones of the east row and set into the stones of the circle. To the right of stone 29, the tall central megalith, the moon briefly gleamed again within the bounds of the circle before finally setting. For a day or two either side of the maximum, the moon would "flicker" through the stones.

Were the avenue lines set out deliberately to give a symbolic descent of the moon into the circle or central cairn? Was this a ceremonial version for the common people of an event which was of great interest to an astronomer priesthood? If so, several more questions arise. Were accurate observations of the sun and moon carried out elsewhere in the area before symbolic incorporation into the greatest monument? Where is the extrapolation device necessary to any lunar observatory?

2.8 THE SOUTHERN EXTREME OF THE MAJOR STANDSTILL

Later in this paper, we shall attempt to demonstrate that several of the less spectacular sites in the area were used to observe lunar extremes. We decided to concentrate at first on the southern extreme of the major standstill; this was partly because of the avenue orientation, partly because it was the most obvious lunar event, and partly because preliminary work suggested that it was indicated for minor sites. Initially, we did this despite strong doubts about prehistoric man's ability to record this event, and despite its occurrence only at the extreme of a cycle lasting more than 18 years. At some sites, we have as yet only surveyed the horizon in this direction.

We wonder, with Hawkins, if this description of Hyperborea from a text by Diodorus in 55 BC (Oldfather 1935) bears a relationship to Callanish-

> "... there is also on the island both a magnificent sacred precinct of Apollo and a notable temple which is adorned with many votive offerings and is spherical in shape ..... They say also that the moon, as viewed from this island, appears to be but a little distance from the earth and to have upon it prominences, like those of the earth, which are visible to the eye. The account is also given that the god visits the island every nineteen years, the period in which the return of the stars to the same place in the heavens is accomplished ...."

From most of the sites in the Callanish area, one or both of 2 ranges of hills are visible. Hills to the southeast make the profile of a sleeping woman (fig. 2.2a) which is well-known to the older inhabitants of Callanish. The hills do not have this appearance from other villages. Sometimes called the Sleeping Beauty, her Gaelic name is less flattering—Cailleach na Mòinteach (the old woman of the moors). The hills forming the Sleeping Beauty are 20 kilometres from Callanish. Roineval is only 12 km away and so it appears to change its position relative to the other hills, as viewed from different sites, due to parallax. Fig. 2.2b shows the mountains of North Harris, of which Clisham is the highest. The V-notch of Glen Langadale is spectacular from many positions around Callanish. Moonrise and moonset at the southern extreme of the major standstill intersect these 2 ranges of hills from any of the sites around Callanish (but see section 2.13)

2.9 EXTRAPOLATION AND STONE SPACING

If an observer wished to observe two successive moonsets with reference to, say, a point on the Clisham range, his maximum sideways movement from one night to the next would be the value which Thom (MLO:ch. 8) has called 4G. This is the equivalent on the ground to the moon's declination decrement between its monthly maximum value and its value exactly one day before or after. (The assumption that two moonsets are exactly 24 hours apart is an acceptable approximation—cf. Wood 1978:117.) As Clisham is 26 km from Callanish, the 4G value would be an impossibly large 1600 metres. This appears to render accurate lunar observation from any Callanish site to the 2 distant ranges of hills very difficult.

Any group of lunar observatories must include an extrapolation device for discovering the moon's true monthly maximum position from the observed rises and sets, which may differ by some hours from the true maximum. We studied different extrapolation methods published by Thom (MLO:ch. 9) and by Wood and Penny (1975). We recognised that parts of the ground plan at Callanish I could be used for a similar geometry. Although not identical, it has more superficial similarities to features at Merrivale than to the Caithness rows. We found that, taken separately from the remaining stones at the site, half of the west side of the avenue, plus the east row, could be used as an extrapolation sector for values of 4G up to 45 metres. Extension of this method to include the full length of the avenue suggested that extrapolation could be carried out here, using a rope and stake geometry, for any 4G value up to 90 metres (see fig. 2.2c). This is equivalent to a horizon distance of up to $1\frac{1}{2}$ kilometres for the major standstill and about 4 kilometres for the minor. Thus we began to take more interest in indicated lunar events where near horizons partially obscured the distant ranges already described. Although we have not yet determined 4G values for all the observation lines studied, a proportion of those calculated do give values approaching 90 metres. This supposes that ropes were brought to the extrapolation device at Callanish I from the smaller observing sites.

This interpretation of the Callanish avenue is a tentative hypothesis. As Wood (1978) has said of his interpretation of Merrivale, it is an "interesting

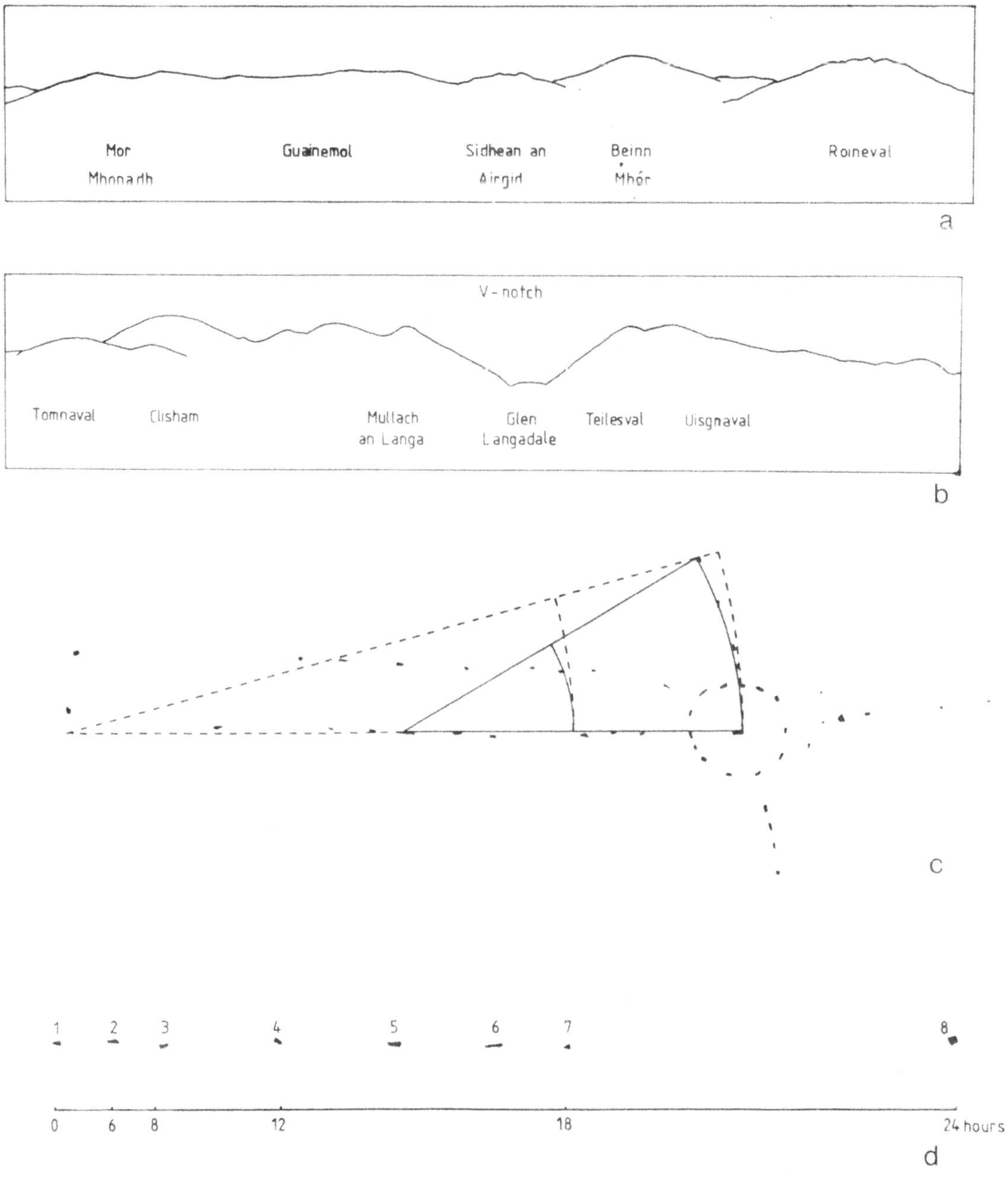

Fig. 2.2 (a) The range of hills in S.E. Lewis known to the people of Callanish as the Sleeping Beauty (Mor Mhonadh = knees, Sidhean an Airgid = face), with Beinn Mhor and Roineval. See fig. 2.10c and profiles 2.7b, 7c, 8a, 8b, 9a, 9c, and 10a. (b) The North Harris hills (the Clisham range). See fig. 2.10c and profiles 2.6c, 7a, 8c, 8d, 8e and 9b. (c) A possible extrapolation device using two of the stone rows at Callanish I. (d) The East side of the avenue at Callanish I, stones 1-8. If the length of this row is taken as a value for 4G, there is a correlation between stone positions and distances representing hours from the moment of maximum monthly declination of the moon.

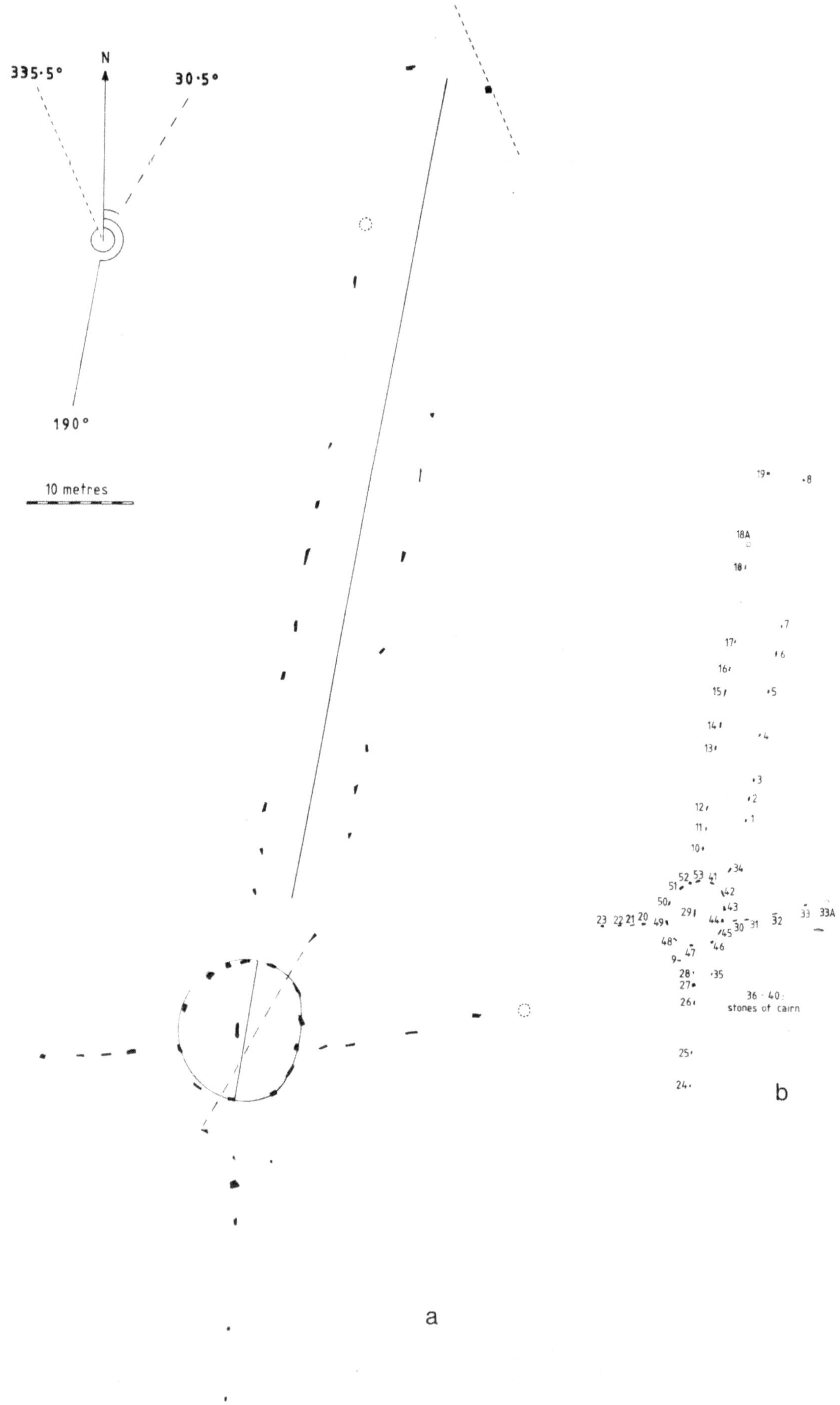

Fig. 2.3 (a) Astronomical indications at Callanish I. Long-dashed line: alignment stone 9-stone 34, see profile 2.6a. Short-dashed line: attitude of stone 8, see profile 2.6b. Continuous lines: midline of avenue and circle axis, see profile 2.6c. Azimuth 218°: midwinter sunset as seen from N end of avenue (over 18A?), see profile 2.6d. (b) Callanish I. Stone numbering system after Somerville 1912b.

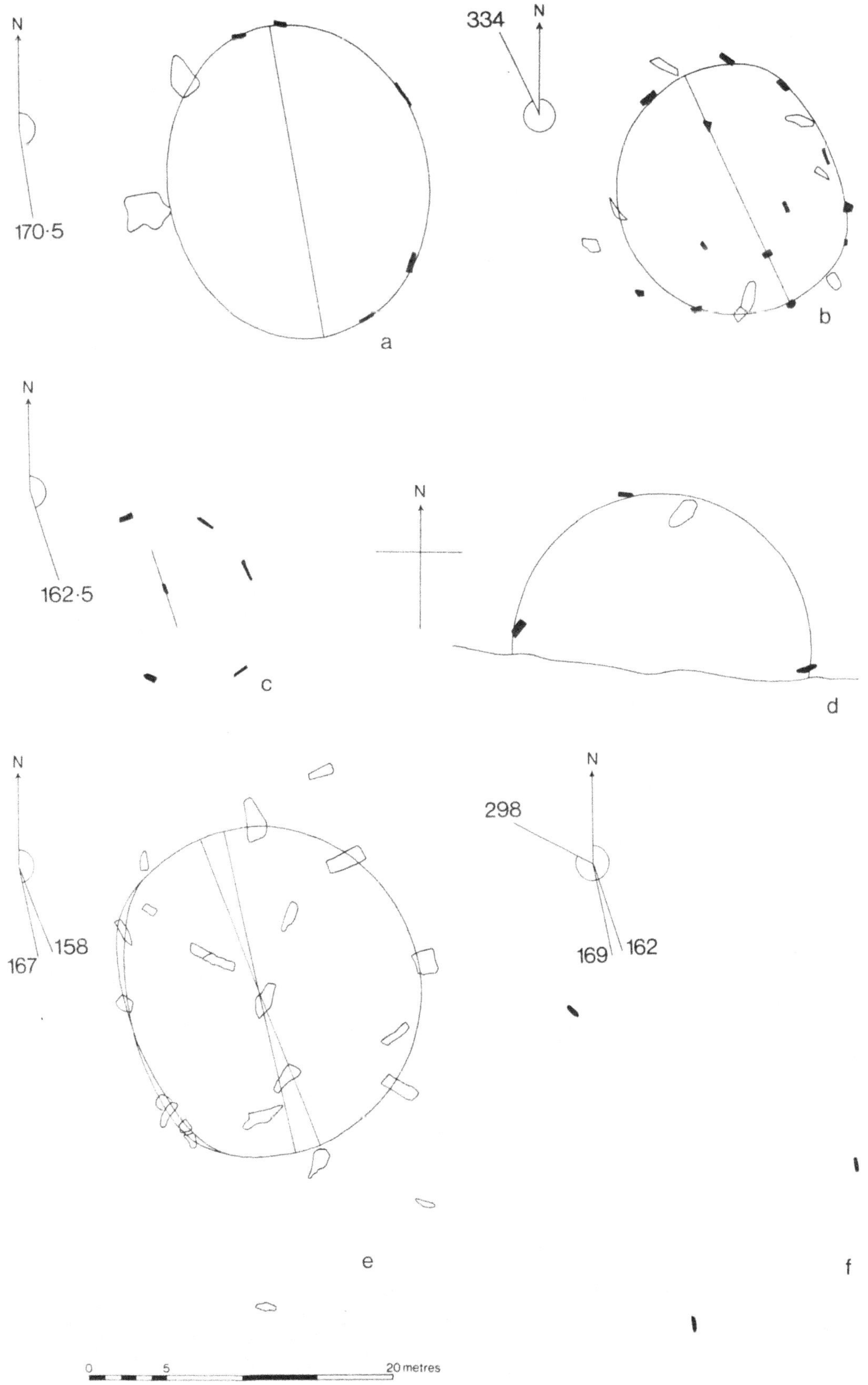

Fig. 2.4 (a) Cnoc Ceann a' Gharaidh, Callanish II. Profiles 2.6f, 6g. (b) Cnoc Fillibhir Bheag, Callanish III. Profile 2.7a. (c) Ceann Hulavig, Callanish IV. Profile 2.7b. (d) Cleiter, Callanish VIII. Profile 2.7f. (e) Na Drommanan, Callanish X. Profile 2.8a. (f) Cliacabhaidh, Callanish XVI. Profile 2.9a.

yet unproven speculation" and "further work on this and other stone rows will clarify the situation eventually".

The presence of the east side of the avenue seemed extraneous to this theory. We realised that the spacing of the stones along the row from stone 1 to stone 8, was reminiscent of the stake spacing which might be expected on a lunar observation line. So, taking the distance along this row of 69.61m as equivalent to a 24 hour declination decrement for the moon, we used the equation

$$y = kt^2,$$

where y = distance along the line in metres
t = time in hours from the maximum
and k = a constant for the particular observation line, given by $k = 4G/(24)^2$; in this case $k = 0.121$,

in order to find the distances along the row equivalent to fractions of 24 hours. The results are listed in table 2.4; see also fig. 2.2d.

Six of the eight stones matched exactly with twelfths of a day, the other two representing periods of $14\frac{1}{2}$ and $16\frac{1}{2}$ hours from the maximum respectively. We wonder whether there is any other evidence to suggest that Megalithic Man may have been able to subdivide the day.

We do not have any evidence to suggest that the line of stones 1-8 was actually used as an observation line. Its length gives a suitable 4G value for the distance from stone 8 to its horizon notch, but the line does not run at right angles to the line of sight to the notch. Might the line of stones 1-8 represent a ceremonial formalisation of the type of stake pattern which emerged on lunar observation lines in the Callanish area?

2.10 TWO POSSIBLE OBSERVATORY SITES

2.10.1 Site XI—Airigh na Beinne Bige

The most northerly of the sites which we have studied—Airigh na Beinne Bige, or Callanish XI—is situated on a hillside plateau with a magnificent view over the whole Callanish area, with the southerly ranges of hills beyond. The large erect stone is 1.5 m tall x 0.5 m x 0.5 m. It contains metal wedges, as though an attempt was made to split it for lintels, and we have been told that other stones from the site were used in this way. Surprisingly, it was never documented until it was recorded by the Glasgow University survey.

With surveying help from Ron Curtis, we have made various investigations at this site and have identified at least 20 features including other stumps, fallen stones and socket holes. We found a neolithic hand axe in the turf alongside one prone slab. We have no doubt that this is a more complex site than yet appreciated. We have attempted to fit our finds onto some kind of ring, or to find extra stones to complete arcs of a ring, but these have never been successful. It would appear that this site is not, as was first thought, the remains of a large ring. By the erect stone are a fallen stone and a socket hole, fairly close together near the rising slope limiting the

Table 2.4 A comparison between stone spacings in the east side of the avenue at Callanish I and distances corresponding to certain declination decrements. The 4G value assumed is such that the distance from stone 1 to stone 8 corresponds to a 24-hour decrement.

Stone no.	Distance along row from stone centre to stone centre (metres)	Time from maximum declination (hours)	Distance calculated from time interval (metres)
1	0	0	0
2	4.32	6	4.35
3	8.25	8	7.73
4	17.20	12	17.40
5	26.15	$14\frac{1}{2}$	25.41
6	33.73	$16\frac{1}{2}$	32.90
7	39.43	18	39.16
8	69.61	24	69.61 (assumed)

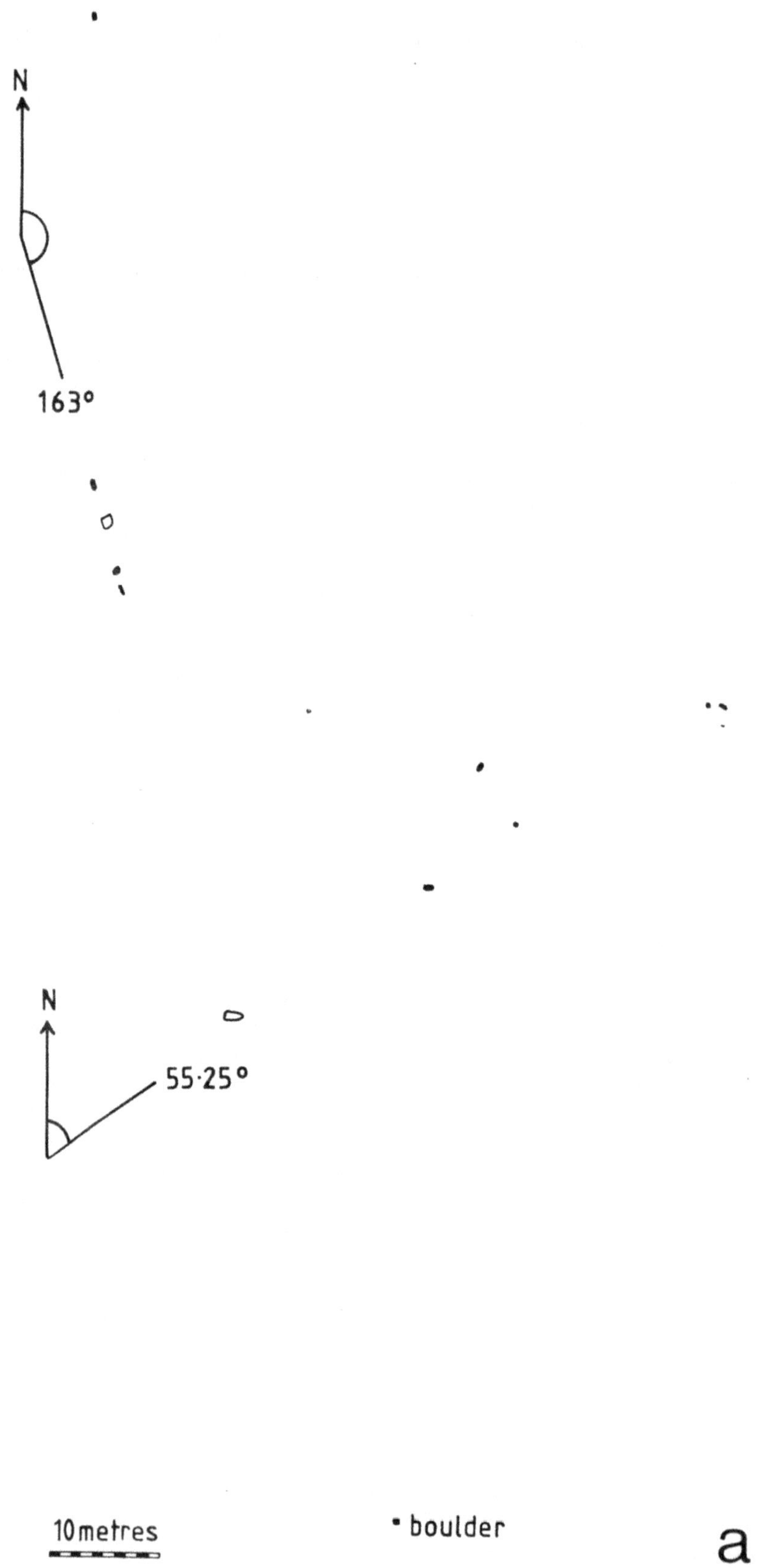

Fig. 2.5 (a) Airigh nam Bidearan, Callanish V. Profiles 2.7c, 7d.
(b) Airigh na Beinne Bige, Callanish XI. Profiles 2.6e, 8b, 8c.

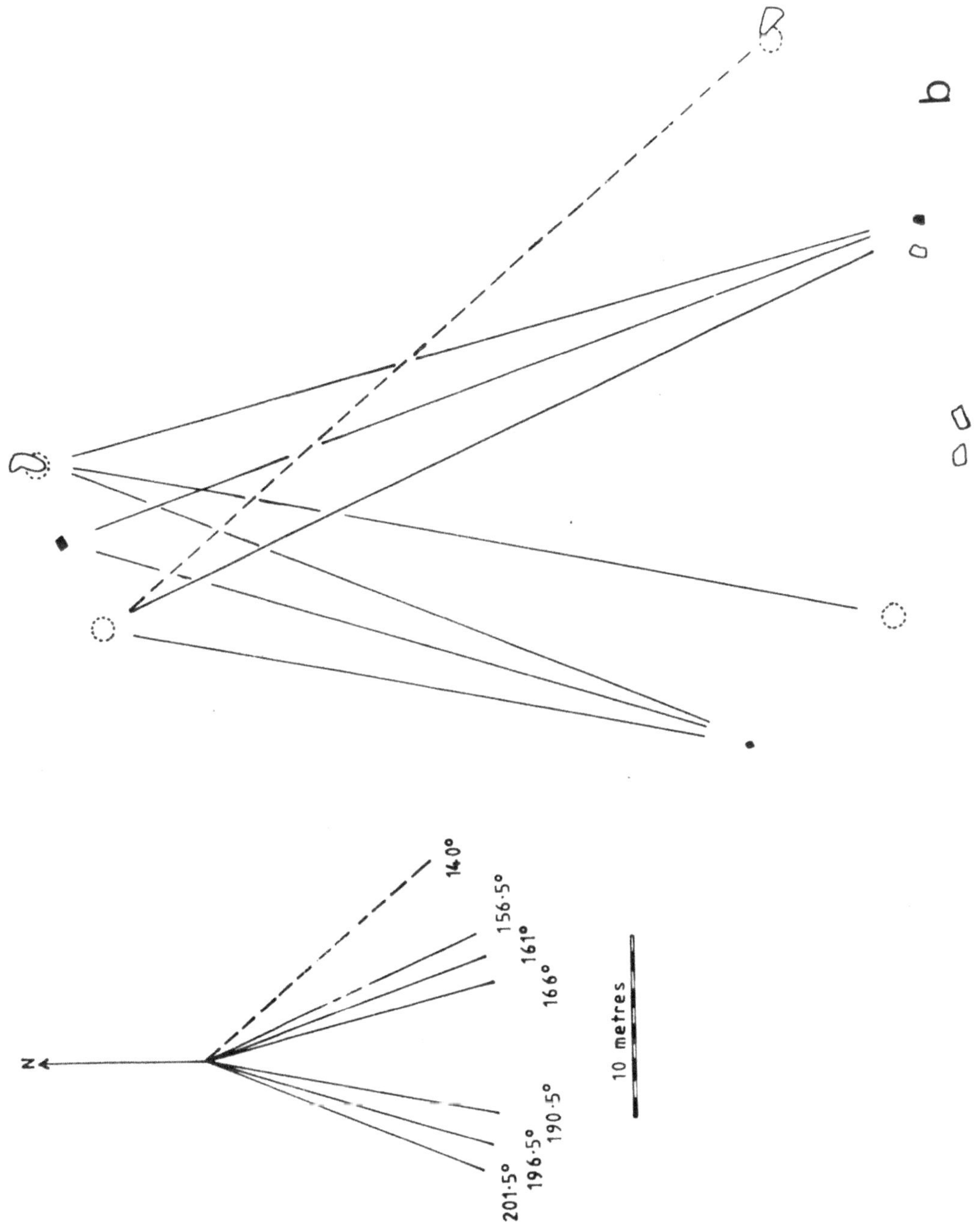

b
N
140°
156·5°
161°
166°
190·5°
196·5°
201·5°
10 metres

plateau to the north. There are similar features along the edge of the plateau where it slopes down to the south.

We erected surveying poles by each fallen stone or socket and tried using the lines from one stone to another as viewing lines (fig. 2.5b). All the lines shown indicate parts of the horizon which were related to the moon at the southern extreme of the major standstill, apart from the dashed line, which indicates midwinter sunrise around 1550 BC. The alignment indicates a small but prominent hillock in the moors, below the true but featureless horizon. Calculation shows that the two stones, the subhorizon feature and the rising midwinter sun were in approximate alignment (fig. 2.6e), but with no obvious feature on the true horizon this cannot be considered an accurate indication.

Fig. 2.8b shows the Sleeping Beauty range as seen from site XI. Three of the alignments shown on the plan, each about 50 metres long, indicate different parts of the range. While one alignment indicated the upper limb of the June full moon (but only when 12 hours off the monthly maximum) it will be seen that the other two alignments seem to have indicated the maximum perturbation, one showing the upper limb (September first quarter moon) and the other showing the lower limb (March third quarter moon).

Another four alignments indicate sections of the Clisham range (fig. 2.8c). We find it difficult to believe that the presence of the main Callanish site in line with the right slope of Glen Langadale is anything more than a coincidence. The two parallel alignments indicated the moon skimming Clisham and the notch to its right at maximum perturbation (March third quarter moon). The other two alignments show two positions where the moon's upper limb may have reappeared after setting at minimum perturbation (June full moon).

To summarise, site XI may well have been an observatory for southern lunar events. Apart from December, when the southerly moon was new and could not be observed, the site seems to have included alignments for use at any season. The remains on the ground are ambiguous and incomplete and might repay detailed archaeological investigation.

2.10.2 Site V—Airigh nam Bidearan

Most of the few plans of site V, Airigh nam Bidearan, show only five stones, which stand about 1 metre high. More stones were shown by Sharbau (c. 1860). We have identified Sharbau's extra stones and more possible ones. With Ron Curtis, we have surveyed all these stones. However, the site is covered with peat to an average depth of 60 cms, eroded down to clay in a few gullies. Even by probing it is very difficult to tell if a small stone visible on the surface is exposed bedrock or the tip of an almost buried standing stone. Inevitably, smaller stones may remain unseen beneath the peat. It is unlikely that the true nature of site V will ever be elucidated without major peat clearance under archaeological supervision. With as many as 30 possible stones at the site, it is easy to imagine alignments running in all directions. We suspect that several astronomically significant alignments do exist at the site, but here we limit ourselves to two. The plan (fig. 2.5a) does not show all the stones which have been surveyed.

The most obvious feature of site V is the alignment of three erect stones pointing to Mor Mhonadh, the knees of the "Sleeping Beauty" (see section 2.8). We believe that other fallen slabs, and a large natural boulder, continue the alignment to the south. Thom's horizon profile (MSB:128; MLO:69) from this alignment was drawn from the O.S. maps and has been rejected for this reason from his more recent lunar material (Thom and Thom 1978b). The "Sleeping Beauty" is not very recognisable and, in fact, very little of the range shown in Thom's figures is visible when looking along the alignment, due to the immediate hillslope—as also noted by Cooke et al. (1977:121). Nevertheless, their profile also omits the near horizon. They concluded that only one of Thom's horizon foresights was visible; this was the only foresight in the whole Callanish complex confirmed as accurate by their rigorous methods. Our own horizon survey (fig. 2.7c) was taken from a position within the alignment of three erect stones, near the prone slab. A heathery slope, less than 50 metres away, forms much of the horizon. The observer's position along the alignment is critical. Further north, the near slope obscures more of the distant range. Further south, more of the Sleeping Beauty may be seen. It is clear that the moon never fully rose when viewed in this way. The alignment clearly indicated the relevant section of horizon, and the large boulder is also in line. The minimum perturbation is drawn, June full moon. We have omitted the limbs of the maximum perturbation to avoid confusion, but it is obvious that the moon would, at most, have produced a brief gleam at that time. The surprising thing is that, by moving a short distance away from the alignment, the rise could be fully observed.

Just as site XI is ideally situated for observations to the south, so site V has an extensive northern and eastern horizon. There are no prominent ranges here, but there are small but noticeable notches at the azimuths of some of the four northerly lunar events.

The event which at present seems most likely to have been indicated on site is the moonrise at the northern extreme of the minor standstill. One suspected prone slab, also on the recognised alignment, one erect stone and two partially buried stones align with a junction between near and far horizons. This junction is obvious in the field, but unfortunately is not clear on our reduced profile (fig. 2.7d). The latter shows that the lower limb of the moon was indicated, and its position with regard to the junction could have distinguished accurately the maximum and minimum perturbations. Thom and Thom (MRBB:173) include a profile from Callanish V, with moonset at the north extreme of the major standstill indicated by the attitude of an erect stone. However, the stone concerned is leaning, and its dimensions do not warrant the determination of an attitude. Also, the moon cannot have been observed in June as they describe, as it would have been new at that time.

2.11 SINGLE STONE SITES

When John MacCulloch (1819) produced the first measured plan of the main site, he also wrote "solitary stones apparently of a monumental nature, are found in this neighbourhood, as well as on the island of Bernera." A few of the solitary stones remain today, of which only site XII was recorded before we extended the Callanish numbering to site XIX, perhaps rediscovering some of the stones known to MacCulloch.

2.11.1 Site XII—Stonefield

Callanish XII is a single stone which stands 1.3 m tall, 0.7 m broad and only 0.2 m thick. In the 1960s, it was surrounded by a plinth when a housing estate called Stonefield was built around it. Suggestions have been made that the stone was moved during building construction, but we have spoken to, among others, the crofter who exposed the lower part of the stone during peat cutting, and the foreman in charge of the building project. We have established beyond all doubt that the stone was carefully protected during the building and was not moved.

To the archaeoastronomer, the surrounding houses provide an added difficulty in the production of profiles. It was fairly obvious that the flattened face of the stone indicated a section of horizon to the right of the V-notch of Glen Langadale. Further investigation involved surveying both by the stone and at a satellite station, climbing a hillock to view over stone and rooftops, and carefully combining the data from these sources. The profile (fig. 2.8d) shows that the attitude of the single stone at site XII indicated the summit of Uigsnaval and the upper limb of the setting moon, which disappeared behind the summit at maximum perturbation but not at minimum. Teilesval and Uigsnaval are only visible above a much closer horizon between azimuths of 194° and 196° 20'. Uigsnaval gives a 4G value of well over a kilometre, while the near horizon gives 60 metres. Only observations on the nearer horizon could have been extrapolated using the avenue at the main site.

2.11.2 Site XIII—Sgeir nan Each

Callanish XIII is situated on a small tidal island, Sgeir nan Each, in an inlet off East Loch Roag. It consists of a thin prone slab of stone (1.3 m x 0.5 m x 0.1 m), two collapsed cairns, and other features which may represent socket holes. The slab has been trimmed relatively recently, perhaps for intended use as a lintel. Other stones may have been taken from the site for the nearby stepping stones. We accept that the archaeological status of site XIII is less certain than that of our other sites.

Its position, however, is very interesting: site I is clearly visible silhouetted against Clisham, though the view is now confused by the housing of Callanish village. We suspected at an early stage that viewing from site XIII to site I to Clisham would have given the same horizon indications as Thom (MSB:ch.11) recorded for the site I avenue lines (cf. fig. 2.6c). On the profile (fig. 2.8e) we have shown the position of stone 29 and the limits of site I. The profile shows that the moon, setting at the southern extreme of the major standstill, first touched the mountain range above the centre of site I—maximum perturbation, March third quarter moon. The declination slope is so gentle that the setting positions of the moon at maximum and minimum perturbation were several degrees apart in azimuth. The V-notch is obscured by a near horizon and the upper limb at minimum perturbation (June full moon) probably set into the junction between this near horizon and the summit of Teilesval. This near horizon gives a 4G value well within the capabilities of the main site for extrapolation.

With the uncertain nature of site XIII, it is not clear whether the Sleeping Beauty range was indicated from here. However our results[4] show that the maximum and minimum perturbations could have been distinguished using this range.

2.11.3 Site XVI—Cliacabhaidh

Cliacabhaidh, Callanish XVI, includes an erect stone, 1 m tall x 1 m x 0.3 m. It stands 800 m from the main site and is in line with the south row and central megalith, at an azimuth of just over 1°. Local traditions imply an ancient origin of the stone, and we have long suspected that it was associated with the low transit of the major south standstill. The view from site XVI to the main site is now confused by houses and poles. Surveys and calculations have shown that, at major south transit, the moon would have appeared to hang low over the main site, barely one degree above it. Walking from site XVI to site I, the ground first falls and then rises. On the whole of the rising ground, now occupied by houses and crofts, the moon at transit would have been silhouetted behind the stones. At the major standstill, this would have occurred for 2 or 3 days each month, as the top of stone 29 has an altitude of about 3° as seen from this area of land. However, this stone does not indicate the main site by its attitude. To the SE it may indicate the sunrise at 2 of Thom's inter-calendar dates (April and August), but we are not at present considering these. To the NW the attitude indicates a very close horizon. A survey[4] showed that the minor north moonset did occur in the general direction indicated by the stone.

Another erect stone, less than a metre high, is about 20 metres away to the south, but it is in the line of a field wall, so its archaeological status is uncertain. Its attitude indicates an azimuth of 169°. A third stone, to the east of the field wall, is almost buried in peat; systematic probing indicates that it is about 0.97 m tall, 0.75m broad and 0.16 m thick, and surrounded by packers (see plan fig. 2.4f). (The layer of peat around this stone may have protected archaeological evidence which may eventually be recovered.) Its attitude, as determined from its visible top surface and the probe survey, indicates an azimuth of 162°. Unfortunately, the horizon is obscured by a nearby house, and we needed to use surveying poles and a step ladder to determine that it indicates part of the Sleeping Beauty range. The attitudes of the 2 stones are shown on the profile (fig. 2.9a). If the position of the stone in the wall was prehistoric and deliberate, it may have indicated the lower limb, maximum perturbation, March third quarter moon. The more definite megalith indicated first gleam of the minimum perturbation, June full moon. After this same moon set, within 9 hours of the monthly maximum, it flashed again.[4]

2.11.4 Site XVII—Druim na h-Aon Chloich

Early in our research, we asked one of our Gaelic speaking neighbours to go over the 1:10000 Ordnance Survey map with us, to find any place names which might indicate the former presence of standing stones. On searching

Druim na h-Aon Chloich, the ridge of the one stone, we found a prone and broken slab about 1.7 m long when complete. We later found that it was marked on the 1853 OS map as "Stone", and that a very brief account of its demolition during peat cutting had been recorded by A. A. MacGregor (1933): "A standing-stone once stood in the doctor's garden at Garynahine; but it fell on its face some years ago, while peats were being excavated in close proximity to it." The nearest house, which stands alone, was formerly occupied by the local doctor, and the stone is on the moor just outside the garden.

From site XVII the V-notch of Glen Langadale is particularly spectacular and the small stone circle Ceann Hulavig (Callanish IV) is immediately in line with the bottom of the notch. It can be seen clearly with the naked eye and would have been more obvious before the growth of peat. Some workers would not consider an alignment from a fallen slab to a stone circle, but we believe that in this case, a distant circle gave an accurate indication of an event on a conspicuous horizon feature. The profile (fig. 2.9b) shows that this V-notch could have been used to distinguish between perturbations, as the upper limb of the June full moon at minimum perturbation reappeared briefly in the notch.

2.11.5 Site XIX—Buaile Cruaidh

Within Callanish village stands a stone, 0.7 m high, apparently the truncated remains of a taller megalith. Certainly the remains are large enough to have been the base of a massive pillar. Local people accept that it may be ancient. We call it Callanish XIX, and the Gaelic name of the hillock on which it stands is either Buaile Cruaidh or Ceann a Cruaidh. Determining the attitude of a truncated pillar, damaged on several faces, is bound to be a somewhat subjective process. We can only say that we determined the attitude of this stone before calculating the moon's position on the horizon beyond. The attitude is to the right slope of Roineval. The indicated section of horizon (fig. 2.9c) shows the lower limb of the moon rising at the southern extreme of the major standstill, maximum perturbation, March third quarter moon. While the setting of this moon was not indicated on site, it is interesting to note that the moon at maximum perturbation (but not at minimum) disappeared behind Mullach an Langa before reappearing in the V-notch.[4]

2.12 THE STONE RING SITES

Having found so many indications of the southern extreme of the major standstill at the single stone sites, we felt that we should examine the better-known small rings referred to by Professor Thom as sites II, III and IV; and also the prone circle first recorded by RCAHMS (1928) and designated Callanish X by the Glasgow Survey team. We have examined this lunar event from each of these sites.

2.12.1 Site II—Cnoc Ceann a'Gharaidh

Callanish II, Cnoc Ceann a'Gharaidh, consists of 5 tall megaliths and at least 2 prone slabs. Sharbau's map of about 1860, shortly after peat clearance, shows several other internal features now lost among the slowly re-growing peat—cists, cairns, etc.—and a 2m-long prone stone later removed to the proprietor's castle in Stornoway, where it can no longer be located. The five erect stones appear to lie on the perimeter of an ellipse (fig. 2.4a). While we have not found evidence for use of the Megalithic Yard at Callanish, we do believe that all the rings could have been laid out as ellipses or flattened circles. Callanish II and IV are both ellipses of exactly the same proportions, though of different dimensions. There is always a certain amount of uncertainty in the assumed axis of a geometrical figure, especially where only some of the stones remain standing, as here. However, the fact that the axis of the ring at site I is parallel to the lines of the avenue led us to investigate the directions of the putative axes of the other ring sites. Our geometrical figures have similarities with those published recently by Curtis (1979), but were arrived at independently without the restrictions imposed by the use of the Megalithic Yard.

Taking the axis of site II as drawn, it indicates the lower limb of the moon rising over Roineval at the southern extreme of the major standstill, maximum perturbation, March third quarter moon.[4] However, we accept that the uncertainty in the axis makes such precision unreasonable, and would only suggest a symbolic orientation of the axis of this circle upon the major south moon rise. While the corresponding moonset is not indicated on site, it is interesting to note that, as at several other sites, the moon at maximum perturbation (but not at minimum) disappeared behind Mullach an Langa before reappearing in the V-notch, and then set into a near horizon which obscured part of the V-notch.[4]

We have also investigated three inter-site lines from site II. The profile in fig. 2.6f shows Cul a'Chleit (site VI) silhouetted on the horizon. Site VI, from which no horizon surveys have been made as yet, consists of 2 erect stones at opposite ends of a small hillock, and various prone slabs. The profile shows that the southern extreme of the minor standstill occurred at the azimuth of site VI. Within a day of the monthly maximum, the distant site would have been silhouetted against some part of the moon, whether at maximum or minimum perturbation. The distance, 3.35 km, gives a 4G value equal to the length of the east side of the avenue at site I.

Site X, a ring now consisting only of prone stones (see 2.12.3 below), is also on the horizon seen from site II, and would have been very conspicuous with the stones erect. On fig. 2.6g the arrows show the limits of the site. As with the previous profile, the stones would have been silhouetted against the rising moon at least once each month near the major standstill (north extreme). The distance again gives a 4G value within the extrapolation capabilities of the main site.

On the croftland of the village of Breasclete, 3 km north of Callanish II, there is a small ruinous chambered cairn, Cnoc a Phrionnsa (Prince's hillock). It is not on the horizon, and unless it was more conspicuous when complete, it is not a likely horizon indicator. Our profile,[4] however, shows that the moon at the northern extreme of the major standstill, having risen over site X, would have set above Cnoc a Phrionnsa. Could the cairn have been made conspicuous with a coating of white quartz? Was this another symbolic setting of the moon into a burial cairn, as perhaps occurred at the main site? Cooke et al. (1977) considered a line through 2 stones at site II indicating this same section of horizon, but which just missed a lunar declination.

2.12.2 Site III—Cnoc Fillibhir Bheag

Callanish III, Cnoc Fillibhir Bheag, consists of nine erect stones in an outer ring, 4 erect stones within the ring, and probably seven fallen stones. Our documentary and site investigations have enabled us to draw a composite plan (fig. 2.4b) which is more complete than any previously published. Thom (MSB:126) gives a plan representing the geometry as a double ellipse, but we believe that a flattened circle fits the stones of the outer ring more effectively. Two of the stones within the circle lie on its axis.

As the axis misses moonrise to the SE, we investigated the opposite azimuth of 334°, already surveyed by Cooke et al. (1977) on the basis of an alignment of 3 erect stones. Cnoc a Phrionnsa is 6 degrees left of the alignment. Our profile[4] shows the moon at major north standstill setting into a small notch. The upper limb is indicated, and its position with regard to the notch could distinguish accurately between the perturbations. This is the third ring which may have a symbolic alignment to a lunar major extreme position.

Although they were not indicated on site, we drew profiles of the two southern ranges. Inevitably, the moon rose slowly out of the Sleeping Beauty, with several degrees of azimuth between the perturbations.[4] When setting (fig. 2.7a), the moon disappeared behind Mullach an Langa and reappeared in the V-notch at both maximum and minimum perturbations. Once again we find that when viewed from within the circle, the floor of the V-notch was obscured by a nearer horizon; and that the upper limb of the moon at maximum perturbation set into this near horizon.

2.12.3 Site X—Na Drommanan

Na Drommanan (the ridges) has been incorrectly named Druim nan Eun (ridge of the birds), and even mis-spelt Druim nan Eum, in several publications.

The site consists of prone stones only, with sockets and packing stones. The stones may never have been erected, or they may have fallen from their erect positions. At present, there is no evidence on which to decide between these two alternatives. As far as we know, the only archaeological examination of the site was that carried out in 1914 by the Royal Commission (RCAHMS 1928). There are about 20 slabs at the site. We believe that the best geometric shape for the site is a flattened circle type A, the same shape as at sites I and III, although the sites differ. However, we feel that there is at least 9° possible variation in the direction of the axis, as shown in fig. 2.4e. Even so, the profile (fig. 2.8a) shows that site X is the fourth stone ring in the area which may have a symbolic axis alignment to a lunar extreme. In fact, the 9° variation which we have allowed covers all the parts of the horizon intersected by the rising moon at southern extreme of the major standstill. When the moon set a few hours later, it flashed in the Glen Langadale notch only at maximum perturbation.[4]

2.12.4 Site IV—Ceann Hulavig

The site at Ceann Hulavig consists of five erect stones. The central area was cleared of peat last century, leaving a wet hollow within the ring, and revealing a smaller upright stone (fig. 2.4d). The stones lie on an ellipse, not drawn, the axis of which does not align with a lunar event, but gives an approximate indication of midsummer sunset over the main site. The small erect stone inside the ring stands within the possible cairn remnants, and is on the axis of the ellipse. It does not align with the axis, but has an attitude of $162^{o}.5$. This indicates the slope just left of an obvious horizon junction, where a nearby hill hides one end of the Sleeping Beauty. The profile (fig. 2.7b) shows that this junction marked the rising major south moon at minimum perturbation, June full moon. Most of the ring sites are situated on ridges, but not on the highest points. From site IV, the moon set into the slope immediately south of the site, which obscures the Clisham range. The stones could have been erected a few metres up the slope, when the range would have been fully visible.

When the moon was observed rising at site IV, it could have been observed setting a few hours later from site XVII, using site IV as a foresight for the moon's dramatic reappearance in the V-notch. Was site XVII, less than 2 km from site IV, a complementary outlying stone?

2.12.5 Other sites

All the sites dealt with so far are within a few kilometres of the main site. We have omitted several sites—VII, IX and XVIII, because we are doubtful if they are prehistoric structures; XIV, because we suspect that it is solely an equinoctial foresight related to the main site; and sites VI and VIIIA, purely because of lack of time.

2.13 SITES ON GREAT BERNERA

Callanish VIII and XV, however, are sites on the island of Great Bernera, which may be considered part of the Callanish complex only because they are intervisible with some of the other sites. They are not within the immediate vicinity of Callanish, and so could not use the same two ranges as horizon markers for the southern extreme of the major standstill. We investigated this event from these two sites.

2.13.1 Site XV—Airigh Mhaoldònuich

Airigh Mhaoldònuich, site XV, is the largest of the single megaliths (3.5 m x 1.3 m x 0.4 m). It was reported to us by the crofter on whose land it now lies. He told us that in his youth, he deliberately chipped away the upper end to make supports for a fence. When complete and erect, this stone must have stood about $3\frac{1}{2}$ metres high. Perhaps this was one of the solitary monumental stones on Bernera mentioned by MacCulloch in 1819. The site has wide views, expecially to the south across Loch Barraglom, the narrows separating Great Bernera from Lewis.

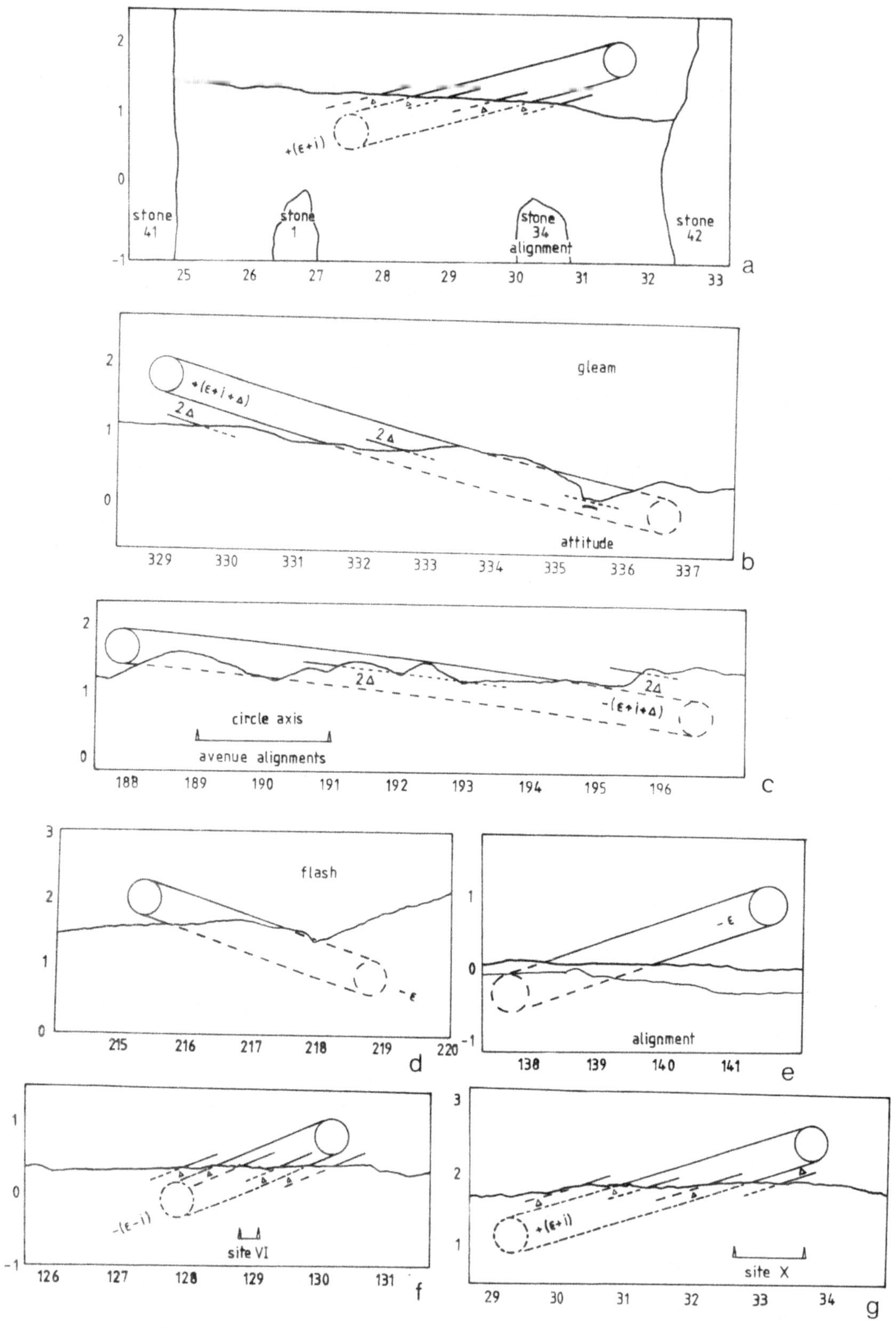
a
stone 41
stone 1
stone 34
alignment
stone 42
+(ε+i)
b
gleam
attitude
+(ε+i+Δ)
2Δ
c
circle axis
avenue alignments
−(ε+i+Δ)
2Δ
d
flash
−ε
e
alignment
−ε
f
site VI
−(ε−i)
g
site X
+(ε+i)

Fig. 2.6

Fig. 2.6 (a) Site I, horizon indicated by alignment stone 9-stone 34 and flat NW face of stone 34 (long-dashed line in fig. 2.3a). On figs. 2.6a-9c inclusive, 10a and 10b, the vertical scale is degrees apparent altitude and the horizontal scale is degrees azimuth. (b) Site I, horizon indicated by attitude of stone 8, flat NE face (short-dashed line in fig. 2.3a). Short line at 335.5 shows former ground level below peat. (c) Site I, Clisham range from by circle. Two sides of avenue indicate the azimuth range 189°-191°. Midline of avenue and circle axis, azimuth 190° (continuous lines on fig. 2.3a). The hills are not visible along these lines. c.f. Thom 1967: 123; and fig. 2.8e. (d) Site I, horizon including "midwinter sunset notch" from N end avenue. Indicated by stone 8-socket 18A? (e) Site XI, horizon (and sub-horizon feature?) indicated by alignment (dashed line in fig. 2.5b). (f) Site II, profile including site VI on horizon. (g) Site II, profile including site X on horizon (all stones now prone).

Fig. 2.7 (a) Site III, profile of the Clisham range showing gleam of moon at both maximum and minimum perturbation. (b) Site IV, horizon indicated by attitude of central stone (see fig. 2.4d). (c) Site V, horizon including "Sleeping Beauty's knees". Slope of near hillside continued below horizon to left of azimuth 164° to include natural boulder on alignment (see fig. 2.5a). Movement along the alignment will render more or less of the Sleeping Beauty range (and of the rising moon) visible. (d) Site V, horizon indicated by line of stones (see fig. 2.5a). (e) Site XV, detail of moon gleam from fig. 2.10b. (f) Site VIII, southern horizon. The moon just appears, near transit, at minimum perturbation. The slope of Teahaval, at az. 174° in this profile, also appears at az. 194°-195° on fig. 2.7e.

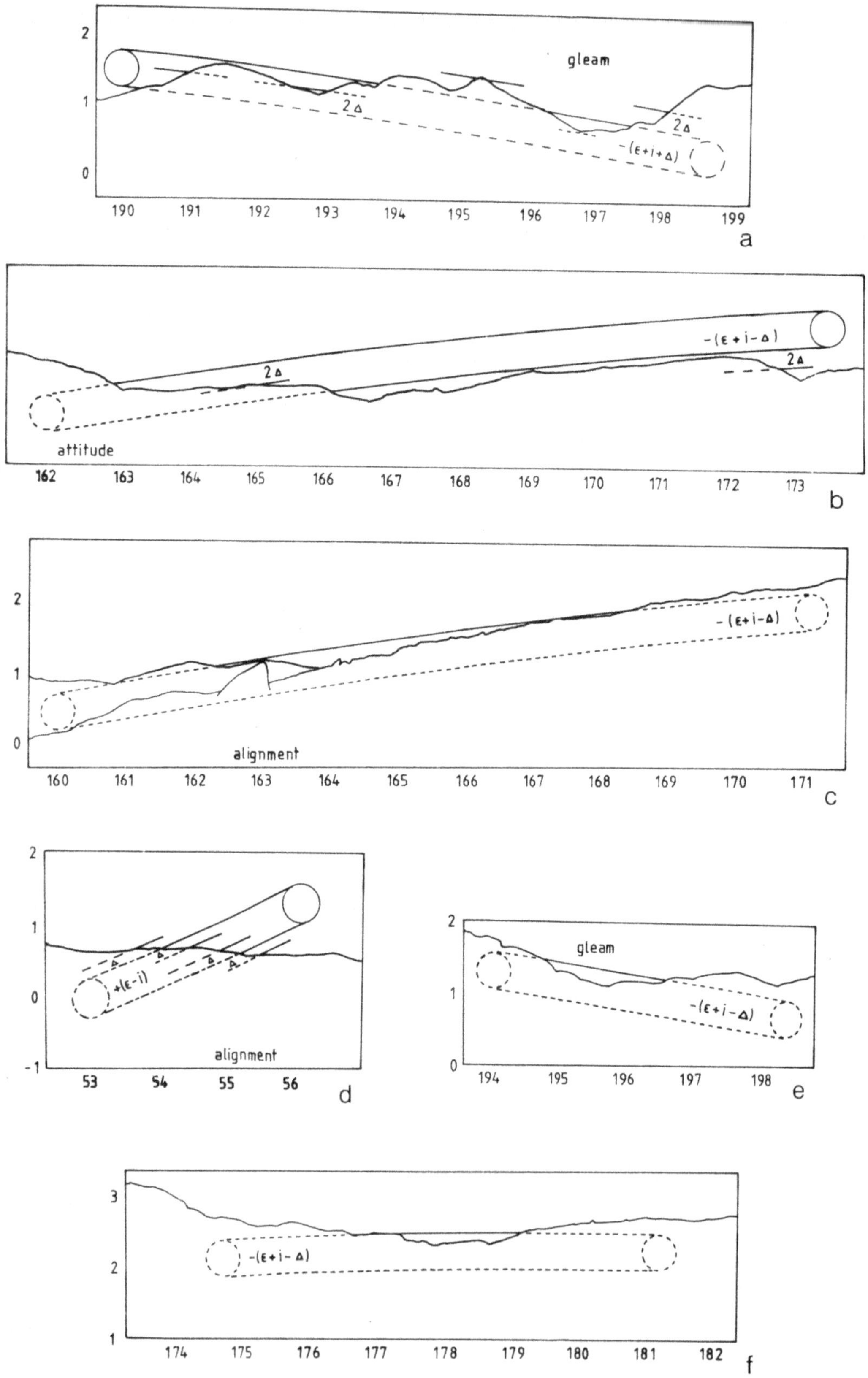
gleam
2Δ
2Δ
-(ε+i+Δ)
a
-(ε+i-Δ)
2Δ
2Δ
attitude
b
-(ε+i-Δ)
alignment
c
+(ε-i)
alignment
d
gleam
-(ε+i-Δ)
e
-(ε+i-Δ)
f

Fig. 2.7

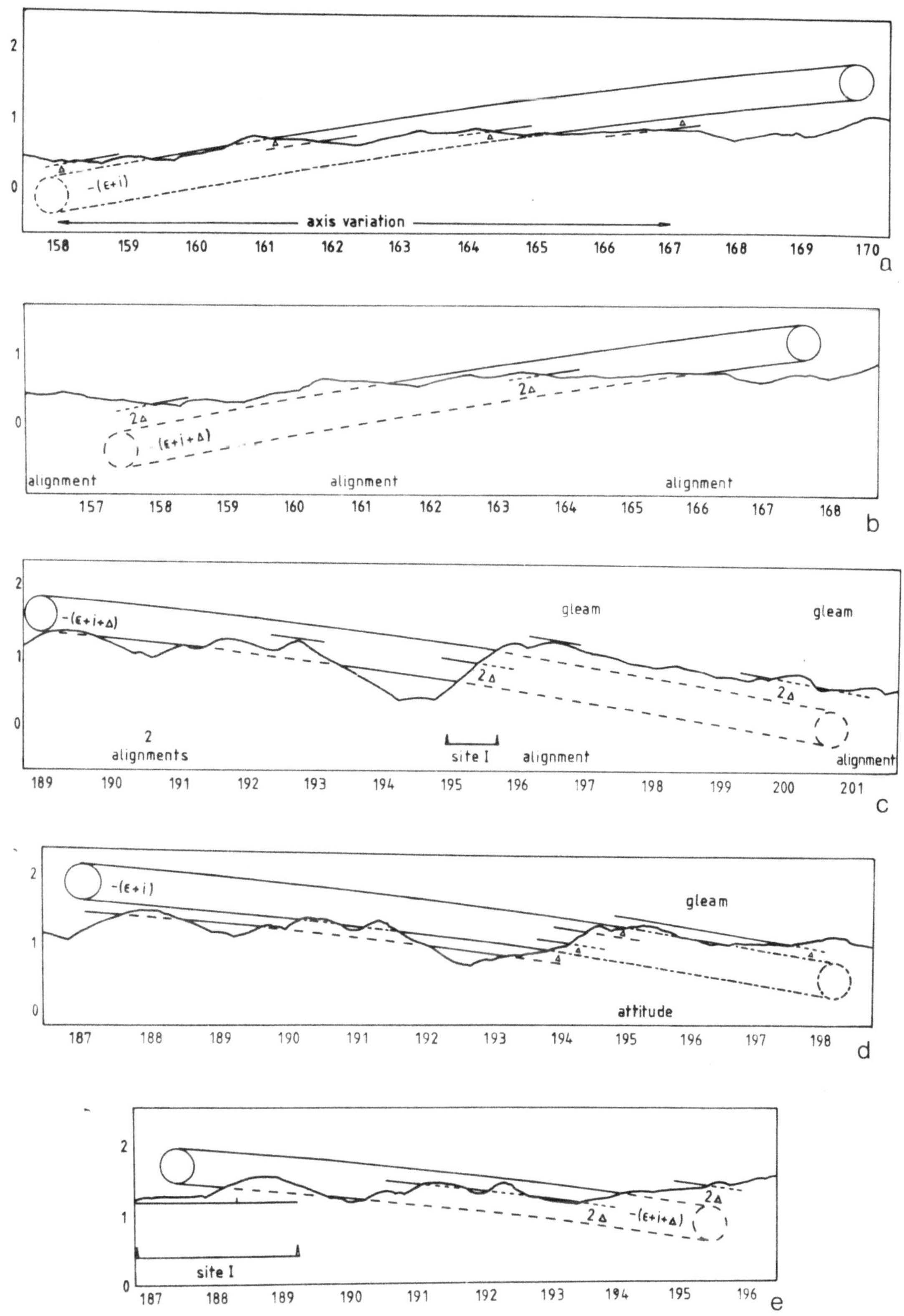

Fig. 2.8 (a) Site X, Sleeping Beauty range. Possible geometric axis with 9° variation (see fig. 2.4e) covers whole range of declinations for -(ε+i) rise. (b) Site XI, Sleeping Beauty range with indications from 3 alignments at the site (see fig. 2.5b). (c) Site XI, Clisham range with position of site I below horizon and indications from 4 alignments at the site (see fig. 2.5b). (d) Site XII, horizon indicated by attitude of flattened stone. (e) Site XIII, Clisham range and site I. The position of stone 29 at the centre of the circle is shown on the line representing site I. (cf. fig. 2.6c).

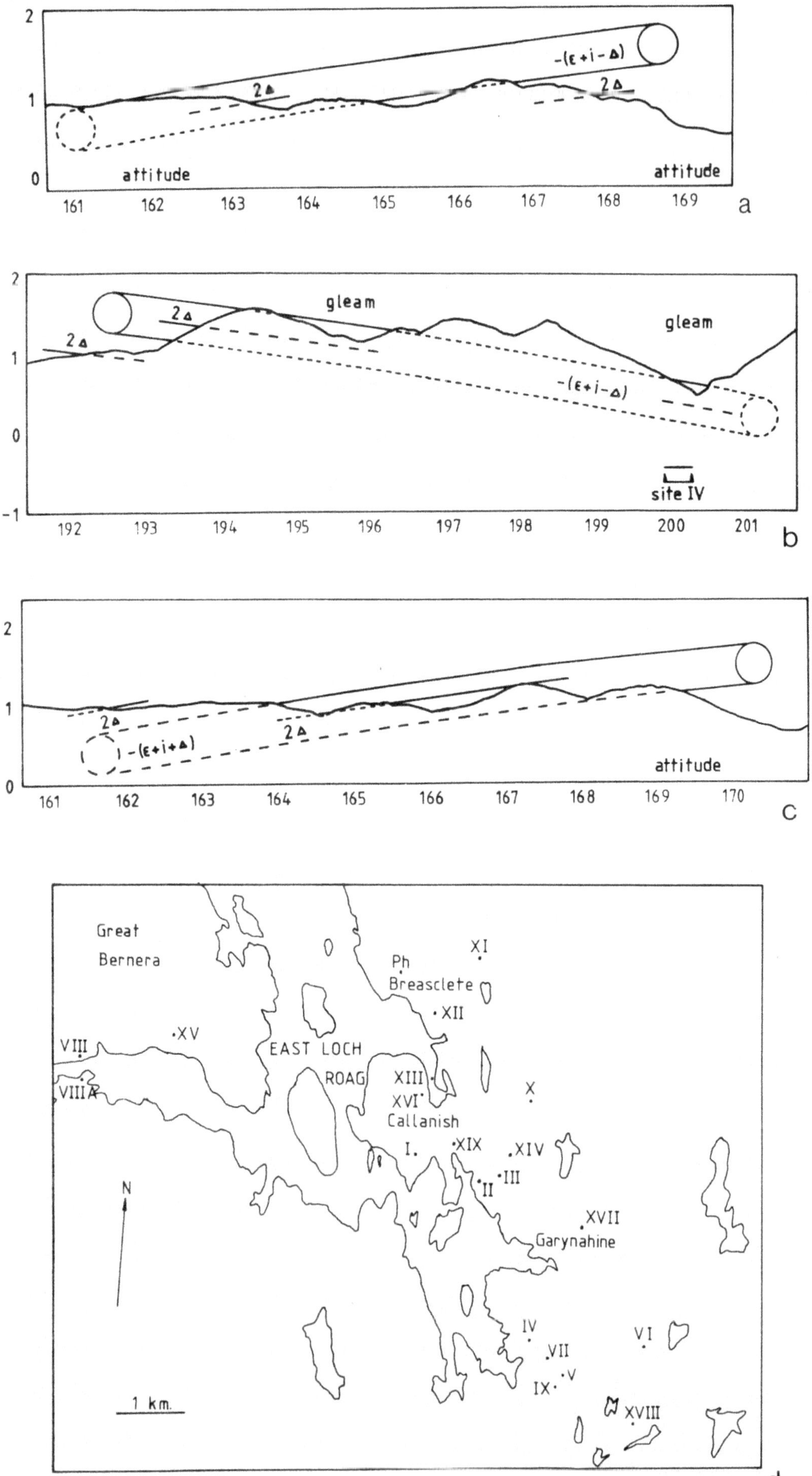

2
1
0
$-(\epsilon+i-\Delta)$
2Δ
2Δ
attitude
attitude
161
162
163
164
165
166
167
168
169
a
2
1
0
-1
gleam
gleam
2Δ
2Δ
$-(\epsilon+i-\Delta)$
site IV
192
193
194
195
196
197
198
199
200
201
b
2
1
0
2Δ
2Δ
$-(\epsilon+i+\Delta)$
attitude
161
162
163
164
165
166
167
168
169
170
c
Great
Bernera
Ph
Breasclete
XI
XII
XV
VIII
VIIIA
EAST LOCH
ROAG
XIII
XVI
Callanish
X
I.
XIX
XIV
III
II
N
XVII
Garynahine
IV
VII
VI
IX
V
XVIII
1 km.
d

Fig 2.9

Fig. 2.9 (a) Site XVI, Sleeping Beauty range indicated by the attitudes of two stones at the site. (b) Site XVII, Clisham range with site IV immediately below the V-notch. (c) Site XIX, Sleeping Beauty range, with right slope of Roineval indicated by the attitude of the stone. (d) Map of the land around East Loch Roag, Isle of Lewis, showing the position of the sites in the area. See tables 2.2 and 2.3 for site names. Ph = Cnoc a Phrionnsa.

Fig. 2.10 (a) Site I, profile from N end avenue. $-(\varepsilon + i + \Delta)$. (b) Site XV. $-(\varepsilon + i - \Delta)$. (c) Map showing relative position of the Callanish sites and the ranges of hills which may have been used for observing the rise and set of the moon at southern extreme of major standstill (the hypothetical "Callanish diamond"). Four sites of dubious status, not used in this paper, are omitted, though they are in the "diamond". (d) Summary of indications found so far.

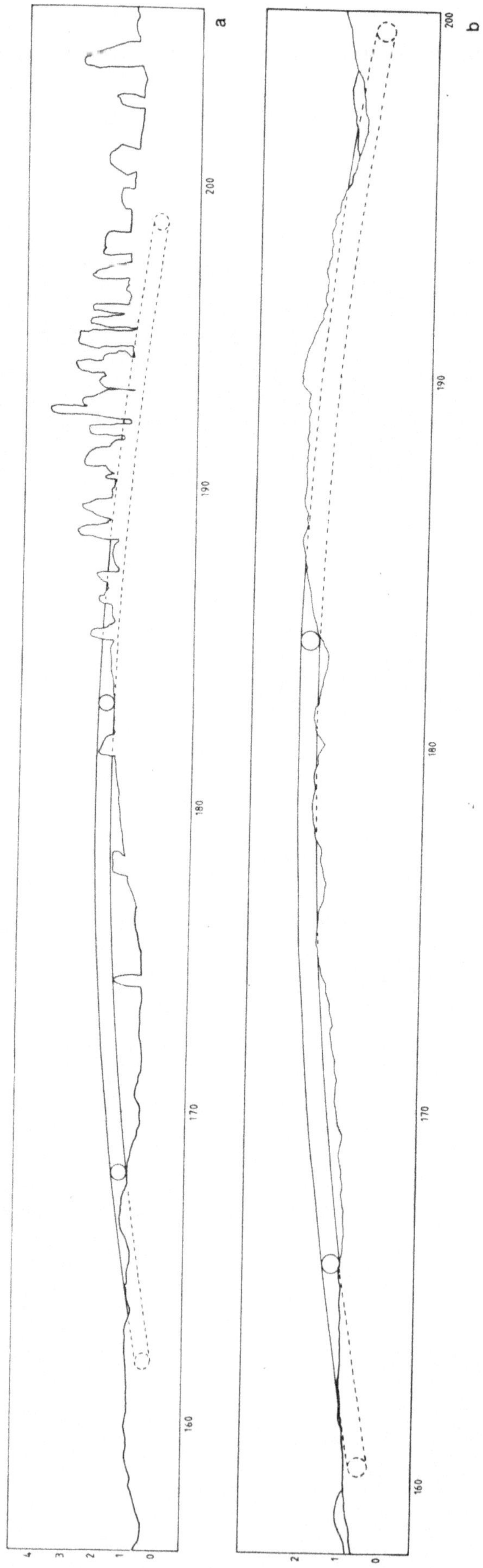

Fig. 2.10

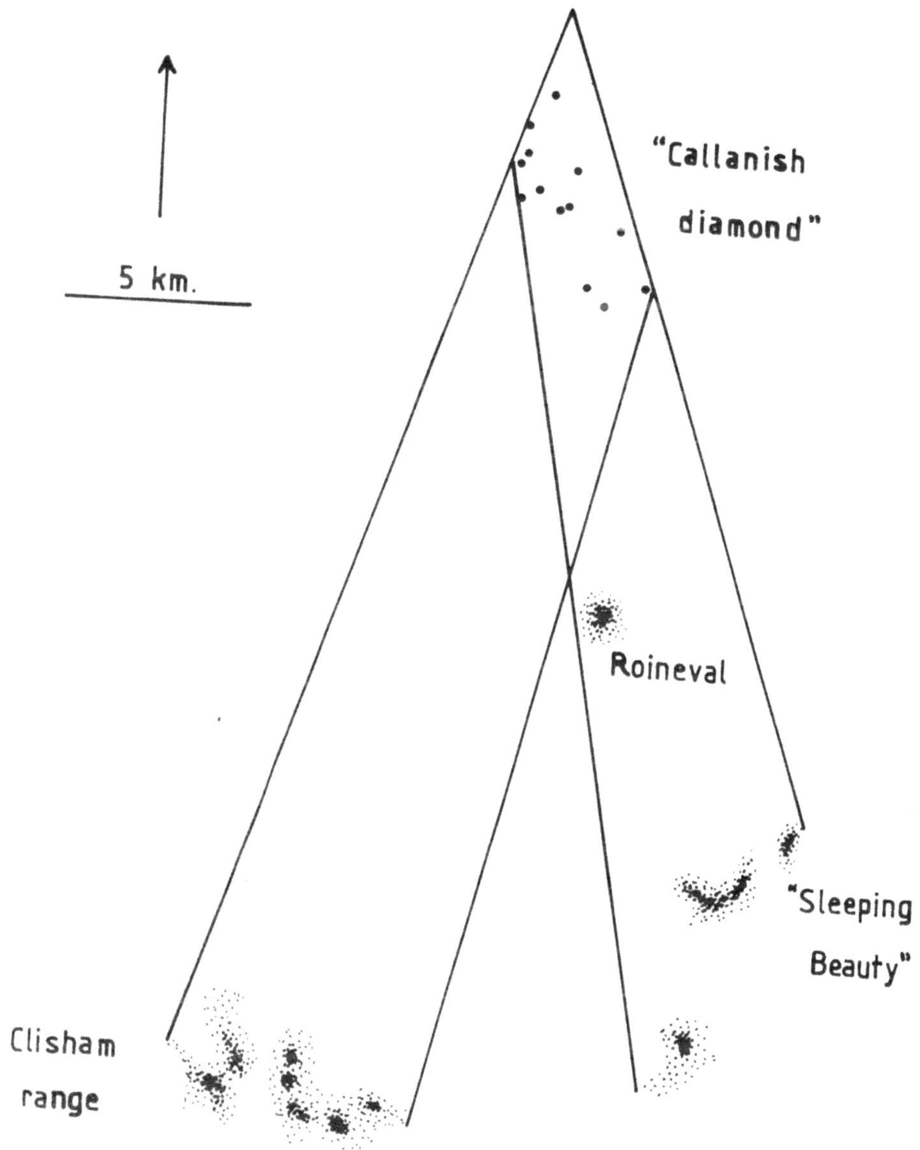
"Callanish
diamond"
5 km.
Roineval
"Sleeping
Beauty"
Clisham
range

c

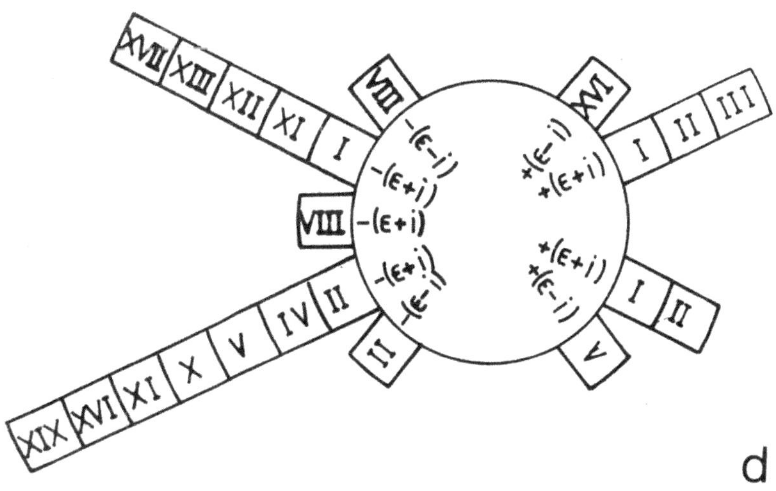
XVII
XIII
XII
XI
I
VIII
XVI
I
II
III
-(ε-i)
-(ε+i)
+(ε-i)
+(ε+i)
VIII
-(ε+i)
-(ε+i)
-(ε-i)
+(ε+i)
+(ε-i)
II
IV
V
X
XI
XVI
XIX
II
V
I
II

d

Having surveyed the southern horizon from this site and plotted the path of the moon at the southern extreme of the major standstill (figs. 2.10b and 2.7e), we found three features similar to those at several Callanish sites:

(i) a distant summit behind a near horizon (at azimuth 162°—not clear on the reduced profile);

(ii) a slope which the moon crept up so that it never fully rose at the maximum perturbation; and

(iii) a notch in which the upper limb of the moon reappeared.

It will be seen that completely different ranges of hills are involved. The June full moon, minimum perturbation, rose with its upper limb behind the two distant summits. Perhaps this acted as a warning of a significant moon-set, as the moon flashed in a notch on the same occasion. At the maximum perturbation, the lower limb of the moon hardly ever lost contact with the horizon and did not flash in the notch (unless it was more than 8 hours away from its monthly declination maximum). The near horizon in front of the distant summits is only $1\frac{1}{2}$ kilometres away and gives a 4G value suitable for extrapolation at Callanish I, which is only a few kilometres away by water through sheltered sea-lochs. Although the stone at site XV is fallen and we cannot determine its former attitude, we strongly suspect that it was a lunar backsight established for observing the flash in this notch. (We also have some evidence that it may have been used as a horizon foresight from sites IV or V—its very size would have made it more suitable as a foresight than a backsight.)

2.13.2 Site VIII—Cleiter

From site XV, there is a fine view of the bridge which now connects Great Bernera with the rest of Lewis. On the cliff immediately north of the bridge is one of the most unusual standing stone sites in Great Britain, Callanish VIII. This site is published as Barraglom, Great Berneray, and Bernera Bridge (a name we have used in the past (Ponting & Ponting 1977a)), but the local inhabitants call the area Cleiter.

The one fallen and three erect stones lie on a semicircular arc (fig. 2.4c). The line drawn approximately along the diameter is the edge of a sheer cliff. There is some evidence of deliberate terracing of the cliff-top, and no reason to suppose that half the site has fallen away. This is not a Guardianship site and it has been despoiled by the erection of a power pole, the digging of a water main through the middle and, within the last year (despite our protests) by a stock-proof fence. Just across the Bernera Narrows, a $1\frac{1}{2}$m single stone was displaced about 10 years ago when a new road was built to serve two crofts (site VIII A—see table 2.2). Nevertheless, this is still a fascinating site. Somerville (1912a) noted the semi-circular shape and recorded the astronomical significance of various hilltop boulders, of which large numbers are visible from this site. When the winter sun is low over the Bernera Narrows, it is a remarkable experience, which can also be seen at the main site, to find that one has a double shadow cast on the stones, one by the sun and one by the sun's reflection on the water.

The flat face and curved top half of the most westerly stone seem to indicate a particular mountain in the Uig range of western Lewis. Our profile[4] shows that this provided a rough indication of the setting moon at the southern extreme of the minor standstill.

The semi-circular site seemed to demand that one looked southwards across the water; admittedly a subjective judgement, but in any case the northern horizon is the immediate hillside. Two of Somerville's boulders are placed almost due south, just visible between azimuths of 180° and 181° on our profile (fig. 2.7f). During the southern extreme of the major standstill, at maximum perturbation, the transit of the moon remained below the horizon if it occurred within 9 hours of the monthly maximum. However, at June full moon, minimum perturbation, the upper limb would have appeared for a very short time. Behind the site is a steep slope and the moon's transit could have been seen at any stage simply by moving a little uphill; the choice of site seems to have deliberately limited the amount of moon seen. In our more imaginative moments, we cannot help wondering if a semi-circular site was designed for observing the moon at a time when no more than half the moon rose above the horizon.

2.14 CONCLUSIONS

In the results presented in this paper, the southern extreme of the major standstill figures prominently (see fig. 2.10d), but this is an inevitable result of our surveying priorities to date (see section 2.8). On a small-scale map (fig. 2.10c) we have shown the hills (figs. 2.2a & 2.2b) which may have acted as reference points for the rise and set of the moon at this event. We have also shown the positions of the Callanish sites—excluding the two on Bernera. The lines represent the limits of possible viewing lines to the hills. There are two possible explanations of the diamond-shaped pattern which emerges -

(a) the sites were all built in this area in order that lunar observation could have taken place against these two ranges of hills; or
(b) lunar events appear to have significance in the Callanish area, purely because of its coincidental relationship to these two ranges.

Determining which of these two alternatives is the true one may be impossible, but statistical analysis may eventually determine if lunar observation and recording actually occurred at Callanish.

We believe that if lunar observation was a preoccupation of prehistoric Man, the evidence for it should still exist in the many sites of the Callanish area. Conversely, if lunar observation cannot be proved at Callanish, it must throw doubt on the whole field of megalithic astronomy.

Archaeoastronomers tend to be polarised between those who find statistical evidence the only valid way of confirming the astronomical use of megaliths, and those who are also prepared to accept the accumulating evidence for specific observations at specific sites.

Clive Ruggles is working on the statistical analysis of declinations occurring on small sections of horizon which are indicated at many sites throughout Britain, the type of indication considered acceptable having been fixed by pre-determined criteria. When complete, our own corpus of data will be of a different, and perhaps complementary, type. We intend to survey the entire

360° horizon from each of the Callanish sites, and to plot in various different types of possible indications. It should be possible to determine whether there is a statistically valid correlation between indications and solar and lunar events, and perhaps to find the preferred type of indication used by any Callanish astronomers.

By the time that this work is complete, archaeological studies should have provided more data, including dates for human activity in the area, perhaps also for the erection of the stones. We should like, in particular, to see more archaeological interest in (and official protection of) some of the minor sites in the area, which may be just as important for our understanding of the Callanish complex as the main site.

We have attempted to categorise our initial lunar results in various ways: rises and sets; which perturbation was indicated; upper or lower limb of the moon; phase of the moon; months in which a sighting line was usable; whether or not the moon gleamed again after setting; and the type of indication on site (see table 2.5). We hope that we shall be able to detect patterns in these results as our data continue to accumulate. However, from our results so far we have developed a working hypothesis to guide our continuing work, without, we hope removing our objectivity.

The basic assumption of this hypothesis is that the Callanish sites formed an inter-related complex, with the function of making and recording lunar observations. The major standstill may have been of particular interest to the Callanish astronomers, as it would have been a spectacular event at this northern latitude. There is some evidence that the minor standstill may also have been recorded, and there are a few possible solar lines which would have enabled the annual calibration of a calendar.

Several isolated erect flattened slabs indicate specific lunar events in an accurate manner, often distinguishing clearly between perturbations. Sites V and XI contain large numbers of stones but are not stone rings. We suggest that both could have been lunar observatories, site XI being concerned with southern extremes, while site V may have recorded six of the eight lunar extremes.

The better known ring sites mostly do <u>not</u> include accurate indications, but appear to incorporate symbolical alignments to solar and lunar extremes in their basic ground geometry. The fact that three of the ring sites can be interpreted as flattened circles (Thom's type A), while the other two are ellipses of identical proportions, supports our hypothesis that all the sites are related.

At the main site, we have shown that accurate solar and lunar observations could have been made from the north end of the avenue, and we wonder if this was the original observing area at this site. However, such a complex site requires a more complex explanation (see below).

At least four sites within the Callanish complex appear to have an indication of the moonrise at the southern extreme of the major standstill, minimum perturbation: $-(\epsilon + i - \Delta)$ (see table 2.5). These may all be related to a "regleam" of the moon on the same night at a nearby site. At site V, the

Table 2.5 Summary of lunar indications

Site	Event	Rise/Set	Perturbation	Indication
I	major S	set	max	alignments
I	major N	rise		alignment
I	major N	set	max	attitude
II	major N	set		inter-site
II	major N	rise		inter-site
II	minor S	rise		inter-site
II	major S	rise		axis
III	major N	set		axis, alignment
IV	major S	rise	min	attitude
V	minor N	rise		alignment
V	major S	rise	min	alignment
VIII	minor S	set		attitude
VIII	major S	transit	min	axis
X	major S	rise		axis
XI	major S	rise	max	alignments
XI	major S	set	max, min	alignments
XII	major S	set		attitude
XIII	major S	set	max	inter-site
XVI	major S	rise	max, min	attitudes
XVI	minor N	set		attitude
XVII	major S	set	min	inter-site
XIX	major S	rise	max	attitude

upper limb of the moon rose from the Sleeping Beauty, only to set into the immediate hill-slope before the full orb had risen. This event is indicated by an alignment and a large sub-horizon boulder. (At maximum perturbation, the moon did not rise at all.) At site IV, at almost exactly the same time, the moon rose from a junction between the Sleeping Beauty range and a nearer hill (indicated by a stone attitude); at maximum perturbation, the lower limb rose much closer to due South, out of another nearby hill. A few hours later, the moon whose $-(\epsilon + i - \Delta)$ rise could have been observed from IV or V, could have been observed setting from site XVII, less than 2 km away. The moon showed a spectacular regleam in the Glen Langadale notch. The rise events observed from IV or V could have acted as a "warning" of the gleam at XVII, which did not occur at maximum perturbation.

At site VIII, as the moon's declination decreased towards $-(\varepsilon + i)$, the moon's path became shorter and shorter above the hills to the south. When the extreme was reached, the moon did not rise at all if at maximum perturbation, while at minimum perturbation, the upper limb appeared for less than 2° of azimuth. This may have acted as a warning for the regleam of the moon about an hour later as viewed from site XV.

Similarly, the $-(\varepsilon + i - \Delta)$ moonrise indicated by the attitudes of two of the stones at site XVI may have acted as a warning for observation of the moonset at sites I, II, III, XII or XIII.

Owing to the infrequency with which the moon would have risen or set <u>precisely</u> at the moment of monthly maximum declination, an extrapolation device was necessary at any lunar observatory. However, the two ranges of hills which intersect with the moon's path at the southern extreme of the major standstill give unmanageable values for 4G of over a kilometre. In many cases the moonset, or less frequently the moon rise, occurred on a much nearer horizon, or on a junction between a near and a far horizon. This often gives a 4G value of more manageable proportions. Our suggested extrapolation device involving two rows of stones at the main site could be used for 4G values of up to 90 metres, corresponding to horizon distances of 1.5 km for the major standstill, or a little over 4 km for the minor standstill. Several of the putative observing sites do give 4G values approaching 90 m.

The east side of the avenue suggests a means of determining the time from a horizon event to the monthly maximum; or, more likely, may be a formalisation of the kind of stake pattern which often emerged at observing sites.

A complex monument such as Callanish I must be expected to have fulfilled several functions. In addition to possible use for observation and extrapolation, it is to be expected that it also had a ceremonial purpose. The symbolic lunar alignment of the avenue may have enabled "astronomer-priests" to involve the general population in their astronomical interests. Perhaps there were great occasions at the major standstill, with the supreme moment being the setting of the low June full moon into the circle itself.

Despite the 40" per century decrease in the obliquity of the ecliptic, a number of the lunar phenomena described in this paper for 1550 BC will still be observable in the 1980s. In particular, at the major standstill due late in

1987, the moon will still (just) set behind Mullach an Langa and reappear in Glen Langadale, as viewed from Druim na h-Aon Chloich (Callanish XVII). At the main site, observers standing at the north end of the avenue will be able to see (clear skies permitting!) the June full moon setting through the stones of the circle, although the moon's final disappearance will occur between the stones of the west row. To return to the words of Headrick (1808)- "these circles were intended (as) rude astronomical observatories ... they might serve these purposes at this day, to one who has bestowed a little attention on the positions of the stones".

NOTES

1. Apparently it was the accuracy of the orientation of the S row at Callanish I which stimulated Thom to begin his megalithic studies following a visit to the site in 1934.

2. During the course of our work, we discovered two important errors in the University of Glasgow group's plans (1978). At site XI the position of the single erect stone is plotted wrongly; and at site II the orientation of the grid lines from the circle centre is in error by 19°. David Tait, the leader of the survey team, has acknowledged the two errors (1980), and the map is now sold with an erratum slip explaining how to make allowance for them. While having caused us delays, they do not detract from the fact that much of our work would have been almost impossible without the basic data provided by the survey.

3. The fifth stone in the east row, designated 33A by the authors (Ponting, Ponting and Curtis 1977:32), was rediscovered by us as the result of a documentary search and probe survey. It was uncovered by the recent excavation (Ashmore 1980).

4. This profile is not included amongst the figures in this paper. Anyone interested in a complete set of profiles should contact the authors direct.

ACKNOWLEDGEMENTS

Thanks must go to David Tait of Glasgow University Geography Department for permission to use the site plans prepared by his team as the basis of our plans of sites I, II, III, IV, VIII, and X in this paper; and also for his unfailing helpfulness in any queries concerning the survey. Thanks also to Ron Curtis for initiating us into the use of the theodolite, for the triangulation of sites XII-XIX, and for co-operation in various other ways over the last four years. Patrick Ashmore, Trevor Cowie and Sjoerd Bohnkce have kindly allowed us to use material from their interim reports on recent archaeological work at Callanish. We have kept up occasional correspondence with Alexander Thom, Gerald Hawkins and Aubrey Burl, all of whom have been most helpful and encouraging in our research. Other correspondents who have helped us iron out particular problems associated with the material in this paper include Richard Brinckerhof, Edwin Krupp, Gordon Moir, Clive Ruggles (and his colleagues John Cooke, Roger Few and Guy Morgan) and John E. Wood.

On the Isle of Lewis, Iain Macarthur and John Maciver have helped us with Gaelic place-names; Anne Maciver allowed us to use the hall of Breasclete School to photograph our 5 m long profiles; and Douglas Leadbitter loaned photographic equipment. Last, but by no means least, our thanks to the crofters of Callanish and the surrounding townships for their co-operation in allowing us on their land, giving us all sorts of local information, and tolerating our somewhat obsessive interest in the old stones.

REFERENCES

Ashmore, P., 1980. Callanish 1980 Interim Report. Edinburgh: Scottish Development Department, Ancient Monuments Division.

Bohncke, S., 1979. Palaeobotanical study of a subboreal and subatlantic peat profile from Callanish, Lewis (unpublished report). Falkirk: Central Excavation Unit.

Burl, A., 1976. The stone circles of the British Isles. London: Yale University Press.

Burl, A., 1979. "Callanish" (letter). Curr. Arch., 6, 191.

Callender, H., 1854. Notice of the stone circle at Callernish. Proc. Soc. Ant. Scotl., 2, 380-384.

Cooke, J. A., Few, R. W., Morgan, J. G. and Ruggles, C. L. N., 1977. Indicated declinations at the Callanish megalithic sites. J. Hist. Ast., 8, 113-133.

Cooke, J. A., Few, R. W., Morgan, J. G. and Ruggles, C. L. N., 1978. Pers. comm.

Cowie, T., 1979. Callanish (Leobag) 1979 Interim Report. Falkirk: Central Excavation Unit.

Curtis, G. R., 1979. Some geometry associated with the standing stones of Callanish. Hebridean Naturalist, 3, 29-40.

Freeman, P. R. and Elmore, W., 1979. A test for the significance of astronomical alignments. Archaeoastronomy (supplement to J. Hist. Ast.), 1, 86-96.

Hawkins, G. S., 1963. Stonehenge decoded. Nature, 200, 306-308.

Hawkins, G. S., 1965. Callanish, a Scottish Stonehenge. Science, 147, 127-130.

Hawkins, G. S., 1971. Photogrammetric survey of Stonehenge and Callanish. National Geographic Society Research Reports—1965 projects, 101-108. Washington D.C.

Hawkins, G. S., 1973. Beyond Stonehenge. London: Arrow Books.

Hawkins, G. S., 1980. Some background information on Dr. Jean-René Roy's paper on Callanish. J. R. Ast. Soc. Canada, 74, 10-11.

Headrick, J., 1808. Editorial notes in G. Barry, History of the Orkney Isles. London.

Henshall, A. S., 1972. The chambered tombs of Scotland, Vol. 2. Edinburgh.

James, H., 1867. Stonehenge and Turusachan in the Island of Lewis. Southampton.

Kerr, J., 1873. The cruciform sun temple at Callernish. The Builder, July 12th.

Lockyer, N., 1909. Stonehenge and other British stone monuments astronomically considered. London.

MacCulloch, J., 1819. A description of the Western Islands of Scotland. London.

MacCulloch, J., 1824. The Highlands and Western Islands of Scotland. London.

Macfarlane, W., 1907. Geographical collections. Edinburgh.

MacGregor, A. A., 1933. The haunted Isles. London.

Mackenzie, W. M., 1904. Notes on certain structures of archaic type in the Island of Lewis. Proc. Soc. Ant. Scotl., 38, 173-204.

Moir, G., 1978. Standing stones and astronomy in the Outer Hebrides. Northumberland Archaeological Group Newsletter, 2, 13-23.

Moir, G., 1979. Hoyle on Stonehenge. Antiquity, 53, 124-128.

Oldfather, C. H. (translator), 1935. "Library of History", Book II, by Diodorus of Sicily. Cambridge (Mass.).

Ponting, G. H. and Ponting, M. R., 1977a. The standing stones of Callanish. Stornoway.

Ponting, G. H. and Ponting, M. R., 1977b. "The stones at Callanish" (letter), The Scots Magazine, 107, 548.

Ponting, G. H. and Ponting, M. R., 1981. The Stonehenge of the Hebrides. In press.

Ponting, M. R., Ponting, G. H. and Curtis, G. R., 1976. Callanish—stone circles and related structures. In C. Lythe (ed.), Discovery and excavation in Scotland, 1976. Edinburgh.

Ponting, M. R., Ponting, G. H. and Curtis, G. R., 1977. Callanish—stone circles and related structures. In C. Lythe (ed.), Discovery and excavation in Scotland, 1977. Edinburgh

Roy, J.-R., 1980. Comments on the astronomical alignments at Callanish, Lewis. J. R. Ast. Soc. Canada, 74, 1-9.

R.C.A.H.M.S., 1928. Inventories of Ancient Monuments, Vol. 9: Outer Hebrides, Skye and the Small Isles. Edinburgh: H.M.S.O.

Sharbau, ?., c. 1860. Manuscript plans and drawings of the Callanish sites. National Monuments Record, mss. nos. RCD/13/13-16.

Somerville, B., 1912a. Prehistoric monuments in the Outer Hebrides and their astronomical significance. J. R. Anthrop. Inst., 42, 15-52.

Somerville, B., 1912b. Astronomical indications in the megalithic monument at Callanish. J. Brit. Ast. Assoc., 23, 83-96.

Tait, D. A., 1980. Pers. comm.

Thom, A., 1954. The solar observatories of Megalithic Man. J. Brit. Ast. Assoc., 64, 396-404.

Thom, A., 1966. Megalithic astronomy: indications in standing stones. Vistas in Ast., 7, 1-57.

Thom, A., 1967 ("MSB"). Megalithic sites in Britain. Oxford: Oxford University Press.

Thom, A., 1971 ("MLO"). Megalithic lunar observatories. Oxford: Oxford University Press.

Thom, A. and Thom, A. S., 1978a ("MRBB"). Megalithic remains in Britain and Brittany. Oxford: Oxford University Press.

Thom, A. and Thom, A. S., 1978b. A reconsideration of the lunar sites in Britain. J. Hist. Ast., 9, 170-179.

Toland, J., 1726. The history of the Druids. (New edn., London, 1814.)

University of Glasgow, Department of Geography, 1978. A map of the standing stones and circles at Callanish, Isle of Lewis, with a detailed plan of each site. Glasgow.

Wood, J. E., 1978. Sun, moon and standing stones. Oxford.

Wood, J. E., 1979. Pers. comm.

Wood, J. E. and Penny, A., 1975. A megalithic observatory on Dartmoor. Nature, 257, 205-207.

Worsaae, J. J.A., 1846. Manuscript field notebooks (unnumbered). Copenhagen: Nationalmuseet.

3. WISE MEN IN ANTIQUITY?

Euan W. MacKie

3.1 GENERALITIES

Dr. Edward de Bono's book on practical thinking (1976) makes interesting reading for archaeologists, and indeed for all concerned with the problems of interpretation in the historical sciences. With the aid of an ingenious experiment he analyses the way the human mind works and identifies 'five ways to be wrong', 'four ways to be right' and 'five ways to understand'. Among the ways to be right—which means ways in which one can convince oneself one is right—is what he calls the 'village Venus' or 'unique rightness' method, a mental process which he believes to be particularly common among scientists and academics. If one has lived one's whole life in a remote village, cut off from contact with other people, then the village Venus must be the most beautiful girl in the world because one cannot imagine anyone more beautiful. In the same way a scientist or scholar who cannot imagine, or who has not heard of, any explanation which will fit a given body of evidence as well as the one he has thought of (or, one might add, has been taught), is capable of being fully convinced of its unique rightness. Consciously he tells himself, and believes, that it is right because it fits all the facts; but actually its rightness derives solely from the lack of rival explanations and thus ultimately from a profound lack of imagination.

It may be that British archaeological theory was a victim of the 'village Venus' effect for a very long time in so far as its explanations of the nature of social changes in prehistory were concerned. The old view of a simple, insular, rural and tribal society evolving through the Mesolithic and Neolithic periods and the Bronze and Iron Ages, occasionally presented with technological advances by intruders from abroad (who ultimately obtained them from Bronze Age urban civilisations of the Mediterranean and the Near East) lasted for so long that it seemed right beyond question to almost everyone. Until about 1964 there was simply no respectable alternative view which was held by people familiar with the intricate details of the hard evidence. However during the 1960s people like Roy Hodson, Colin Renfrew and a few others advanced an entirely new view of British prehistory. At the beginning it was a matter of pointing out that immigrations had previously been assumed on far too little evidence (Hodson 1964): but later, and using as a clinching argument the correction of radiocarbon dates by the tree-ring calendar, this deduction was taken much further (Renfrew 1968; 1970; 1973). The resulting adjustment backwards by several centuries of the C-14 dates appeared to make, for example, the Atlantic European megalithic monuments and the emergence of copper-working in the Balkans—two dramatic and important cultural and technological developments—earlier than the civilised societies of the Mediterranean world and the Near East which had, until that time, been thought of as the ultimate sources for such innovations. Instead of these diffusionist explanations, accepted without question hitherto, Renfrew offered the alternative hypothesis of independent development among the indigenous European societies. He thus released a swelling torrent of questioning, re-assessment and re-interpretation of the archaeological evidence which is still continuing. Even prehistoric immigrations which seemed beyond argument—like that of the round-headed Beaker people in the Chalcolithic period in the middle of the 3rd millennium BC–are being doubted (Burgess & Shennan 1976), although this will probably prove to be an example of an excess of reforming zeal (MacKie 1981: ch. 6). However whether the non-diffusionist view will ultimately prevail or will have to be drastically modified is beside the point. What is clear is that, for the first time in decades, a radical new outlook has emerged which has caused a long overdue re-assessment of the primary archaeological evidence in may fields, the questioning of old assumptions and the framing of alternative hypotheses. The 'village Venus' of British archaeology —diffusionism—suddenly seemed less attractive beside her new rival!

Renfrew's new hypothesis did not radically alter our picture of the structure of British and European Neolithic society, which was still envisaged by most archaeologists as rural and tribal, usually egalitarian but with chiefdoms sometimes being invoked to explain the exceptionally massive monuments of the fourth and third millennia BC in southern England. However in 1977 I offered a drastic and detailed reassessment of this view in the more limited field of Neolithic Britain (MacKie 1977a) with a more tentative and generalised extension to the whole of Atlantic Europe (MacKie 1977b). Much new information had been accumulating over the previous decade about the sophisticated geometrical and astronomical skills possessed by the builders of the stone circles and standing stones, to such an extent that it had begun to seem incompatible with traditional archaeological views; and my work was an attempt to help British archaeology, in Aubrey Burl's words, in the task of 'finding a Megalithic society to suit its monuments' (Burl 1976). It is the purpose of this article to review some of the important evidence, to assess some of the reactions to this attempt and, by doing so, to outline the present position of the debate.

My attempt at re-interpretation was stimulated largely by the work of Professor A. Thom (1967; 1971; Thom & Thom 1978) which suggested that the stone circles and standing stones of the Late Neolithic period and the Early Bronze Age embodied a completely unsuspected and high level of geometrical and astronomical knowledge. Having tested by excavation the interpretation as an advanced solar observatory of one standing stone site, at Kintraw in Argyll, and found it supported (MacKie 1974), it seemed to me that these extraordinary and widespread activities could best be explained if one supposed that British Neolithic society was not a simple homogeneous one but was stratified, with a professional priestly upper class either ruling the rest (a theocracy) or having a privileged position like the Druids of Iron Age times (Piggott 1968; Burn 1969). Such a full-time class of priests, astronomers and wise men, being supported by the agricultural population, would certainly have had the time to develop the astronomical and geometrical expertise, detected by Thom; but the obvious question was–is there any other evidence for such a stratified society apart from the controversial work of Thom himself? I came to believe that there was and that, for example, many of our most prominent Neolithic sites and artefacts could be completely re-interpreted to favour this view.

Three of the massive Late Neolithic earthworks in Southern England—the giant henge monuments of Durrington Walls and Marden in Wiltshire and Mount Pleasant in Dorset—had recently been excavated and they show a unique combination of ceremonial and domestic features. This is hard to explain in terms of the traditional view of Neolithic society, but it fits well with the idea that these henges were the prehistoric equivalents of Early Christian monastic sites, or perhaps of the Maya 'ceremonial centres' in Central America—that is that they were the living and training places of learned orders of astronomer-priests and wise men, the activities of which are everywhere to be seen in the standing stones and stone circles. The results of the excavation of the three earthworks mentioned, as well as of the stone village at Skara Brae in Orkney, were reviewed in detail with this alternative hypothesis in mind and were found to fit it as well as, and probably better than, the traditional view; it was gratifying to find that the excavator of the henge sites considered this an interesting and useful exercise (Wainwright 1977). The traditional view of course fails to account for the evidence of astronomical and geometrical knowledge at the standing stone sites except at a fairly low level (Renfew 1973: 237ff.). Moreover the settlement sites mentioned (some of them widely separated) are linked by their associated pottery, Grooved ware, and the new explanation was the first ever which has accounted in social terms for the remarkable distribution of that pottery and of what was once called the 'Rinyo-Clacton culture' (Warren et. al., 1936; Piggott 1954: 321).

Since the book was an avowed attempt to overturn a long-held view of prehistoric British society and to substitute an entirely different one, the first chapter included a long section on the nature of archaeological evidence and of the type of deductions which can be made directly from it. It was concluded that all kinds of social explanations of prehistoric archaeological evidence—including the new one on offer and all the traditional ones, no matter how established—do not and cannot derive directly from such evidence, but have to be produced by drawing analogies with present day or historically recorded societies with similar technologies (fig. 3.1). This is an absolutely fundamental point which is sometimes overlooked in British archaeology and, while it means that my new picture of a priest-dominated, stratified society in Neolithic Britain can never be finally proved in the true scientific manner, so must it also mean that the traditional views are equally lightly anchored to the hard evidence and should be blown away without trouble if a better explanation emerges. The real anchor of every social theory in prehistoric archaeology is the zeal with which it is believed in by archaeologists, especially when it is a 'village Venus'. The only satisfactory way of assessing such social hypotheses and theories is by noting how much of all the available evidence they account for and by testing them if at all possible.

3.2 THE EVIDENCE

We may now briefly review some of the more important evidence which bears on the new interpretations, beginning with the three sites where I have carried out excavations or fieldwork myself.

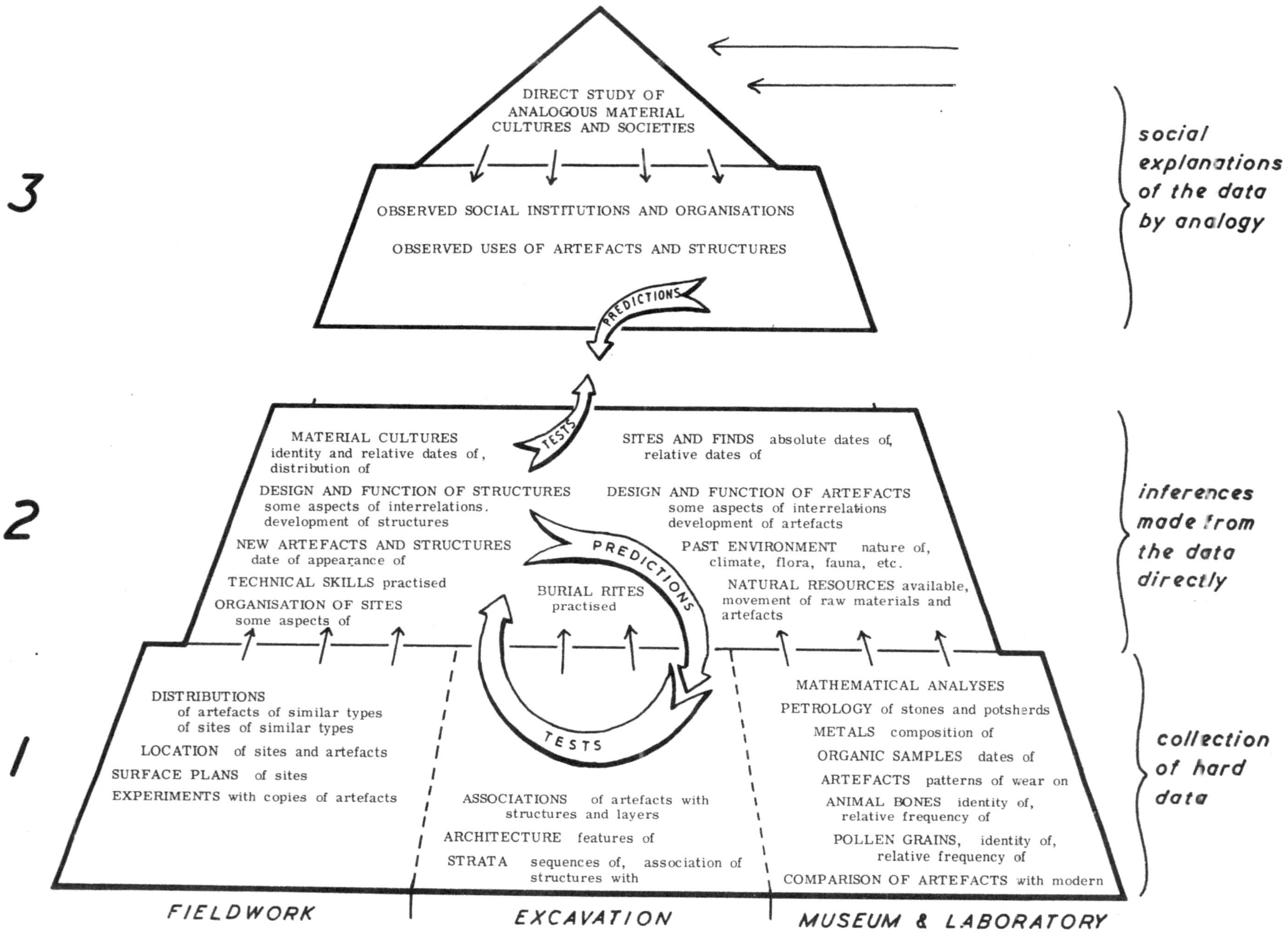

Fig. 3.1 The nature of archaeological induction.

3.2.1 Kintraw

Here in mid Argyll was carried out in 1970 and 1971 the first experimental test of Thom's astronomical hypothesis (MacKie 1974; 1977a: 84-92). The site was claimed as a highly accurate solstitial observatory, in theory capable —because of the very long line of sight between its standing stone observation point and its mountain peak foresight on Jura (28 miles)—of picking out the sunset on the shortest day of the year (Thom 1967: 154; 1971: 36). However the local topography means that it was impossible to have established the site from the field in which the stone stands; another observing point is needed slightly higher up on the hillslope north-east of and behind the stone. A search was made for traces of a level stone platform at this point—where there is an unobstructed view to Jura—and one was found. In the absence of potsherds, flints and charcoal, recourse was had to a simple statistical study of the orientation and dip of the stones composing the platform, that is to what is known as petrofabric analysis (Bibby 1974). The result of this demonstrates clearly that the platform is much more likely to be a man-made than a natural formation. In fact there can be little doubt of this when one considers how one end of the stone pathway (that at which the midwinter sunset could once have been seen) is laid behind a revetment composed of two massive contiguous boulders chocked into position on the edge of a steep drop. These boulders form a conspicuous and solid marker for the observation point. The main gap in the evidence is any independent clue to the date of the platform and boulder revetment but there is really very little doubt that it is artificial.

This has to be stressed because of the importance of the discovery, which in effect has provided a decisive and independent verification of Professor Thom's astronomical hypothesis, or at least of that part of it which postulates that about 4,000 years ago simple but highly accurate devices for working out the exact days of the solstices had been invented, and were being used. These alignments were observing instruments, capable of providing useful information, not just crude orientations for ceremonial purposes like the famous midsummer line through the Heel stone at Stonehenge. Their existence must, I believe, clearly transform our ideas about the organisation and capabilities of British Neolithic communities. They must surely imply the existence of a professional class of observers and their existence must, moreover, make more likely that of the more controversial and sophisticated lunar alignment at the site, proposed by Thom (1971: fig. 4.4).

It is hardly surprising that the discoveries at Kintraw have caused some interest and dispute. British archaeologists and astronomers who do not believe in Thom's theories (or who, perhaps, cannot accept their implications) naturally tend to assume by a process of deductive reasoning (from the general to the particular) that there must be something wrong with the evidence of Kintraw. However, I have not yet seen anything in print on the subject. In America the followers of the catastrophic theories of Dr. I. Velikovsky (1950) also reject the Kintraw site deductively, but this time because it fits with a uniformitarian view of the solar system and therefore seems to contradict Velikovsky's ideas that the Earth was shaken by great convulsions in the 15th and 8th centuries BC. Attacks on the significance of the site have appeared in Kronos, the propaganda magazine of the movement (Cardona

1979; McCreery 1980b). One simple point can be made about this first criticism of the petrofabric evidence.

McCreery has searched through the literature and he illustrates formations from several natural petrofabric diagrams which are strikingly similar to those obtained from the Kintraw platform. Thus he seems to have destroyed the main plank in my argument about Kintraw—that the diagrams obtained from the hill platform are so markedly different from those obtained from natural deposits, and similar to one from a known man-made pavement (in a hillfort on Sheep Hill, Dunbartonshire), that the structure must be artificial. The problem here of course is lack of field experience. For example, great stress is laid on the remarkable similarity between a diagram obtained from an outwash of gravels in a glacial terrace in Illinois and the two from Kintraw, it being said that this kind of evidence, since it could equally show that the gravel deposit was artificial, is thus useless. However, the Kintraw hill platform is 47 m above sea level and on the steep slope of an isolated hill in a situation where no gravel terrace could be; the local raised beach is lower down and nearer the sea. The comparison is thus quite invalid since it is surely obvious that only natural deposits which could occur <u>in the same situation</u> as the Kintraw platform can legitimately be compared with it. However it must be admitted that some of his other comparisons are more plausible.

It should be stressed again that the artificiality of the boulder revetment and platform on the hillside at Kintraw does not depend entirely on the petrofabric evidence. It was clear for example that there was no such pavement behind a large boulder in a very similar situation on the edge of the gorge and 50 m upstream. It was certainly my impression during the excavations that the level stone pavement had been deliberately laid behind two massive boulders and that these had previously been carefully chocked into position with their pointed ends touching and at the edge of a steep drop down into the stream gorge. I laid less stress on these features in the original report because I hoped that the artificiality of the whole structure could be confirmed with quite independent evidence, as it was.

In the spring of 1979 the standing stone at Kintraw, which had been leaning for many years, finally fell down; the ground around its base was subsequently excavated and the impression of the butt in the subsoil found (Cowie 1980). The stone was later re-erected into its original position and concreted in. As seen from the hill platform the upright stone now points a little more directly at the 'solstitial' mountain notch on Jura.

3.2.2 The Cultoon circle, Islay

In 1974 and 1975 the ruined stone circle near Cultoon Farm on the Rhinns of Islay, Argyllshire (fig. 3.2), was excavated and, although the original reason for undertaking the work was simply to investigate the site prior to its reconstruction, by good fortune the results provided an excellent test through archaeological fieldwork of Thom's geometrical hypothesis. This maintains that the stone circles were built around precisely-drawn geometrical figures —true circles, flattened circles, ellipses and egg-shapes, some of which must have been constructed on the bases of right-angled (Pythagorean) triangles— set out on the ground using a standardised unit, the so-called 'megalithic yard' (MY) of 0.829 m (2.72 ft or 2 ft. 8.64 ins) (Thom 1967: chs. 5-7).

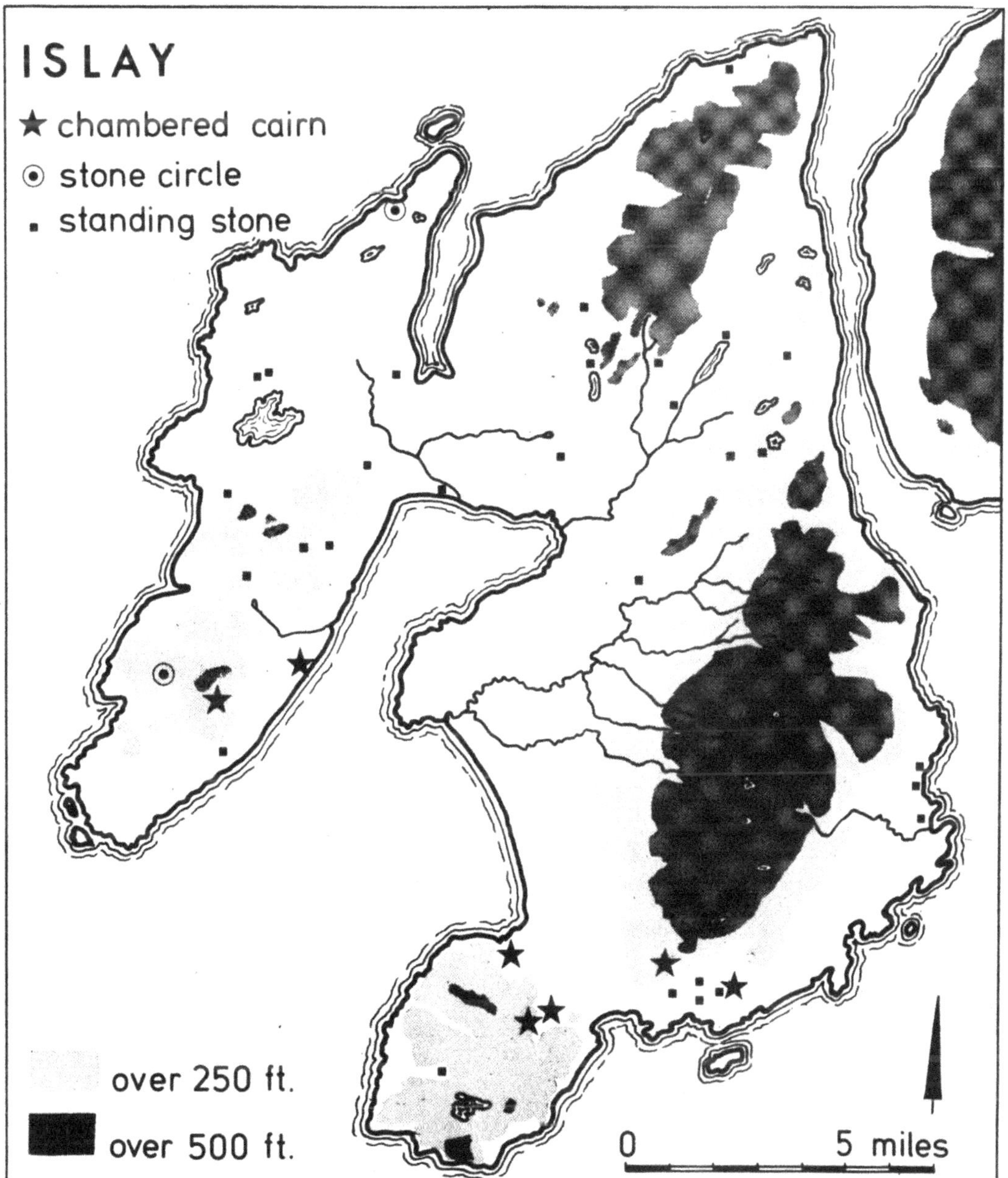

Fig. 3.2 Map of Islay showing chambered cairns and standing stone sites.

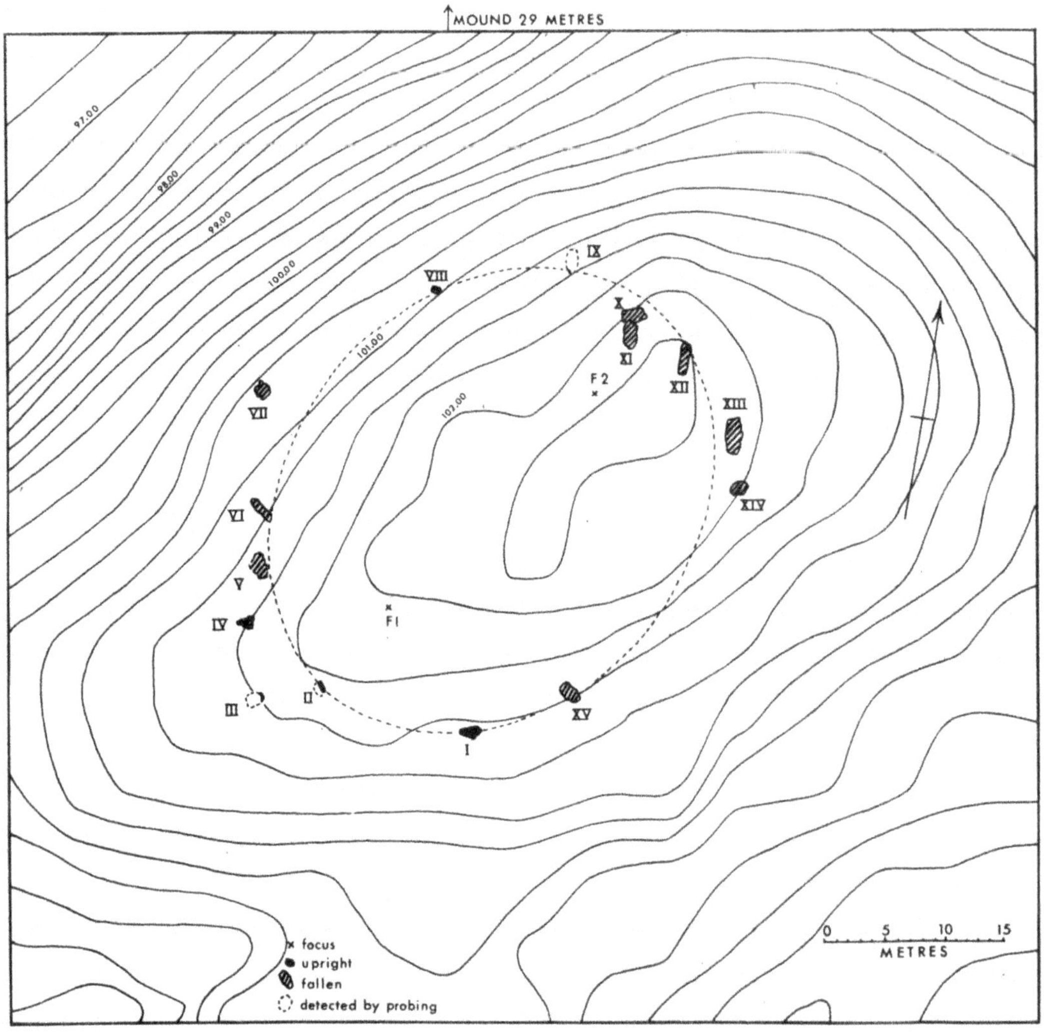

Fig. 3.3 Contour survey of Cultoon showing the original ellipse suggested. The contours are at 0.5 m intervals.

Pl. 3.1 Cultoon: View of stones X, XI, XII, XIII and XIV after the removal of the peat, showing the rubble-covered prehistoric ground surface.

Pl. 3.2 Cultoon: Stones X and XI showing the large, paved unused socket for no. XI partly underneath it. The butt of no. X is still in its socket.

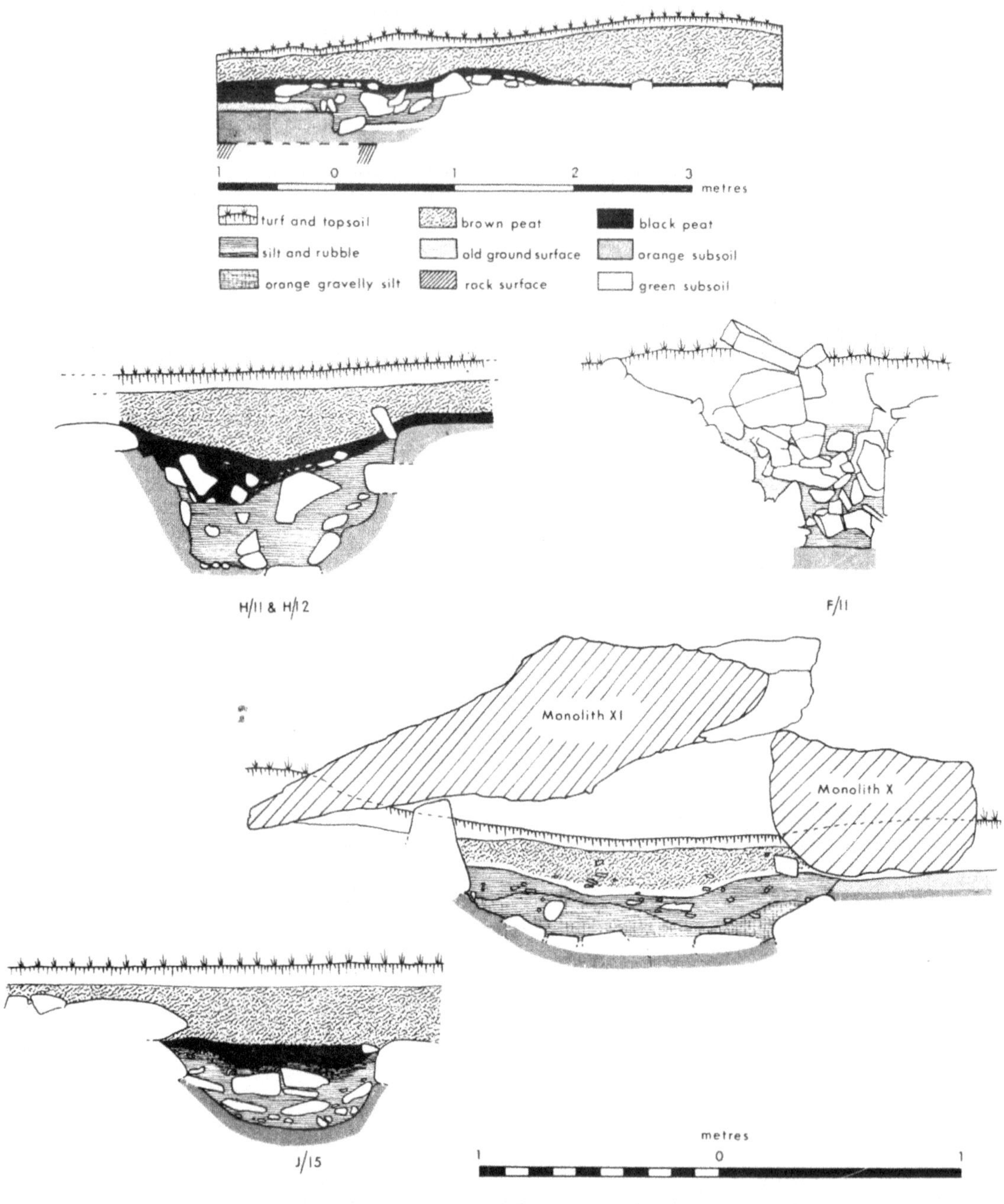

Fig. 3.4 Sections of stone sockets at Cultoon.

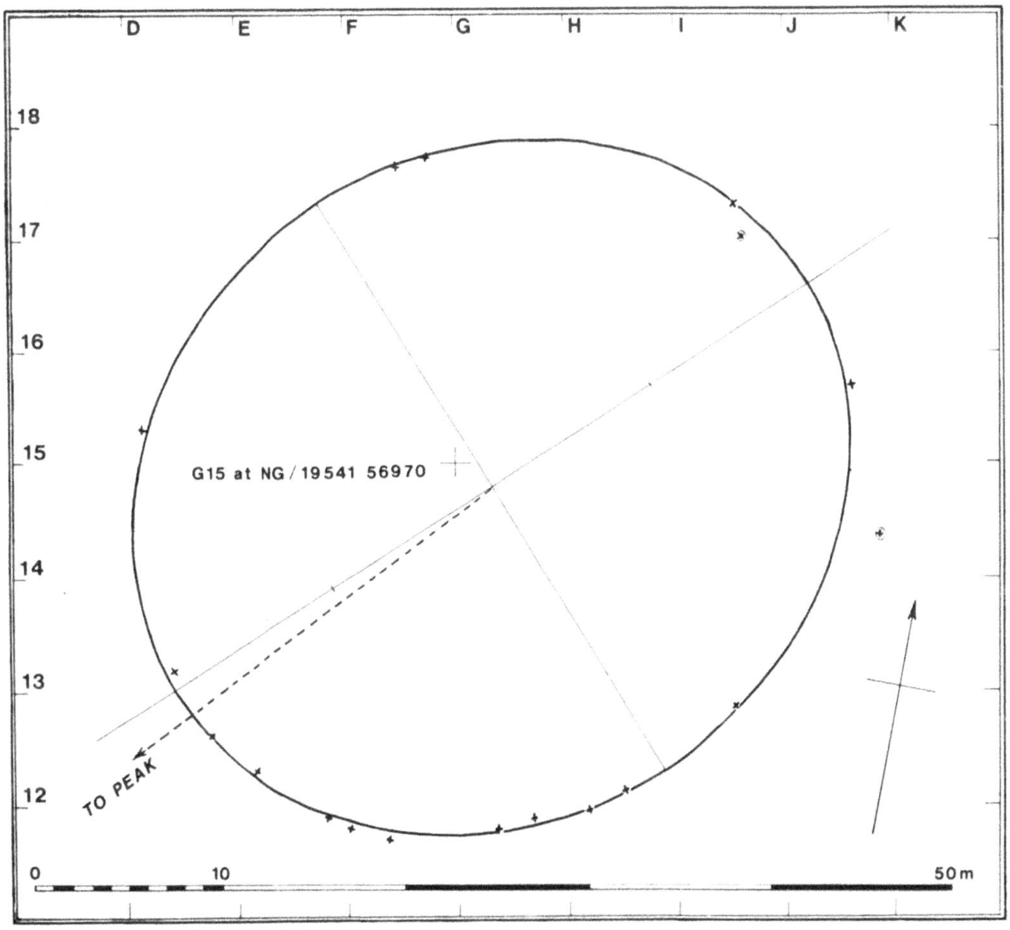

Fig. 3.5 Plan of Cultoon showing the computed best-fitting ellipse fitted against the measured positions of stones and sockets (Xs). The two dotted circles represent the sockets not included.

One possible objection to this idea is that, without excavation (and the great majority of the sites have not been excavated), one cannot be sure that the individual stones are now in the same position that they were 5,000-4,000 years ago; any geometrical figures based on their present positions may therefore be open to doubt (Hogg 1977; Wainwright 1977). Statistical techniques were utilised by Thom to take account of this possibility to some extent but the objection sounds reasonable. It was interesting therefore to find that the Cultoon excavations showed rather neatly that the argument can be reversed.

Before work started the stones, which are mostly lying flat, fitted very poorly round a 'megalithic' ellipse based on a 3:4:5 triangle, and with axes of 40 and 50 megalithic yards, some of them lying several metres away from the supposed perimeter (fig. 3.3). Only two stones still stood, though the butt of a third was found still in position but buried under the peat which covered the whole site to a depth of up to 40 cm. The many prone monoliths were found lying directly on the old ground surface below the peat, which had evidently only begun to grow after they had reached their present position (pl. 3.1). A radiocarbon date of 765 ± 40 bc (SRR-500) was obtained for the base of this peat layer.

Further examination of the prone monoliths showed clearly that, with one exception, they had never been set up; several of them lay next to sockets but in situations which showed that they had not fallen out of them (pl. 3.2). Others had no sockets beside them and there were pits without adjacent stones. All the excavated pits except one had been dug and then abandoned, gradually to fill again with rubble and water-washed silt; all this had happened before any peat grew on the site (fig. 3.4). The stone sockets were not quite full when the peat did start to grow and they were thus identified by circular shallow depressions of peat in the old ground surface (pl. 3.3). Unfortunately none of the sockets produced any material that could be dated by radiocarbon. One socket was found by chance which had been deliberately filled with rammed rubble and one stone had been re-erected close beside another lying beside its empty stone-hole: they are shown as dotted circles on the plan (fig. 3.5). This last however fell down again on top of its prone neighbour and was the only example on the site of a fallen stone with its butt still in a socket.

Thus the excavation showed that the site was unfinished; work on it had evidently been abandoned when it was only partly built. However the three standing stones and the fourteen empty sockets allowed the original shape planned for the site to be exactly determined. These seventeen positions (see table 3.1) were analysed for the quality of their fit to a geometrically drawn ellipse with the help of a computer programme devised by Dr. G. I. Crawford of the Department of Natural Philosophy of the University of Glasgow, and the plan (fig. 3.5) shows the calculated best fitting ellipse matched against the actual stone and socket positions. One fact immediately revealed was that the re-erected stone did not lie on the perimeter of the ellipse but some distance from it; this confirmed that it was a later alteration.

The computed best-fitting ellipse has a long axis of 40.716 m a short one of 35.310 m and an eccentricity of 0.498; the standard deviation of the positions from this perimeter is 29.4 cm, a remarkably small figure when set against the overall size of the ring. For the standard deviation to be only 1.4%

Table 3.1 Details of positions of stones and sockets at Cultoon, measured on the ground from peg G/15 with 0° being grid north. The 'difference' column shows the distance of each point from the perimeter of the calculated ellipse

	Feature	distance (m)	angle	difference (cm)
1.	Socket in I/17, under stone X	20.70	48.25°	3.5
2.	Socket in J/15, next stone XIII	22.20	79.25°	36.5
3.	Socket in I/12	20.07	129.80°	10.9
4.	Socket in H/12, next stone XV	19.60	151.83°	6.6
5.	Socket in H/11	19.70	158.33°	9.8
6.	Socket in G/11	19.20	167.25°	37.4
7.	Stone I	19.45	173.17°	13.8
8.	Socket in F/11	20.20	190.00°	49.1
9.	Socket in F/11	20.10	196.17°	34.4
10.	Socket in E/11	19.90	200.08°	12.3
11.	Stone II (stump)	19.50	213.67°	25.4
12.	Grave marker slab in D/12	19.60	222.67°	1.6
13.	Socket in D/13	18.80	234.58°	43.0
14.	Shallow socket in D/15	17.10	276.33°	26.2
15.	Socket in F/17	16.10	349.25°	7.6
16.	Stone VIII	16.35	354.83°	11.4
	Not included are			
17.	Socket at base of stone XI	19.89	52.75°	-
18.	Deliberately filled socket in J/14	23.65	99.25°	

Table 3.2 Possible planned dimensions of the Cultoon ellipse

	computed	'Pythagorean' 1	'Pythagorean' 2	Equilateral
triangle	-	$12^2+21.91^2=25^2$	$12^2+22^2=25.06^2$	$12\frac{1}{2}^2+21.65^2=25^2$
semi major axis	20.358 m	20.725 m	20.775 m	20.725 m
semi minor axis	17.655 m	18.163 m	18.238 m	17.948 m
F^1-F^2	20.275 m	19.896 m	19.896 m	20.725 m
eccentricity	0.498	0.480	0.479	0.500
perimeter	119.57 m			
perimeter in MY	144.23 MY	147.60 MY	147.55 MY	146.74 MY
perimeter in M Rods	57.69 Rods	59.04 Rods	59.02 Rods	58.70 Rods

(1 Rod = $2\frac{1}{2}$ MY)

even of half the short axis is an astonishing tribute to the accuracy with which the plan of the site was set out on the ground and a clear proof, if any was needed, that the builders planned the stone ring around a precisely drawn ellipse. Moreover the eccentricity of the calculated ellipse is almost exactly 0.5—further confirmation that the geometry underlying the site is real. Thus we see how, far from the excavation of this stone circle showing that its original plan is a less good fit to a sophisticated geometrical figure, exactly the reverse has happened.

For interest, in table 3.2 the computed best-fitting ellipse may be compared with three possible "megalithic" ellipses. In each case the 'triangle' given is a right-angled one, half of the isosceles triangle of the ellipse itself, which is made by one end of the short axis and the two foci (fig. 3.5). A loop of rope forming this triangle, and hooked round two pegs at the foci, can be swung tightly round to draw the ellipse on the ground. The dimensions of the triangles are given as $\underline{a}^2 + \underline{b}^2 = \underline{c}^2$ where $\underline{a}$ is half the distance between the foci, $\underline{b}$ is half the short axis and $\underline{c}$, the hypotenuse, is equal to half the long axis.

Since neither of the two Pythagorean triangles have integral sides they are perhaps less likely interpretations—even though one may doubt that the circle builders knew the difference (especially for the second) since we assume that they worked empirically and did not actually know Pythagoras' theorem. The ellipse based on the equilateral triangle with sides of 25 MY has the advantage of having an eccentricity of 0.5—a neat proportion strongly suggested by the computed dimensions of Cultoon—and of requiring to be set out only in a very simple way, with a loop of rope 75 MY long. It is impossible to accept that the Cultoon ring should fit by chance so closely to an accurately drawn ellipse and extremely difficult to believe that this ellipse should also by chance be so close in size to one with a long axis of 50 MY and an eccentricity of a half.

It is also noteworthy that each axis of the computed ellipse is almost the same proportion of the equivalent one in the 'ideal' figure having a long axis of 50 MY, the short one being 98.37% and the long 98.23%. This, and the fact that the eccentricity of the computed figure is almost exactly a half, suggests that either the stone positions were systematically dug slightly inside the marked out elliptical perimeter or—more likely I suggest—the acutal unit of length employed at Cultoon was slightly shorter than the 'standard' megalithic yard of 0.829 m. If one divides the axes of the computed ellipse by their 'ideal' lengths in MY (50 and 43.30) one gets figures for the unit used of 0.814 m and 0.816 m respectively; the average of 0.815 m may well be close to the length of the actual rod employed at this site. It is hard to see otherwise, at a site where the circumstances are almost ideal for determining the exact points plotted by the prehistoric surveyors, why the eccentricity of the ellipse should be so close to 0.5. This inference of course supports the idea that the megalithic yardsticks varied a little from region to region and were not rigidly standardised throughout Britain. However the accuracy involved surely means that pacing out the dimensions is a completely inadequate explanation.

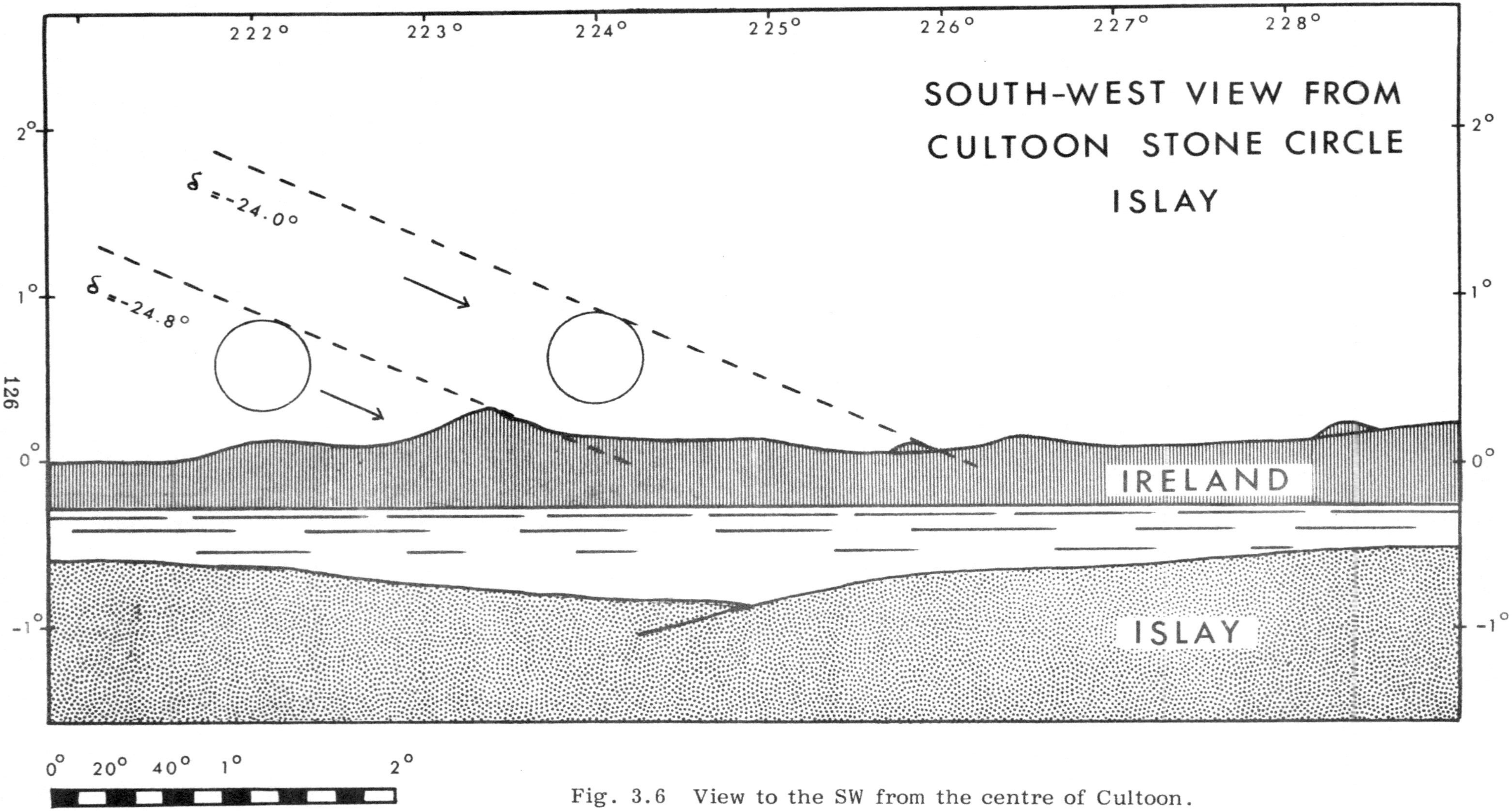

Fig. 3.6 View to the SW from the centre of Cultoon.

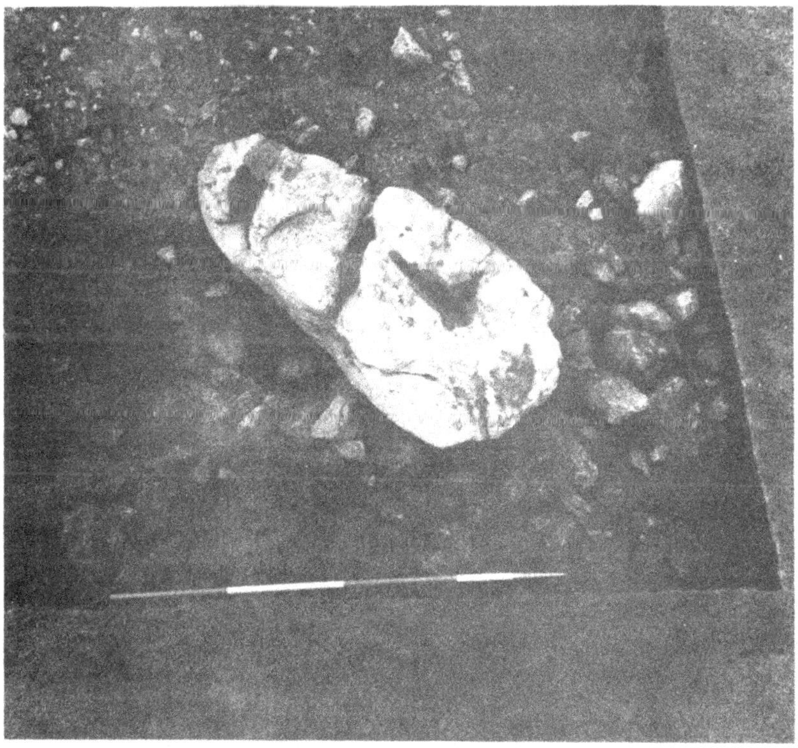

Pl. 3.3 Cultoon: Stone XV with half its unused socket showing (the dark, peat-filled half circle next to the pole).

Pl. 3.4 Brainport Bay: View from behind the 'observation boulders' north-east towards outcrop with notch and paved areas. Stones 1 and 2 set in sockets.

The computer, when it discovers the ellipse which best fits the positions measured, also determines the direction of the figure's axes in relation to the site grid. The nature of the computation means that this direction is only accurate within a degree or two, and indeed it varies slightly with each calculation; however all the results agree that the long axis points almost exactly NE-SW.

In the south-west, on exceptionally clear days (rare now in Islay!), the coast of Ireland can be seen and on this two or three prominent peaks are visible, at least 50 miles away. Conditions for observing these peaks and for calculating their azimuths with the aid of the Sun have not been ideal on any of the occasions I have been there, and the azimuths are in fact determined fro from the 8-figure grid references of three neighbouring Ordnance Survey triangulation stations and from that for the Dubh Ardtach lighthouse, some miles to the north. Fig. 3.6 represents the horizon profile in the south-west as seen from Cultoon and shows how the Sun setting behind the main peak would have had a declination of -24.8°, and one of -24° when setting behind the smaller, more distant peak to the right. If visibility was regularly good the latter peak should have been able to pinpoint the day of midwinter sunset at some period towards the end of the 3rd millennium BC.

Cultoon has thus proved to be a remarkable and important site in several ways, and it is a pity that no exact dates for its abortive construction are available. The point is that its excellent fit to a geometrical construction—and the closeness of the best fitting ellipse to an exceptionally neat and simple one in terms of megalithic yards—emerged from an objective calculation by a machine and can by no stretch of the imagination be said to have been imposed on to the site a priori. Indeed, as we have seen, before excavation the site seemed to be rather unpromising in this respect. Moreover the shape computed by the machine indicated, again quite independently of any preconceived ideas, that the long axis of the ellipse pointed almost exactly towards the only place on the horizon where really distant peaks are to be seen; one of these peaks could have served as a precise midwinter solstice indicator in late Neolithic times.

Thus this stone circle could have been a carefully designed observing station planned, among other things of course, to mark midwinter's day. Here for once geometrical design and astronomical function are neatly linked. One last point is that the contour plan of the site (fig. 3.3) shows that the ring was not built symmetrically around the summit of the low rise, as one might have expected if its position had not been determined by some more abstruse factors. It is offset towards the west-south-west for some reason.

3.2.3 Brainport Bay, Loch Fyne, Argyll

Loch Fyne is one of several long sea lochs in Scotland which are aligned north-east to south-west and about half way down its western shore, near the village of Minard, is a short promontory which forms the sheltered Brainport Bay on its north-eastern side. Near the outer end of this bay, and only a few metres back from the shore, is a rocky outcrop which in the past has been modified and added to with paving and revetted platforms (pl. 3.4; fig. 3.7). The site has been explored since 1976 by Col. P. Fane Gladwin and

Pl. 3.7 Brainport Bay: Stone 1 as it was found.

Pl. 3.8 Brainport Bay: View north-east along the alignment from the back platform, seen in the foreground.

members of the mid Argyll Archaeological Society (Fane Gladwin 1978), and I am grateful to the excavator for ready permission to descrine the site here in advance of full publication. The new plan (fig. 3.7) was made by myself in June 1980 and is complete except for the back platform (described below).

There are three main parts to the site, namely the main outcrop with its V-shaped cleft and paved areas, a pair of isolated boulders about 9 m to the south-west and, 25 m further in the same direction, what seems to be a large, circular artificial platform on rising ground. There is a cup-marked rock and an unexcavated cairn nearby and at least two groups of fallen standing stones have been located within a quarter of a mile of the main site.

Until 1977 the main site (the front outcrop and its associated platforms) was thought to be an ancient settlement of some kind, though no clear traces of dwellings or midden deposits were found. Artefacts recovered included prehistoric flints—scrapers, cores and waste flakes—as well as later material such as iron implements and some slag. One radiocarbon date of ad 974 ± 74 (GU-1000) has been obtained for charcoal from a hearth area in a depression which seems to have been broken through the paving in square T43. Since that time the possibility of the site being concerned in some way with the mid-summer sunrise has been considered and this hypothesis has explained well many features which appeared incomprehensible before.

On top of the main outcrop is an area of paving which narrows as it leads back to a notch or cleft between two projecting rocks at the south-west end of the raised area; this cleft appears to be natural though the discoveries made at the rear boulders (below) make one wonder if it may be partly artificial.

Towards the front (north-eastern) end of the paved top of the outcrop a socket was found and, lying 4 m away on the south-eastern edge of the outcrop, a long, narrow stone 1.3 m in length which fits well into it (pl. 3.6). Beyond, or south-west of, the V-shaped cleft is a small semi-circular paved platform with a much larger, slightly lower one beyond it (pl. 3.4). Two more long stones, measuring 1.1 and 1.35 m in length, were found buried on the surface of the larger platform and a socket for one of them is immediately behind the cleft.

When viewed from the two rear boulders the rock outcrop presents a striking silhouette against the north-eastern sky (fig. 3.8; pl. 3.4). The cleft appears V-shaped from this angle and the fron stone, inserted in the socket, can be seen through it looking like the foresight of a rifle beyond the cleft.

On midsummer morning in 1977 I was at the site before dawn and, stationed between the two rear boulders, saw the Sun appear dramatically through thick haze exactly over the cleft and the foresight stone, then marked by a stake (pl. 3.5). The observing position chosen was the obvious one and a firm pebbled area between the boulders appeared to be an artificial floor. The front stone is 1.60 m (5 ft 3 ins) high and convenient to look over.

Further observation and measurement, some of which was carried out by Dr. A. S. Thom, showed that the notch visible against the rising solar disc is that between Beinn Dubhchraig and Beinn Oss, two peaks nearly 30 miles away from Tyndrum; this is theonly fragment of the horizon visible from the

Pl. 3.5 Brainport Bay: The same view at dawn, June 21st 1977 with wooden post in front socket and ranging pole between 'observation boulders'.

Pl. 3.6 Brainport Bay: The 'observation boulders' from the north-west, showing the paved area between them and the angular stratification of the front boulder.

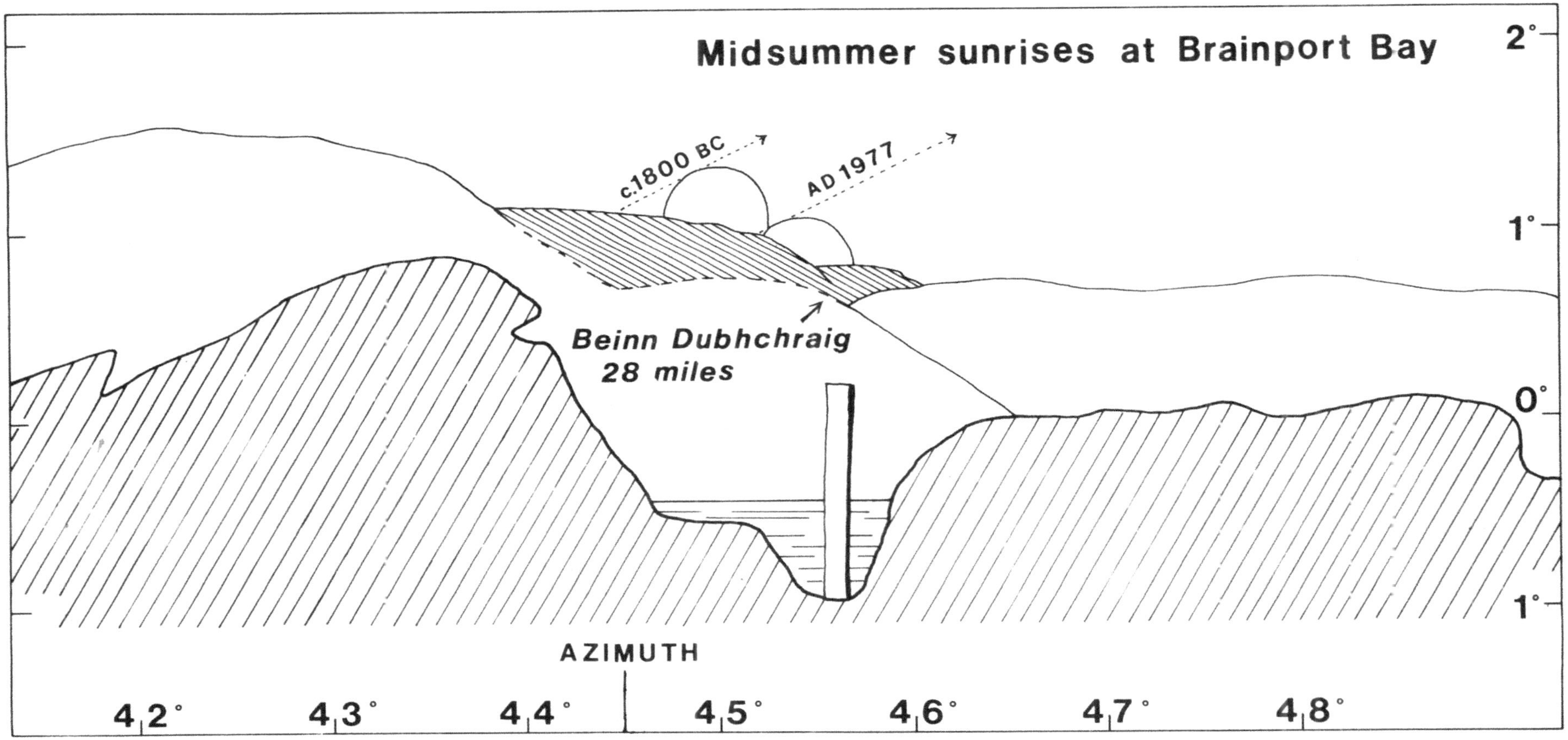

Fig. 3.8 Midsummer sunrises seen from Brainport Bay.

site which is more than about seven miles away (fig. 3.8). The horizon profile shows the present rising position of the Sun and its calculated one of about 4,000 years ago. It will be noted that this distant foresight—so clearly indicated by man-made features—is not of the same kind as that visible from other standing stones claimed as solstitial observatories. It does not, from this position, show a left hand slope rising at about the same angle as the Sun's ascent and against which the very gradually changing declinations around the solstices could have been exactly observed (Thom 1967; MacKie 1974). However the notch between the two peaks (clearly visible on the photograph) could have been used to mark the sunrise 15 days before and after midsummer (at a time when the Earth's axis was tilted 23° 54' from the vertical, instead of 23° 27' as at present—at about 1800 BC on uniformitarian assumptions) and in this way the day of midsummer could have been exactly discovered and predicted.

The whole design and arrangement of the Brainport Bay site suggests as much a ceremonial concern with the midsummer solstice as an astronomical or calendrical one. This is particularly true of the back platform (not shown on the plan), which is approximately circular and apparently at least partly artificial; a fairly large number of people could have stood on it and had a magnificent view of the dawn (pl. 3.5). At its front edge, and facing directly north-east down the alignment, is an artificial projection—a modified rock outcrop (pl. 3.8)—and it is difficult not to stand on it without imagining oneself a High Priest of the Sun, or some similar dignitary, leading a dramatic dawn ceremony in ancient times. This is fanciful of course but the site does suggest these thoughts and no-one who has not actually seen the Sun flash into view over the marks in the grey stillness of a clear midsummer morning there can have the faintest idea of how awesome the spectacle is.

At the same time the site has most of the important features of an accurate standing stone solar observatory of the kind first identified by Professor A. Thom—namely an artificial structure, including what seem to have been standing stones, pointing clearly to the solstitial sunrise which occurred behind a really distant mountain peak or notch, thus allowing the apparatus to function as an accurate calendrical instrument if required. There are also strong indications of other useful astronomical alignments radiating from the site but more investigation of these is needed. At Brainport Bay the artificial nature of the site is so clear that it is just not reasonable to raise doubts about the deliberate arrangement of the various features as an alignment towards the the north-east. Some of the paving might perhaps be later but the important features—the small once standing stones, the two sockets which seem designed for them, the V-shaped rock cleft and the rear 'observation boulders'—all seem clearly arranged to point to the midsummer sun at dawn.

Of great importance is the pair of rear boulders, with a flat rubble surface between them, which serve so conveniently as an observation point; from between them the 'rifle barrel' sighting effect to the north-east is seen most dramatically (pl. 3.5). The rubble between them looks like a deliberately laid floor; it was not found, except for a scatter of quartz and other pebbles immediately in front of the larger boulder, on the phyllite bedrock round about. However there are clear indications, first observed by Mr. S. Rankine, that the entire front boulder has been moved and tipped through 90° or more in a

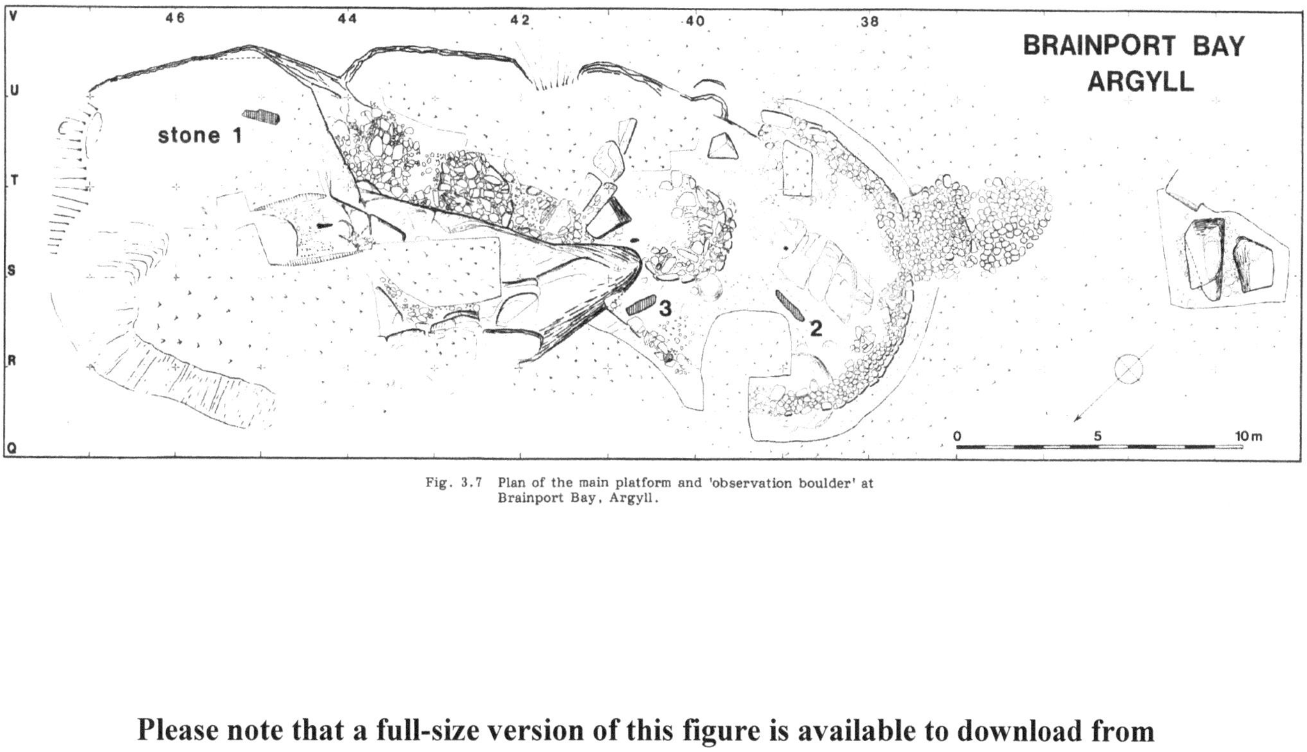

Fig. 3.7 Plan of the main platform and 'observation boulder' at Brainport Bay, Argyll.

Please note that a full-size version of this figure is available to download from www.barpublishing.com/additional-downloads.html. The original foldout has been reduced in size to match the A4 format of this book, the image is therefore not as clear as the original foldout. The image might also be cut off or not complete in the printed book. Please refer to the original foldout via the download for the original content.

vertical plane to reach its present position. It appears to have been attached at one time to an outcrop close by and the vertical side of this shows the strata of the metamorphosed rock turning sharply through an angle of about 90°, the result of some ancient heating and twisting of the sediments. The angle of these layers forms an arrow pointing upwards and to the right.

The same bent strata can be seen, more clearly, on both sides of the front 'observation boulder'; but here the angled strata form an arrow pointing down and to the right, the whole rock having evidently been tipped over from its original position (pl. 3.6). Moreover this must have happened after the passage over the site of the ice sheet which left scratch marks on the rocks down by the shore. The only likely reason for turning over the rock in this way was surely to lever it into its present position, where it serves as a marker from which to watch the midsummer sunrise.

Thus the site at Brainport Bay provides, in my view, the most dramatic and decisive evidence yet recovered that prehistoric man was concerned with precise solar observation and that this was most probably in the service of some powerful religion connected with the celestial bodies and for which an exact calendar was important. Further radiocarbon dates for the site are awaited with interest.

3.2.4 Giant henges and ceremonial centres

An important part of my hypothesis about the presence in Late Neolithic Britain of a stratified, possibly theocratic society was the identification of suitable training and residential centres for the religious and learned orders assumed to have existed then. This was achieved by re-interpreting the function of the giant henge monuments of southern England and the two contemporary stone villages known from Orkney—Skara Brae and Rinyo. The discovery of what may be a new example of a giant henge, Meldon Bridge in Peebles-shire, is discussed later (section 3.4); here I wish only to make some further remarks about one aspect of the evidence which I put forward to support the idea that sites like Durrington Walls in Wiltshire were permanently inhabited by a special class of people, probably members of religious orders, who were supported by the rest of the agricultural population (MacKie 1977a: chs. 7-10).

The diet of the inhabitants of Durrington Walls was compared with that of the people living in, or camping seasonally at, the older causewayed camp at Windmill Hill further north in Wiltshire and found to be significantly different from it (MacKie 1977a: fig. 35). The two sites were inhabited simultaneously —in the latter part of the 3rd millennium BC—and of course they stand in the same chalk downland environment and only a few miles apart. Yet the meat eaten at the great henge was very different from that consumed by the presumably peasant visitors to the causewayed camp and, while this cannot by itself tell us that the former were an upper class, this difference in diet matches very well with other sharp differences between the material cultures found on the two sites, and of course between the sites themselves.

It now seems that it is possible to infer from animal bones that prehistoric communities were at different levels in the social scale. In their new book, Baker and Brothwell (1980) demonstrate how traces of diseases can be found in animal bones and suggest that differences in the state of health of the flocks and herds of past communities could be explained by differences in the wealth, and therefore presumably the social status, of their owners. Unfortunately the large quantities of animal bones recovered from the extensive excavations at Durrington Walls and related Late Neolithic sites in the late 1960s (Wainwright & Longworth 1971) were not available for inclusion in this study but, when they are, they should surely be examined in the samey way.

It would seem that we have here an almost perfect experimental test for the hypothesis advanced in my book Science and Society in Prehistoric Britain (MacKie 1977a). If the giant henges of the south, and the related sites like Meldon Bridge in Peebles-shire and Skara Brae in Orkney, really were the residential centres of élite groups of some kind, then the animal bones from these sites may well be from healthier animals than those whose remains have been found on the (admittedly rarer) more ordinary domestic sites like Windmill Hill.

3.3 REACTIONS AND REVIEWS

3.3.1 Reviews of "Science and society in prehistoric Britain"

How has all this work, and the controversial ideas set out in Science and Society (MacKie 1977a) been received? A fair number of reviews by professional archaeologists have appeared and they fall into two groups: there are those which welcomed the new hypotheses, without necessarily agreeing with them completely, and offered constructive criticism; and those who dismissed them as fantastic nonsense, either overtly or by innuendo. Fortunately the latter were a minority but they include some of our best known archaeologists whose views are to be valued, so their views must be examined carefully.

Professor Stuart Piggott is probably the most eminent of those in the latter group with an unusually wide knowledge of European prehistory (Piggott 1965) and it was therefore sad that he found little good to say of the book in his review in Antiquity (1978a) although he was surprisingly short of specific criticisms. One of his major objections was that I misused archaeological evidence; phrases occurred like 'here is some very odd use of evidence...' and 'Despite his protestations Dr. MacKie has not played fair with his non-technical readers but presented them with something of an intellectual confidence trick, building, with a wild disregard for the proper handling of the evidence a grotesque structure of theory upon Thom's thesis.' An elegant literary jibe at the end of the review was thought a suitable way to complete the job of demolition.

It was sadder still to have to come to the conclusion that the confidence trick was on the part of the reviewer, who substituted for specific arguments condemnation by the authority of his reputation and by the trick of implying guilt by association. As already explained, the nature of archaeological evidence was fully discussed in the book and it was made clear that an attempt was being made to apply the methods of natural science to a difficult problem. These methods are well described by Beveridge (1961) and consist in essence of analysing a given body of evidence, of developing an hypothesis to explain it, of defining predictions made by the hypothesis and testing these predictions by experiments so that they are verified or refuted. Two crucial such experiments by excavation (at Kintraw and Cultoon) were described in the book, both of which were tests of fundamental aspects of Thom's theory, and they were the first such to be carried out by an archaeologist. However not a single archaeologist reviewer mentioned either piece of work except Fischer (1978)—though the astronomer Heggie did (1978), as also did Wood (1978: 85) and Hadingham (1975: 109) in their books. This is the more extraordinary since Beveridge's book was favourably reviewed by Stuart Piggot (1962) and recommended to the profession as essential reading: in fact it influenced me strongly in my later work. Particularly critical reviewers and commentators like Piggott (1978a), Daniel (1980) and Jacquetta Hawkes (1977) surely had a duty at least to mention what support archaeological excavation has given to the new hypotheses; but none did.

3.3.2 Kintraw

The results of the Kintraw excavations have in fact caused some interesting contortions in archaeological commentary which might be construed as unintentional compliments to the implications of the work. For example in his comprehensive and authoritative survey on stone circles, Burl briefly discussed the site (1976: 199). He states:

> "The other point is that despite this overall confirmation of Thom's astronomical analyses" [seen in the orientations built into a variety of stone circles] "he may be wrong about Kintraw which he has claimed as one of the most important solstitial sites known to him... He also postulated an artificial sighting-platform cut in a precipitous hillside to the NE of the cairn but as this continues and widens for several hundred feet all around that side of the hill it is much more likely to be natural (but see MacKie 1974: 181). In fact, as at Monzie and Fowlis Wester, the outlier probably belongs to the small kerb-cairn B, and provides a good alignment to the midsummer sunrise."

There are several notable features about these remarks. Thom did imply that the platform, or terrace, had been cut into the hillside and he surmised that the boulder on this terrace marked the primary observation point for setting up the observatory. However I pointed out that the terrace was probably a natural feature (MacKie 1974: 181) and the subsequent excavations on it, next to the boulder, were designed to see if this part of it had been modified to form a viewing platform and they showed that it almost certainly had. This remarkable discovery which shows —exactly contrary to what Burl says —that Thom is almost certainly right about Kintraw, is concealed in a '(but see)' and a bare reference! Moreover the suggested midsummer sunrise line is just not credible: the standing stone is only a few feet from the small 'kerb-cairn' B, the suggested observation point, far too close to give anything but the crudest indication of direction. Worse, the horizon to the NE is less than a mile away, with a great number of irregular bumps and notches on it. As has often been explained, solstitial alignments, to be credible and of any use at all as practical observing instruments, should include horizon markers many miles distant which should be clearly indicated by one or more standing stones.

We seem to have here a desire to emphasise simple, approximate orientations towards astronomical positions in ancient sites—which everyone can accept without mental strain because they imply no more than a general interest in celestial phenomena and a seasonal calendar (Renfrew 1973: 237ff; Fleming in MacKie 1976), or the existence of a ritual need like the one which produces the E orientation of Christian churches, and to play down the existence of immensely long alignments of which Kintraw and Ballochroy are classic examples (MacKie 1974: 176). Clear-cut, distant natural foresights, 18 miles away at Ballochroy in Kintyre and 29 at Kintraw, could have served as accurate observing instruments and they point, if genuine, to a desire for precise measurement and a degree of skill in astronomical observation which must surely indicate the presence of a trained class of observers. These would have been committed to searching for years over the highland terrain for suitable viewing points and to making years of observations after these had been established. This is what the discoveries at Kintraw imply and this is why the site is so important; the kindest thing one can say about the almost universal failure to discuss the evidence properly, or even to mention it, is that it is odd.

A more recent example of misleading comment about Kintraw comes from Dr. G. Moir, in the course of a joint article discussing some mistakes in Professor Thom's archaeological identifications and some ambiguous surveys (Moir 1980; Ruggles & Norris 1980). Moir says:

> "The site at Kintraw (NGR: NM 830050) is one of potential extreme accuracy for the observation of mid-winter sunset, and has been much discussed following MacKie's excavations of the associated 'observing platform' (MacKie 1974; 1977a)."

He gives details of discrepancies between the plan of the two cairns next to the standing stone prepared by Thom and that made by the archaeologist who excavated them, Dr. D. D. A. Simpson; both dimensions and azimuths differ. After several further related comments he says:

> "In view of the importance of Kintraw in discussions of Megalithic Astronomy it seems somewhat surprising that no-one has remarked ... on the differences in the two published site plans."

This is extremely misleading as the last sentence implies quite clearly that the value of the site as an accurate prehistoric astronomical observatory is diminished by these discrepancies (I have had experience of people saying this). Of course it is nothing of the kind. The cairns, their sizes and orientations are quite unconnected with the long solstitial alignment which runs from the standing stone—or, more probably, from the excavated hill platform —south-west to the Paps of Jura 28 miles away. Again Dr. Moir glosses over the main point about the Kintraw platform.

The most incomprehensible of these episodes is still a statement by Dr. Jon Patrick, in a lecture read for him at the 1977 Glasgow conference 'Ceremonial, science and society in prehistoric Britain' (MacKie 1976), that he was unable even with a theodolite to see the mountain foresight from the hill platform at Kintraw (see paper 5 of this volume—eds.). This is a nonsense which Dr. Patrick has never explained; photographs are available, taken from the platform, which show the V-shaped notch on Jura fully in view.

Evan Hadingham (1975) has also written a general survey of stone circles, and his treatment of the site makes a contrast to those just outlined. The site and the platform excavations are fully discussed and the significance of the long accurate alignment—or observing instrument—is clearly grasped. Indeed Hadingham was one of the first to acknowledge in print the plausibility of the idea (MacKie 1974: 188) that the kind of site represented by Ballochroy and Kintraw means "that the demand for an accurate calendar was set up by some specialist group in ancient Britain, perhaps by a ruling aristocracy or priesthood" (Hadingham 1975: 109-115).

3.3.3 Professional priests and wise men

Piggott and several others, in their reviews of Science and Society, accepted that it was interesting and useful to look at the Maya Classic civilisation of the first millennium AD in Central America as a possible model with which to explain British Neolithic society (MacKie 1977a: ch. 10). Here is a striking example of a non-urban, stone-using society, dominated by chiefs and by a professional priesthood, which had developed advanced intellectual skills in astronomy, mathematics and calendar-making. These upper classes lived in special ceremonial centres most of which were isolated groups of stone buildings among the scattered dwellings of the rural population, relatively small and composed solely of temples, priestly residences and 'palaces'. A very few, like Tikal in Guatemala, attained gigantic size and can be regarded as cities. However the Maya also developed a form of writing, which our Neolithic society did not, so one might legitimately doubt the validity of a close comparison with this culture.

What is strange however is that all the hostile reviewers failed to mention a much more convincing piece of evidence much nearer home—the Druids of Iron Age Britain and Gaul. Here was exactly the kind of upper class of priests, wise men, jurists, poets and so on—separated out from the Celtic tribal societies of the time and with immense prestige—that would fit the Neolithic evidence. Moreover they lived in exactly the same areas as, and less than two millennia later than, the stone circle builders. The Druids were highly skilled in astronomy and cosmology as Piggott describes in detail (1968: 122) and they also had a strong tabu against writing down any of their specialised knowledge, spending many years instead learning by heart the huge quantities of verse which doubtless contained their accumulated wisdom.

Even more exciting, in the light of Thom's work, now appears the information, derived from Julius Caesar, that "'their doctrine is believed to have been found existing in Britain—disciplina in Britannia reperta—and thence to have been imported into Gaul and nowadays most people who wish to acquire a thorough knowledge of it, go there to study'" (Piggott 1978b: 52; Burn 1969: 4). This idea of the Druidical orders originating in Britain made no sense in the old diffusionist days when their knowledge was supposed to have been obtained from the Greeks through their colony established at Massilia (Marseilles) at around 600 BC, but is easily explicable if the standing stone sites (confined to Britain and Brittany) were the work of similar, and ancestral, orders of priests and wise men in Neolithic times.

Table 3.3 List of 38 surveyed brochs

1	2	3	4	5	6	7	8	9	10	11	12	12	13
Clickhimin	A/1	HU 464408	A	SB	C	4.615	61	11	+55	0.839	(2.75)	34.98	13.99
Clumlie	A/2	HU 404182	A2	SB	C	5.103	93	12	+29	0.851	(2.79)	38.68	15.47
Jarlshof	A/3	HU 399096	A	SB	C	4.493	20	11	-67	0.817	(2.68)	34.06	13.62
Levenwick	A/4	HU 415197	A2	SB	C	4.173	76	10	+116	0.852	(2.80)	32.30	12.92
Loch of Huxter	A/5	HU 173569	A2	TR	C	4.851	45	12	-103	0.810	(2.66)	36.92	14.77
Borwick	B/2	HY 223167	A2	?	C?	(uncertain)		-	-	-	-	-	-
Burray East	B/3	HY 491968	B	GG	C	5.503	43	13	+114	0.847	(2.78)	41.71	16.69
Gurness	B/8	HY 383268	A	GG	C	5.080	56	12	+106	0.847	(2.78)	31.51	15.40
Howe of Hoxa	B/19	HY 425939	A	SB?	C	4.531	51	11	-30	0.824	(2.70)	34.34	13.74
Achvarasdal Lodge	C/2	NC 983647	A	SB	C	5.036	37	12	+62	0.839	(2.75)	38.17	15.27
Keiss North	C/9	ND 354614	A	SB	C	3.952	25	10	-193	0.790	(2.59)	29.96	11.98
Keiss South	C/10	ND 354612	A	SB	C	5.835	25	14	+32	0.834	(2.74)	44.23	17.69
Keiss West	C/	ND	A	SB	C	5.175	35	12	+201	0.796	(2.61)	39.23	15.69
Ness	C/14	ND 381666	A	SB	NC	(egg shaped)		-	-	-	-	-	9.291
Nybster	C/15	ND 370632	A	SB	C	3.211	51	8	+105	0.803	(2.64)	24.34	9.74
Skirza Head	C/16	ND 394684	A	SB	C	3.322	61	8	-5	0.827	(2.66)	25.08	10.03
Yarrows	C/18	ND 308435	A2	TR	C	4.859	53	12	-115	0.810	(2.66)	36.83	14.73
Backies	D/1	NC 835026	A2	SB	C	4.132	60	10	-13	0.826	(2.71)	31.32	12.53
Carn Liath	D/2	NC 871013	A	SB	C	5.105	94	12	+165	0.857	(2.81)	38.95	15.58
Carrol	D/3	NC 846065	A2	SB	C	4.698	49	11	+138	0.854	(2.80)	35.61	14.25
Dunbeath	D/5	ND 155305	A2	SB	C	4.240	36	10	+95	0.848	(2.78)	32.14	12.85
Kintradwell	D/8	NC 929081	A	SB	C	4.724	56	11	+164	0.859	(2.82)	35.81	14.32
Ousedale Burn	D/9	ND 172188	A	SB	C	3.321	38	8	+11	0.832	(2.73)	25.22	10.09
Sallachadh	D/10	NC 548092	A2	SB	C	4.924	43	12	-50	0.821	(2.69)	37.32	14.93

(contd.)

Table 3.3 (contd.)

1	2	3	4	5	6	7	8	9	10	11		12	13
Armadale Burn	D/23	NC 800625	B	?	C	3.503	34	8	+187	0.876	(2.87)	26.55	10.62
Forsinain	D/30	NC 899509	B	?	NC	5.250	35	-	-	-	-	-	-
Caisteal Grugaig	E/1	NG 867251	A	TR	C	4.686	56	11	+126	0.852	(2.80)	35.52	14.21
Dun Beag	E/4	NG 340386	A	GG	C	5.387	45	13	-2	0.829	(2.72)	40.84	16.33
Dun Fiadhairt	E/5	NG 230507	A	GG	C	4.735	70	11	+190	0.864	(2.84)	36.01	14.40
Dun Osdale	E/9	NG 242464	B	?	C	5.155	58	12	+181	0.859	(2.82)	39.08	15.63
Dun Telve	E/12	NG 829173	A	SB	C	4.919	19	12	-55	0.820	(2.69)	37.29	14.92
Dun Troddan	E/13	NG 834173	A	SB	C	4.280	25	10	-135	0.856	(2.81)	32.44	12.98
Dun Boreraig	E/18	NG 195532	B	?	C	5.072	54	12	-99	0.846	(2.78)	38.46	15.38
Dun Borodale	E/19	NG 555364	B	?	NC	elliptical, c. 10 x 14 MY							15.28
Dun Carloway	F/e	NB 190412	A	TR	C	3.660	59	9	-61	0.813	(2.67)	27.74	11.10
Dun Torcuill	F/9	NF 888737	B	?	NC	elliptical		-	-	-	-	-	17.84
Dun an Sticir	F/8	NF 898777	B	?	NC	elliptical		-	-	-	-	-	17.84
Torwood	G/2	NS 838848	A	SB	C	egg-shaped		-	-	-	-	-	15.44

The columns give the following information: 1 = name of site; 2 = site code (A = Shetland, B = Orkney, C = lowland Caithness, D = highland Caithness and Sutherland, E = west coast and Inner Hebrides, F = Outer Hebrides, G = central Scotland); 3 = O.S. grid reference; 4 = category of information (A = excavated and interior wall exposed most of the way round at ground level, A2 = excavated but much material fallen back in, B = unexcavated so that the internal wallface is visible only several feet above the ground and not completely); 5 = type of broch (GG = ground-galleried, SB = solid-based, TR = transitional form); 6 = shape (C = circular, NC = non-circular); 7 = radius in metres of best-fit circle; 8 = mean error in mm of this circle from the wall-face; 9 = nearest whole number of MY in diameter of best-fit circle; 10 = difference in mm between half that number and the radius of the best-fit circle; 11 = length (m & ft) of the unit near the MY which divides exactly into the diameter of the best-fit circle; 12 = length of the perimeter of the best-fit circle in MY: 13 = the same length in megalithic rods ($2\frac{1}{2}$ MY).

Owen Gingerich (1977) saw that the existence of the Druids provides very strong support for my hypothesis of a theocratic, stratified society in Neolithic Britain since the Iron Age learned orders could easily have been directly descended from earlier ones. This of course was discussed both in the book (MacKie 1977a: 226-228) and before that (MacKie 1974: 189). In this context Professor D. W. Harding (1977) mentioned the extremely important Irish Late Bronze and Iron Age sites of Tara, Dun Ailinne and Emain Macha which excavation is showing to have originated around 600 BC or earlier and to have been an interesting combination of royal and religious centres, quite probably analogous to Durrington Walls and the other giant henge sites in S England 2,000 years earlier. What is more it is steadily becoming clear, as Harding also mentioned, that various aspects of British Iron Age society, including its ritual monuments, are probably derived from a much earlier local tradition; continuity over the 'three ages' (Neolithic, Bronze and Iron) is gradually becoming clearer from traditional archaeological evidence.

There is in fact some evidence which gives unequivocal support to the hypothesis that the learned orders of the Iron Age inherited the specialised knowledge and traditions of the Neolithic period and it has already been discussed twice (MacKie 1975: 85; 1977a: 61). The brochs are circular Iron Age drystone forts probably dating to the period from the first century BC to the second AD and which are found only in the highland and island regions of the far north and west of Scotland. They are skilfully built and strikingly standardised in their design. As a check on the methodology Thom used to discover the existence of the 'megalithic yard' in the stone circles, I took in 1971 accurate measurements of the shape and dimensions of the central courts of 37 brochs (table 3.3). The results were analysed by Dr. G. I. Crawford to discover (1) whether the broch builders had set out the towers accurately round compass-drawn circles and (2) if they had, whether a standard unit of length had been used to measure out the circles. There was no prior expectation that such a unit would be discovered, or what it might be, although I certainly expected that the brochs would prove to be exactly circular, which they did.

The results for the unit of measurement investigation were unexpected and surprising. It appeared that a 'broch yardstick' had been used and that its mean length was 0.839 m, only 1 cm longer than the Megalithic yard. The degree of confidence in the result was 2%, that is there were only two chances in a hundred that this unit had emerged fortuitously from the analyses. Moreover several brochs fitted with extraordinary accuracy (to within a few cms at most points) round other geometrical shapes; two were elliptical and two followed an unusual shape which consisted of half an ellipse with an egg-shape or a half circle resting on it. Unless one assumes that these traditions of measuring and geometry emerged again in the Iron Age and by chance duplicated those of the stone circle builders two millennia earlier, there is no alternative to accepting that there was continuity between the specialised knowledge of the two eras. Again no reviewer mentioned this crucially important evidence, except Fischer (1978).

Stuart Piggott himself went into exactly this problem of continuity between the Iron Age and earlier times in considerable detail in a lecture in Rome in 1975 and, among many other things, discussed the probable derivation of the

Irish Iron Age ceremonial sites mentioned earlier from the English Neolithic henge monuments (Piggott 1978b: 49ff.). He also mentioned favourably there the importance of the Druids in this context and my own ideas of the links these may have had with Neolithic society:

> "What was transmitted from Britain to Gaul in Druidic teaching did not include the Irish type of ceremonial centre, but as MacKie [1971] has recently suggested, a contributary component of insular Celtic religion, again from the same substrate, might well have been something of the calendrical expertise and numeracy which, even if we reduced the claims of the archaeo-astronomers to the most sceptical minimum, must have been an important factor in the construction and purpose of the earlier circular monuments."

The contrast between these sentiments and the extreme hostility of the book review (Piggott 1978a) is baffling. Could it be that, when the inevitable conclusion to be drawn from all these indications—that there were learned orders of professional priests and wise men in Neolithic Britain and it is possible to identify their residential and training centres and their temple-observatories —is stated openly, that is when the foreign girl is paraded through the village square with her charms plainly visible instead of being talked about in whispers, then instinct takes over and the local Venus finds many champions?

3.3.4 The work of Professor Thom

A number of reviewers felt that the first half of Science and Society— which described and analysed Thom's most important discoveries and some of his methods—was too favourably disposed towards them; that is in Piggott's words

> "While the author is at desperate pains to give the appearance of playing absolutely fair with the readers, his treatment of Thom's hypotheses is that of the passionate conviction of disciple rather than that of a disinterested critic of a number of propositions which have not yet received by any means universal acceptance from either astronomers and mathematicians, or from archaeologists."

I would question only the 'passionate' here; my present conviction arose from the results of the fieldwork tests described but it was certainly not uncritical, as Heggie (1978) noted. The sections on the 'delusions of accuracy' which can come through mathematical analyses, on the degree of standardisation of the megalithic yard, on the sites claimed as lunar observatories as well as on the astronomical hypotheses about Stonehenge should make this reasonably clear. The stated purpose of the book was to carry out a further major test in the light of the results of the fieldwork experiments—that is to find out whether the 'traditional' archaeological evidence could be reinterpreted convincingly to provide the kind of Neolithic society the Thomian hypotheses required. What is hard to understand is the persistent failure of some archaeologist critics to assess the archaeological evidence relevant to the Thom and MacKie hypotheses, evidence which their training does equip them to evaluate properly. Instead there is a tendency to rely on generalised condemnations and on the simple acceptance of the criticisms of specialists in alien fields such as that of statistics.

3.3.5 The form of publications

Another criticism voiced by Professor Piggott (1978a) and also by Sharon Gibbs (1979) is that the new hypothesis about the structure of Neolithic society in Britain would have been better set out first in a learned journal rather than in a book. Presumably the reasoning here is that the experts ought to be allowed to assess the new view first and then present a united front to the public, who cannot judge such things for themselves. Indeed Piggott explicitly says this:

> "...the general public is in no position to do this" [assess the book properly]. " Readers may too often belong to Sir Peter Medawar's large population of people, often with well developed literary and scholarly tastes, who have been educated far beyond their capacity to undertake analytical thought";..." worse there will be encouragement (in a book) for the weak-minded, the credulous seekers for magic and ancient mysteries."

One would like to know whether this last point means that he thinks archaeo-astronomical research, and hypothesising based on the results, should be banned because it attracts fringe cults; if so one could equally argue that archaeology itself should be avoided because of the works of, for example, von Däniken, which its discoveries have stimulated. The point about laymen needing guidance is certainly true of course although, while intelligent non-archaeologists clearly do not possess the detailed knowledge of the evidence that professionals do, there is no need to underestimate their capacity for distinguishing logical from illogical argument, or indeed fair from prejudiced reviews. Indeed anyone who maintains that 'the expert knows best' to this extent obviously has an exceptionally clear duty to appraise new work for the public as open-mindedly and thoroughly as possible.

However it does seem fair to maintain that a scholar should give his professional colleagues the first chance to judge a new hypothesis, if only out of courtesy. I have already mentioned that some of the crucial evidence which I found by excavation to favour Thom's hypothesis was published in 1974 in the Philosophical Transactions of the Royal Society and that this included a disquisition on the possible significance of the Druids for our understanding of the builders of the henge monuments and stone circles. All this was first presented in a lecture to a conference called 'Astronomy in the Ancient World', organised in 1972 jointly by the Royal Society and the British Academy, and I also spoke to the Prehistoric Society on the same theme in 1973. The published paper was gratifyingly well received by Professor Piggott (1978a) as explained. In 1976 I planned a conference called 'Ceremonial, Science and Society in Prehistoric Britain' in Glasgow (organised by the Department of Extra-Mural Education) in which various aspects of Neolithic Britain were discussed. There I outlined for the first time in public the theme of the second half of the book—that many important Neolithic sites could be drastically re-interpreted as the residential and training centres of a theocratic élite and could thus provide Professor Thom with the society that his discoveries required. Although this conference was widely advertised only about twenty professional archaeologists came and these included none of the eminent ones who later disapproved so strongly of my hypotheses when they appeared in book form. A summary of what was said at the conference, and of my lecture, appeared in Antiquity (MacKie 1976) so it could be said that a fair amount of warning was given of what was coming.

In this context it is worth explaining briefly how Science and Society in Prehistoric Britain was conceived and how it developed. I was originally approached by Paul Elek in December 1971 with the suggestion that I should write a more popular book on the Thomian ideas than had appeared up to that time and the first draft outlines which I submitted were concerned solely with explaining Thom's work and with considering how plausible it was when set against the traditional picture of prehistoric Britain. It was really only when much of the first part, dealing with the astronomical and geometrical evidence, had been drafted that it became clear that the archaeological implications were much more profound that I had realised. The drastic reinterpretations of several British Neolithic sites which appear in the second half actually evolved as the work was being written so it seemed only fair that they should first appear with the book.

3.4 FRESH EVIDENCE: MELDON BRIDGE, PEEBLES-SHIRE

Fresh evidence is still emerging to give support to the new interpretation. A. H. A. Hogg (1977) pointed out that although the 'ceremonial centres' for the Neolithic theocracy identified in Wessex are quite convincing,

> "elsewhere the evidence does not compel belief in anything more than a few itinerant priest-architects. A national theocracy ought surely to have maintained some ceremonial centres in the now blank areas between Avebury and Skara Brae...... Not the least merit of Dr. MacKie's stimulating book is that it will lead to a new search in these empty regions for the missing great henges, which if discovered will go far to confirm his theories."

Such a centre and one of supreme importance, seems to have been identified in Peebles-shire in SE Scotland, a little further than halfway between Wessex and Orkney.

In 1974 and 1975 Colin Burgess explored a site at Meldon Bridge, discovered by air photography and threatened by a road widening scheme (Burgess 1976). The circular dark patches seen from the air were at first thought to be an alignment of pits and they lay in a triangular area of flat ground on river gravels at the foot of a slope, between the Meldon Burn and the Lyne Water. This site is only a few hundred metres above the junction of the latter river with the Tweed and 7 km W of Peebles.

The excavations showed that the 'pit alignment' was in fact the entrance avenue in a massive timber barrier which cut off the triangular promontory formed by the two water courses referred to, and which enclosed an area of about 50 hectares (20 acres). The site is in a narrow but strategically important part of the Tweed Valley which leads by way of the Biggar gap to the upper valley of the Clyde and so to the west; two Roman forts nearby and two marching camps (one of which partly overlies the Neolithic site) well illustrate this. The palisade was about 430 m long and composed of massive posts, some 25-30 cm in diameter but others up to 60 cm: they stood about 3-4 m apart and in places there were traces of pairs of intermediate lighter uprights which suggested to Burgess that the whole had been a huge fence with horizontal planking. It seems to me unlikely that the site was a purely defensive one since the water barriers on two sides are not too difficult to cross and no clear signs of a gate in the entrance avenue have been spotted (though this has not yet been excavated). In the interior were many pits, some for rubbish and some for uncertain, possibly ritual purposes, and much decorated middle Neolithic pottery was recovered. There was no Grooved ware. A series of radiocarbon dates showed that the enclosure had been built and used near the beginning of the third millennium bc and was thus somewhat older than the earliest phase of Stonehenge.

Meldon Bridge is remarkably similar in its layout to the great Neolithic henge at Marden in Wiltshire where the earthwork formed two sides of the enclosure and the River Avon the third. The early Christian monastic site at Old Melrose, further down the Tweed, shows the same idea and it seems likely that the enclosure near Lyne is another Neolithic 'ceremonial centre' —showing that combination of ritual and domestic features which suggest that it was inhabited by a priestly class. A massive timber barrier, a secondary construction, was found at the giant henge at Mount Pleasant in Dorset. Meldon Bridge is older than the Grooved ware sites which were discussed in my book and seems to show that the learned orders of Neolithic times had a variety of cultural traditions, or at least that they lived among a variety of different communities.

If my interpretation of its social function is correct (and it is not necessarily supported by the excavator), Meldon Bridge is also extremely important in that its date takes the existence of the prehistoric priestly classes further back into Neolithic times and makes more likely their ultimate origin among the builders of the megalithic and other collective tombs of the 5th and 4th millennia BC. This is the theme of the other half of this reappraisal of the British and Atlantic European Neolithic period and was set out in The Megalith Builders (MacKie 1977b); here again the journal Antiquity provided the vehicle for a counter-attack (Fleming 1978) and its editor also took part (Daniel 1980). However the question of the nature of the people buried in the collective tombs and their origin is a vast one and further discussion of it must be held over for another occasion.

Meldon Bridge has also provided a decisive piece of evidence which strongly supports the idea that it really was a class of astronomer-priests who lived in these 'ceremonial centres'. The straight avenue of posts leading up to the enclosure is 27 m long and was studied by Dr. G. Hulme of the University of Lancaster, who found that in both directions it pointed to places on the horizon of astronomical significance. To the SE across the Tweed the avenue is directed exactly at the prominent summit of Cademuir Hill; this peak marks the sun rising with a declination (astronomical latitude) of $-16^{\circ}26'$, a position corresponding exactly to one of the 16 subdivisions of the solar year inferred by Thom to be marked at many standing stone sites. It falls halfway between the autumnal equinox and the winter solstice and again halfway between the latter and the spring equinox (about 8 November and 5 February). In the opposite direction, to the NW, the avenue points to a spot on the nearby hilly horizon where the sun sets at midsummer. Exactly the same pair of alignments are marked by the Castlerigg stone circle in Cumbria (Thom 1967: 150) and it would be hard to think of a clearer demonstration of the nature of the people who lived in these huge enclosures. Meldon Bridge seems likely to provide formidable support for the new hypotheses.

3.5 CONCLUSIONS

Considering first the implications of the tests by archaeological fieldwork of Professor Thom's two major hypotheses outlined earlier, the results have convinced at least one archaeologist of the basic correctness of his astronomical and geometrical interpretations of the standing stone sites, although obviously one can and must argue about many of the details. For example the existence of precise and potentially accurate solstice-observing instruments seems proved beyond doubt, but it is not clear to me yet how much further the technique was developed—how far for example the tracking of the Moon's more complicated motions was taken (McCreery 1979; 1980a; Ellegård 1981)—since no independent tests have been carried out on any of these sites.

Likewise the existence of accurately set out geometrical figures, and the use of a well developed, standardised measuring system based on a traditional unit of length which may well have been imported rather than have been locally derived from a pace (MacKie 1977: ch. 3), seems to be clear in the stone rings but I am not personally convinced that the statistical evidence compels the belief that these measuring rods were exactly the same length throughout the country (MacKie 1977: 36-43; Ellegård 1981).

Thus it seems quite reasonable to think in terms of the existence in Late Neolithic Britain of full time castes or orders of priests and wise men, no doubt practising some celestial religion for which fairly advanced skills in surveying, practical geometry and observational astronomy were required. While it is not impossible that some members of these orders went on to investigate astronomical phenomena for their own sake, this seems difficult to prove while the case rests entirely on statistics—particularly in the absence of any evidence for written records (allowing that the cup-and-ring rock carvings may prove one day to be something of the sort). 'Science' in the modern sense is probably not the word to apply to the activities of our Neolithic ancestors, any more than it could be applied to those of their Iron Age successors, the Druids, and I would now be inclined to omit the word from the title of my book (1977a) if I were writing it again. What remains still seems quite revolutionary enough in terms of the picture of ancient British society which was accepted without question by almost everyone only fourteen years ago.

When reflecting on the second phenomenon—the reaction of professional colleagues to the fieldwork and hypothesising briefly reviewed here—enough has probably been said to suggest that objective assessment of the new evidence has not always been manifest, but of course this can change. Perhaps it is sufficient to quote, in conclusion, the words of one eminent British archaeologist who has mastered both sides of the evidence, both that provided by Thom and by more traditional methods; those who oppose both Thom's discoveries and the hypotheses that flow from them could well reflect carefully on them.

> "It is hardly surprising, therefore, that many prehistorians either ignore the implications of Thom's work because they do not understand them or resist them because it is more comfortable to do so. I have myself gone through the latter process; but I have come to the conclusion that to reject Thom's thesis because it does not conform to the model of prehistory on which I was brought up involves also the acceptance of improbabilities of an even higher order. I am prepared, in other words, to believe that my model of European prehistory is wrong, rather than that the results presented by Thom are due to nothing but chance" (Atkinson 1975: 51).

It seems that we are at a major turning point in Neolithic studies and I have tried in Science and Society in Prehistoric Britain and The Megalith Builders to sketch out an alternative view of prehistoric society in NW Europe which will accommodate all the available evidence, both old and new. This new model of British Neolithic society may or may not turn out to be the best one in the long run; but, I submit, it explains more of the total evidence available at present than any other currently being considered. It deserves more careful and objective consideration than some colleagues have given it up to now.

REFERENCES

Atkinson, R. J. C., 1975. Megalithic astronomy—a prehistorian's comments. J. Hist. Ast., 6, 42-52.

Baker, J. & Brothwell, D., 1980. Animal diseases in archaeology. Academic Press.

Beveridge, W. I. B., 1961. The art of scientific investigation, London.

Bibby, J. S., 1974. Petrofabric analysis. Phil. Trans. R. Soc. Lond., S276, 191-194.

de Bono, E., 1976. Practical thinking. Harmondsworth.

Burgess, C. B., 1976. Meldon Bridge: a Neolithic defended promontory complex near Peebles. In C. B. Burgess & R. Miket (eds.), Settlement and economy in the third and second millennia BC. Oxford: B. A. R. 33, 151-179.

Burgess, C. B., & Shennan, S., 1976. The beaker phenomenon: some suggestions. In C. B. Burgess & R. Miket (eds.), Settlement and economy in the third and second millennia BC. Oxford: B. A. R. 33, 309-331.

Burl, A., 1976. The stone circles of the British Isles. London: Yale University Press.

Burn, A. R., 1969. Holy men on islands in pre-Christian Britain. Glasgow Arch. J., 1, 2-6.

Cardona, D., 1979. The cairns of Kintraw. Kronos, 4(3), 38-55.

Cowie, T., 1980. Excavations at Kintraw, 1979. Glasgow Arch. J., 7, in press.

Daniel, G. E., 1980. Megalithic monuments. Scientific American, July, 78-90.

Ellegård, A., 1981. Stone age science in Britain? Curr. Anthrop., in press.

Fane Gladwin, P., 1978. Discoveries at Brainport Bay, Minard, Argyll: an interim report. The Kist, no. 16, 1-15.

Fischer, U., 1978. Review. Germania, 56, 252-256.

Fleming, A., 1978. Review of "The megalith builders". Antiquity, 52, 58-59.

Gibbs, S. L., 1979. Review. Archaeoastronomy (bulletin, Centre for Archaeoastronomy), 2, no. 2, 21-22.

Gingerich, O., 1977. Review. The American Scholar, 46, 530-532.

Hadingham, E., 1975. Circles and standing stones. London.

Harding, D. W., 1977. Review. Times Higher Educ. Suppl., June 24th.

Hawkes, J., 1977. Review. Sunday Times, Aug. 14th.

Heggie, D. C., 1978. Review, J. Hist. Ast., 9, 61-63.

Hodson, R., 1964. Cultural grouping within the pre-Roman Iron Age. Proc. Prehist. Soc., 30, 99-110.

Hogg, A. H. A., 1977. Review. Times Lit. Suppl., Nov. 4th.

McCreery, T., 1979. Megalithic lunar observatories—a critique, Part I. Kronos, 5(1), 47-63.

McCreery, T., 1980a. Megalithic lunar observatories—a critique, Part II. Kronos, 5(2), 6-26.

McCreery, T., 1980b. The Kintraw stone platform. Kronos, 5(3), 71-79.

MacKie, E. W., 1974. Archaeological tests on supposed astronomical sites in Scotland. Phil. Trans. R. Soc. Lond., A276, 169-194.

MacKie, E. W., 1975. The brochs of Scotland. In P. Fowler (ed.), Recent work in rural archaeology. Stratford-upon-Avon. pp. 72-92.

MacKie, E. W., 1976. The Glasgow conference on ceremonial, and science in prehistoric Britain. Antiquity, 50, 136-138.

MacKie, E. W., 1977a. Science and society in prehistoric Britain. London: Paul Elek.

MacKie, E. W., 1977b. The megalith builders. Oxford: Phaidon.

MacKie, E. W., 1981. Ancient Scotland. Harmondsworth, in press.

Moir, G., 1980. Megalithic science and some Scottish site plans: Part I. Antiquity, 54, 37-40.

Piggott, S., 1954. Neolithic cultures of the British Isles. Cambridge.

Piggott, S., 1962. Review of The art of scientific investigation, by W. I. B. Beveridge (1961). Antiquity, 36, 69-70.

Piggott, S., 1965. Ancient Europe. Edinburgh.

Piggott, S., 1968. The Druids. London.

Piggott, S., 1978a. Review. Antiquity, 52, 62-63.

Piggott, S., 1978b. Nemeton, Temenos, Bothros: sanctuaries of the ancient Celts. In I Celti a la loro cultura nelloepoca pre-Romana e Romana nella Britannia. Rome: Accademia Nazionale dei Lincei.

Renfrew, C., 1968. Wessex without Mycenae. Ann. Brit. Sch. Arch. at Athens, 63, 277-285.

Renfrew, C., 1970. The tree-ring calibration of radiocarbon: an archaelogical evaluation. Proc. Prehist. Soc., 36, 280-311.

Renfrew, C., 1973. Before civilisation. London.

Ruggles, C. L. N. & Norris, R. P., 1980. Megalithic science and some Scottish site plans: Part II. Antiquity, 54, 40-43.

Thom, A., 1967. Megalithic sites in Britain. Oxford: Oxford University Press.

Thom, A., 1971. Megalithic lunar observatories. Oxford: Oxford University Press.

Thom, A. & Thom, A. S., 1978. Megalithic remains in Britain and Brittany. Oxford: Oxford University Press.

Velikovsky, I., 1950. Worlds in collision. London & New York.

Wainwright, G. J. & Longworth, I. H. (eds.), 1971. Durrington Walls, Excavations 1966-1968. London: Society of Antiquaries.

Wainwright, G. J., 1977. Review. Journ. Arch. Sci., 4, 394-395.

Warren, S. H. et. al., 1936. Archaeology of the submerged land surface of the Essex coast. Proc. Prehist. Soc., 2, 178-210.

Wood, J. E., 1978. Sun, moon and standing stones. Oxford: Oxford University Press.

4. A CRITICAL EXAMINATION OF THE MEGALITHIC LUNAR OBSERVATORIES

Clive Ruggles

(with an appendix by R. J. C. Atkinson)

"... our critical examination of our theories leads us to attempts to test and to overthrow them; and these lead us further to experiments and observations of a kind which nobody would ever have dreamt of without the stimulus and guidance both of our theories and of our criticism of them."

Karl R. Popper,
(in Conjectures and Refutations,
London, 1963).

4.1 INTRODUCTION

Professor Alexander Thom has hypothesised that many megalithic sites were constructed primarily to serve as precise astronomical indicators, and in some cases as multipurpose "observatories", by marking significant rising or setting points of the sun, moon or stars. At each stage of his work, the strongest supporting evidence for his theories comes from the analysis of data from many sites taken together. Thus it is to these analyses in their entirety, rather than to the interpretations of a few particular sites, that our criticism must be directed in order to be most constructive.

Thom's astronomical work in recent years (see Thom & Thom 1978a, hereinafter "MRBB": chs. 9-11) has focused largely upon the interpretation of four sites—two near Carnac, the Ring of Brogar in Orkney, and Stonehenge—as possible lunar observatories. However it is three earlier, more broad-based analyses that provide the foundations and motivation for this work, each one seeming to establish the existence of lunar indications of a greater precision than did those preceding it. Thus the idea of megalithic lunar observatories essentially rests upon a cumulative structure of evidence, which can be divided into three levels as follows:

Level 1: The analysis of 262 potential indications at 145 sites suggests the existence of deliberate lunar (amongst other astronomical) indications of a rough nature, set up to a precision of about half a degree to record the extremes of the moon's movements in its 19-year cycle (Thom 1967a, hereinafter "MSB": table 8.1 & fig. 8.1).

Level 2: The further analysis of those potential indications from the above sample which might be construed as lunar, about 40 of them, suggests that the upper and lower limbs of the moon were preferentially observed; this increases the inferred precision to about 10 minutes of arc (MSB: table 10.1 & fig. 10.1).

Level 3: The analysis of 40 distant horizon features at 23 sites suggests the use of distant natural foresights to observe to great precision (3 arc minutes or better) and record the tiny additional perturbation in the moon's motion (Thom 1971, hereinafter "MLO": table 7.1 & fig. 7.1).

Reference will later be made (section 4.5.1) to a further level of analysis which has appeared in the most recent work by the Thoms.

At each of these levels the critic must ask: does the data set, taken as a whole, force us to accept the hypothesis of deliberate lunar indications and foresights, or can such putative lunar alignments as have been presented quite adequately be explained away by other means? The key questions are these:

(i) Is each item of data archaeologically and intrinsically (i.e. in the broad terms of the hypothesis) suitable for inclusion in the analysis?
(ii) Is each item of data correct?
(iii) Is the overall selection of data for inclusion free of subjective bias?

Having re-examined, and perhaps revised, the data set, we must then ask

(iv) Does the data set force us, statistically speaking, to accept the lunar hypothesis?

What follows is directed towards answering these questions.

Section 4.2 describes the methods used by the author for collecting independent site data. After a short discussion of the problems involved in a re-examination at Level 1, section 4.3 presents a full reassessment of the evidence at Level 2. Section 4.4 does the same for Level 3. Section 4.5 deals with some further relevant topics, and a general summary of the conclusions is presented in section 4.6. Appendix 4.1 contains a reference list of those sites which are considered at one stage or another in the paper. Appendix 4.2, by Richard Atkinson, consists of archaeological comment on several of the sites.

It is assumed that the reader is familiar with the meanings of "azimuth" (clockwise bearing from true north) and "declination" (see, e.g., MSB: 14). It has not been found necessary to include any astronomical terms apart from these, or to get involved in any discussion of the lunar motions, except to a small extent in section 4.5. The concept of a lunar "extreme" declination is used without elaboration up until then: it is entirely equivalent, but employed in preference, to Thom's term "standstill" (MLO: 18).

The words "accuracy" and "precision" have definite and distinct connotations in some disciplines. Throughout the present paper "accuracy" is used in relation to what we can actually measure, and "precision" to what we are trying to infer. (This is elaborated in section 4.3.5.) This nuance, it is hoped, will aid formality without impeding the progress of the casual reader.

4.2 METHODS FOR COLLECTING INDEPENDENT DATA

Fifty-five sites, listed in appendix 4.1, are relevant to the discussions in this paper. All but one of them have been visited by the author, mostly during the summer of 1979. Somewhat over half have been partially or completely re-surveyed; in other cases horizon surveys were deemed unnecessary, or were rendered impossible by persistent bad weather, crops, or afforestation around the site.

The instrument used for surveys during 1979 was a Kern DKM-1 microptic theodolite reading to 10". Four months of continual use enabled theodolite adjustment errors to be regularly monitored. A few sites were surveyed in previous years, using different instruments, as part of a project (mentioned in section 3.1) following on from the work of Cooke et al. (1977). At four sites inadequate previous surveys were augmented in 1979, and the horizon profiles at one site (Unival) were completely resurveyed in 1979 in order to check consistency.

Plate bearing zero (PBZ) is the true azimuth of the zero graduation on the horizontal circle of the theodolite. When weather permitted, this was determined from observations of the sun, timed using a calibrated quartz crystal wristwatch. A series of 12 observations typically gave PBZ to ±10". In other cases PBZ was determined from sightings of three or more Ordnance Survey triangulation stations. At two sites (Brogar and Unival) both methods were available and provided a consistency check.

Where it was not possible or convenient to place the theodolite exactly on the assumed sightline, a parallax correction was applied to transform the measured profile into that seen from the assumed observing position. Complete profiles were constructed by fitting measured horizon points (relocated at least once) on to photographs taken with a 200 mm lens from the assumed observing position.

Surveys of the sites themselves, using theodolite and steel measuring tape, were usually only strictly necessary in order to determine the centres of circles or rings, but they were often undertaken anyway for the sake of completeness.

Many of the horizon surveys are not directly relevant to the conclusions of this paper, but will be included in other work. Those that are have been divided into two classes. At those sites where the survey is considered reliable (marked "S" in tables 4.1, 4.3 & 4.4) the quoted azimuths, altitudes and declinations have a probable error of around ±1'. In the other cases (marked "E") either (i) the probable errors are greater, owing to an excessively large parallax correction; or (ii) there is a possibility of a gross systematic error in the azimuths, since PBZ was deduced from only 1 or 2 sun observations, or from unreliably identified triangulation stations.

4.3 A CRITIQUE OF THE ROUGHER LUNAR INDICATIONS

4.3.1 Level 1, Level 2 and the overall selection of sites

Thom (MSB: table 8.1) presents a selection of 262 potential indications at 145 sites. A declination is quoted in each case accurate to $0^{o}.1$. This is either the mean declination indicated by a structure on the ground or, where there is considered to be an indicated foresight such as a distant notch or mountain top, the declination of the supposed foresight. The results are displayed graphically (MSB: fig. 8.1) by representing each indication as a gaussian probability hump of fixed height and width centred upon the quoted declination; a wider spread of probability is used in a few cases where the indicated declination was determined less accurately. The humps are plotted cumulatively so as to produce a sort of histogram showing the preferred declinations. We shall henceforth refer to such a graph as a "curvigram", to MSB: fig. 8.1 as the "master curvigram" and to MSB: table 8.1 as the "master source list". Noticable peaks occur in the master curvigram at declinations corresponding to the rising and setting points of the sun both at the solstices and at regular intervals of 1/16th year between them (the so-called "calendar declinations"); at the four lunar extreme declinations; and at certain other declinations interpreted by Thom as stellar.

In fig. 4.1a we have focused upon the lunar lines by taking four declination intervals centred upon the mean lunar extreme declinations ($-29^{\circ}.95$, $-19^{\circ}.58$, $+17^{\circ}.94$ & $+28^{\circ}.17$) and superimposing gaussian humps representing all indications listed in the master source list which contribute to the graph in these intervals. They are listed in table 4.1. A standard deviation of $0^{\circ}.1$ has generally been assumed, corresponding to a probable error of $\pm 0^{\circ}.07$; $0^{\circ}.2$ has been used for those less accurate lines marked "±". These values may be slightly different from those employed by Thom, which are not explicitly stated; but they do appear to match the specimen humps shown in the master curvigram. It is the clear build-up of lines within about $0^{\circ}.5$ of one or other mean lunar extreme declination (i.e. between about $-0^{\circ}.5$ and $+0^{\circ}.5$ on the graph), over and above the background "noise", which provides the evidence for deliberate lunar indications precise to about half a degree (Level 1). Fig. 4.1a also shows a sharp peak to the left of the main build-up of humps. Several (but not all) of the lines contributing to it come from the same one of the four superimposed declination ranges and have an explanation in terms of the alleged solar calendar.

In reassessing Level 1, it is key question (iii) —whether the selection of data for inclusion in the master source list was free of subjective bias—which presents the most problems. The reason is that in order to answer it we must look beyond the sample of sites presented by Thom. In MSB he does not state his criteria for the inclusion of sites in the site lists. Indeed, the sequence of site reference numbers (A1/2, etc.) in a general list (MSB: table 12.1) indicates the omission of many sites which were present in an original catalogue. In an earlier paper (Thom 1955: 284) a shorter reference list contained "all the available data", but MSB: table 12.1 comprises (only) those sites which were surveyed and "contribute to the material of this book" (MSB: 135); we must ask why other sites which were visited were found unsuitable for surveying and inclusion. "Legitimate" reasons, that is reasons which would not introduce any overall bias, are unrelated to the possible astronomical function of the site; its being in too bad a state of repair, for example, or weather conditions being too bad for a survey when the site was visited. If on the other hand (say) possible indications at each site were first examined

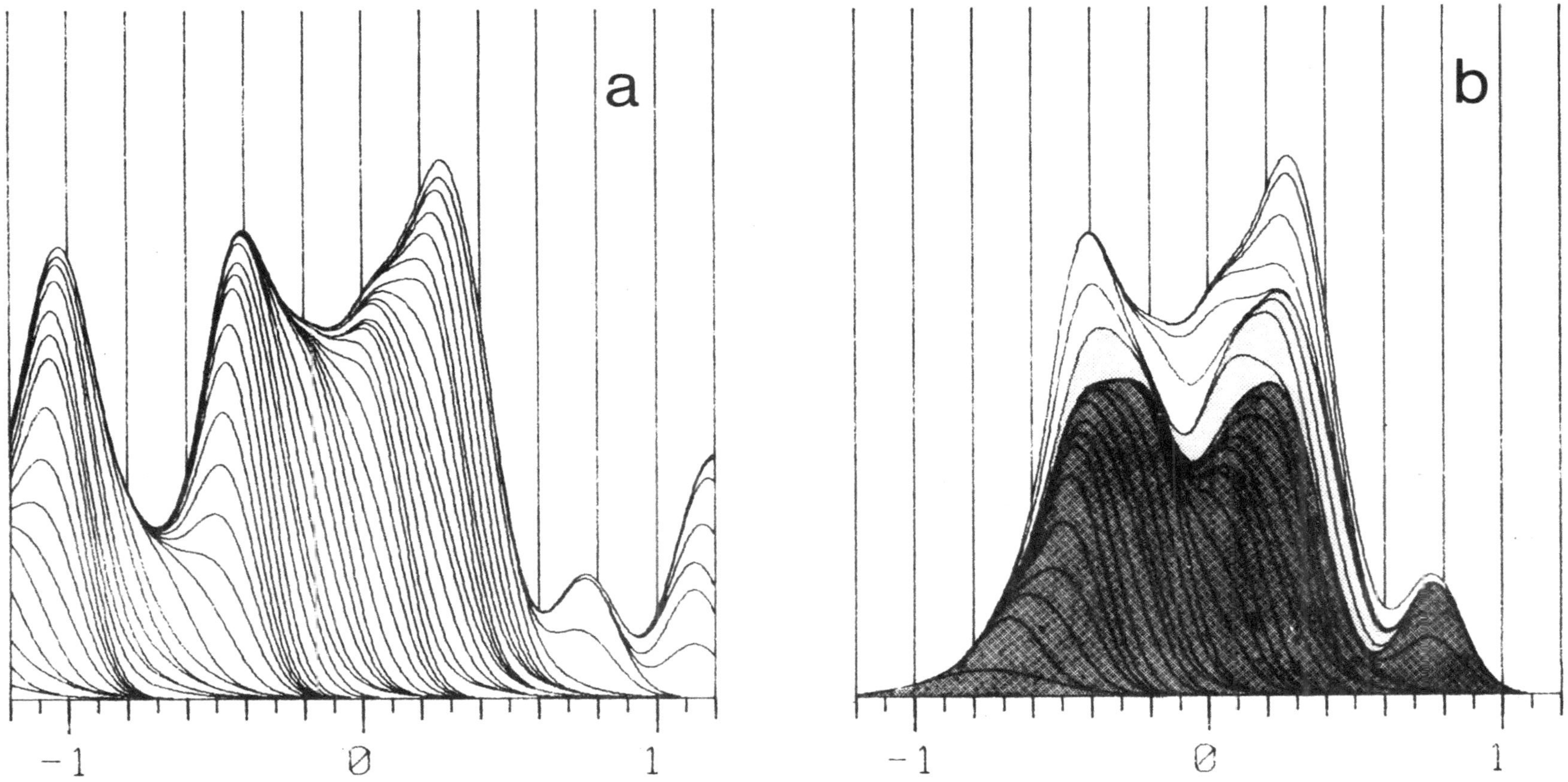

Fig. 4.1a Indications listed in Thom's master source list falling within 1°.5 of a lunar extreme declination. See section 4.3.1.

Fig. 4.1b Indications listed in Thom's master source list falling within 0°.8 of a lunar extreme (see section 4.3.2), with archaeological status (see section 4.3.3) marked by differential shading.

roughly with a magnetic compass, and if those sites with indications in astronomically "interesting" directions were preferentially surveyed subsequently, large-scale bias would have been introduced. The only way to check for possible bias is to build up a body of data independent of Thom's by systematically surveying complete samples of known sites in certain areas—not only those listed by Thom, but other archaeologically similar ones as well.

A further problem attaches to the selection of potential indications at each site: although rigorous criteria are presented in Thom's earlier paper (Thom 1955), when it comes to the master source list subjective judgements are involved (MSB: 96) and unperceived bias may have entered. The fact that several potential astronomical indicators, such as stone alignments, appear in the general site list but are not carried through to the master source list reinforces this doubt. It can be eliminated only by the use of a rigorous selection methodology, the development of which was the theme of a recent paper (Cooke et al. 1977).

The independent examination of some 200 sites in western Scotland has been undertaken in an attempt to resolve these questions and provide an adequate reassessment of the claims at Level 1. This work is being prepared for separate publication (Ruggles, Morgan, Few, Cooke & Burch, in preparation).

The conclusions of Level 2, in contrast, depend upon the precise values of those few indicated declinations falling near to one of the four mean lunar extremes, rather than to the actual number of such lines in the overall sample. If astronomical indications were deliberate but a precision of about half a degree was the best that was achieved, we would expect a concentration of humps building up to an overall peak at about the origin. Instead, fig. 4.1a shows two definite peaks, one at about $-0^{\circ}.4$ and the other at about $+0^{\circ}.25$. As the middle of the graph represents the centre of the moon's disc at its extremes, and the moon's semidiameter is about $0^{\circ}.25$, the two peaks are suggestive that the upper and lower limbs of the moon were preferentially indicated, increasing the inferred precision to about 10 minutes of arc. These conclusions can not be prejudiced by subjectivity in the selection of potential data prior to an accurate survey, since rough compass measurements could not adequately determine the exact values of any declination within one of the four general "target" areas. Wilful selection, after the careful reduction of survey measurements, could admittedly have biased the data; but as Thom twice refers to the unexpectedness of the double peak (MSB: 165; Thom 1967b: 95-96) such selection has clearly not occurred. Thus we may adequately reassess Level 2 on the basis of the sample of lines provided by Thom. This is done in the remainder of section 4.3, addressing in turn each of the key questions of section 4.1. We shall briefly return to Level 1 in section 4.3.6 in order to make a general comment.

4.3.2 The data set at Level 2

We begin by identifying the exact data set which provides the evidence at Level 2.

We need to isolate those lines with conceivable lunar significance from others more obviously explicable in terms of alternative astronomical hypotheses such as the solar calendar. Thus we need to include the central heap and double peak in fig. 4.1a but to exclude the left-hand peak. A convenient demarcation is $+0^{\circ}.8$. Thirty-eight declinations quoted in the master source list fall within $0^{\circ}.8$ of one of the mean lunar extremes. These are the central group of alleged indications listed in table 4.1–those with a site reference number given in column 1. Thom missed out one line (site 11) from his equivalent short list (MSB: table 10.1), apparently by mistake. He also gives different declinations in his two lists for all three lines at site 49 (one of which was interpreted as stellar, not lunar, in the master source list), and for the one at site 10, where there is a gross $0^{\circ}.9$ discrepancy. In each of these cases we have adopted the value quoted in the short list. We also note that only two lines (sites 47 & 50) are marked in the short list as less accurate, whereas a further four (sites 6, 12, 16 & 51) have "±" marked by azimuth, altitude or declination in the master source list. We have taken all 6 lines to be less accurate.

The 38 selected lines are plotted in fig. 4.1b, which resembles Thom's equivalent figure (MSB: fig. 10.1). The slight disparities are accounted for by the addition of a hump centred at $-0^{\circ}.37$, four humps being relegated to less accurate status, and a possible small difference in the assumed standard deviations. In addition, our humps were plotted in a different order.

It is worth re-emphasizing that only the double-peaked shape of the build-up of humps, rather than the actual number amassed, is relevant to the conclusions of Level 2. We wish to know whether this characteristic shape will remain when the data have been reassessed.

4.3.3 Archaeological status of the sites

In this section we attempt to classify the 38 alleged indications in terms of archaeological status. A number of them, designated class C, involve sites which are either of doubtful authenticity or else are particularly dubious contenders for reasonably precise astronomical indicators. Further lines, designated class B, are also considered to be of doubtful status but not as seriously so. After all these have been singled out, the typological and spatial coherence of the remaining (class A) sites is briefly examined. (It should be noted that this classification is entirely distinct from Thom's own [MSB: 96].)

There are 3 sites whose authenticity is in serious doubt:

Site 23: Kell Burn. This site, described by Thom (MLO: 70) as "an alignment of small stones", was visited by the author early in 1976, when the stone marked as standing on Thom's plan (Thom 1966: fig. 10) was identified as a tiny pointed stone about 0.2 m high atop a low mound, and the other stones as merely small boulders, downhill from the pointed stone in both directions and not intervisible with it. These boulders may have been the remains of an oval enclosure wall noted by RCAHMS (1924: no. 224) and more recently by the Ordnance Survey (Archaeological Records, Southampton & Edinburgh–hereinafter "OSAR": NT 66 NW no. 2). By 1979, improvements associated with the construction of a nearby reservoir and a new farm access road seemed to have destroyed all but the pointed stone.

Table 4.1 Table of rougher lunar indications

Column headings

1. Site reference number
2. Thom's site reference
3. Site name
4. Brief site description
5. Nature of indication
6. Indicated foresight?
7. Azimuth quoted by Thom
8. Altitude quoted by Thom
9. Declination deduced by Thom
10. Difference from relevant lunar extreme
11. Standard deviation used in figs. 4.1a, 4.1b & 4.1c
12. Archaeological status
13. Intrinsic status
14. Accuracy in azimuth measured or estimated by author
15. Status of re-survey
16. Comments
17. Azimuth measured by author
18. Altitude measured by author
19. Declination deduced by author
20. Difference from relevant lunar extreme
21. Standard deviation deemed appropriate

<u>Key to Column 4 (Site description)</u>

A	Alignment
C	Circle or ring
M	Menhir
Q	See text (section 4.3.3)
RC	Recumbent circle
T	Chamber tomb or cairn

For fuller details see appendix 4.1

<u>Key to Column 5 (Nature of indication)</u>

A	Alignment
CA	Circle centre to (and along) alignment
CC	Circle centre to circle centre
CHH	Circle centre to excavated stone holes
CO	Circle centre to outlying menhir
F	Flat side of menhir
FS	Flat side of menhir to distant site
MM	Menhir to menhir
SO	Site to outlying menhir
SS	Site to distant site ("inter-site")

<u>Key to Column 12 (Archaeological status)</u>

A	Reasonable
B	Not unreasonable but somewhat dubious
C	Very dubious

For details see section 4.3.3

<u>Key to Column 13 (Intrinsic status)</u>

As for Column 12, plus

Z	Necessarily dismissed or excluded

For details see section 4.3.4

<u>Key to Column 15 (Status of re-survey)</u>

S	Surveyed
E	Surveyed, but measurement errors large
U	Unsurveyed
P	Previously surveyed and published (Cooke et al. 1977)

<u>Key to Column 16 (Comments)</u>

a	Alignment not quite straight
c	Exact position of circle/ring centre uncertain
f	One or more menhirs are fallen; original position uncertain
g	Results are indicative of a gross error by Thom or the author
h	Foresight well below horizon
i	Thom's quoted declination seems to be that of a presumed indicated foresight
l	Outlying stone holes not quite in line with monument centre
m	The two faces of the menhir point in somewhat different directions
n	There exist several equally plausible foresights in the possibly indicated azimuth range
o	Outlier close
q	Foresight is a cairn: hence larger angle subtended
s	Stones close together
t	Horizon (now) hidden by trees, hence direct (re)measurement impossible
w	Third menhir noted by Thom not found
y	Third menhir noted by Thom not original

For details in representative cases see section 4.3.5

<u>Columns 7-11, 14, and 17-21</u>

All angles quoted are in degrees

Table 4.1

1	2	3	4	5	6	7	8	9	10	11	12	13	14	15	16	17	18	19	20	21
	W11/2	Y Pigwn				233.3	1.0	-21.0	-1.42	0.2										
	N2/1	Learable Hill				61.6	2.4	16.6	-1.34	0.1										
	W9/4	Castell-Garw				214.3	0.0	-31.2	-1.25	0.2										
	L1/7	Long Meg				65.1	3.4	16.7	-1.24	0.1										
	G4/14	Cauldside				59.5	0.3	16.8	-1.14	0.1										
	G7/4	Loupin Stanes				201.2	1.5	-31.0	-1.05	0.1										
	H1/6	Cul a'Chleit				304.9	-0.2	16.9	-1.04	0.1										
	H4/4	Rueval				303.8	-0.1	16.9	-1.04	0.1										
	W5/1	Moel ty Ucha				298.6	0.6	16.9	-1.04	0.1										
	A10/3	Ballimore				228.2	1.8	-20.6	-1.02	0.1										
	A2/6	Carnasserie				169	2.4	-30.9	-0.95	0.2										
	S3/1	Stanton Drew				211.4	1.7	-30.9	-0.95	0.2										
	H3/11	Unival				304.1	-0.2	17.0	-0.94	0.1										
	B1/8	Sheldon of Bourtie				55.9	0.0	17.1	-0.84	0.2										
	M2/6	Tiraghoil				59.9	1.5	17.1	-0.84	0.1										
51	L6/1	Boroughbridge	A	A	-	151.2	0.4	-30.7	-0.75	0.2	A	B	±0.5	U	a					0.2
49	L1/6	Burnmoor (E to D)	3C	CC	-	311.9	5.5	27.5	-0.67	0.1	A	A	±0.4	S	c(x2)	310.4	5.4	26.7	-1.47	0.2
29	A2/8	Temple Wood	C,6M	CO	-	135.0	3.7	-20.1	-0.52	0.1	A	A	±0.2	U	h					0.1
33	A6/4	Knockrome	M	F	IF	203.4	1.0	-30.4	-0.45	0.1	A	A	±2.6	S	m	203.1	1.0	-30.5	-0.55	0.7
46	G3/3	Laggangarn	2M	SO	-	150.3	0.0	-30.4	-0.45	0.1	C	Z								
24	M4/2	Balinoe	M	F	IF	190.2	2.8	-30.4	-0.45	0.1	A	A	±6.0	S	mn	191.3	2.5	-30.6	-0.65	1.0
55	S1/2	Altarnun	C,2A	CA	-	63.5	1.5	17.5	-0.44	0.1	C	C	±0.1	U						0.1
34	A6/6	Craighouse	A	A	IF	228.0	2.6	-20.0	-0.42	0.1	A	A	±3	U	tf					1.5
11	H1/5	Airigh nam Bidearan	A,2M	SS	-	332.8	0.3	27.8	-0.37	0.1	B	B	±0.1	P		333.0	0.2	27.8	-0.37	0.1
29	A2/8	Temple Wood	C,6M	A	-	206.1	0.3	-30.3	-0.35	0.1	A	A	±0.3	U	ti					0.2
36	A10/5	Stillaig	M	FS	IF	325.5	0.8	27.9	-0.27	0.1	A	A	±0.2	E		325.3	0.6	27.5	-0.67	0.1
9	H1/1	Callanish I	C,5A	A	-	190.6	1.3	-30.2	-0.25	0.1	A	Z								
50	L1/11	Kirksanton	2M	MM	-	210.8	0.5	-30.2	-0.25	0.2	A	A	±0.9	S	sw	212.6	0.0	-30.0	-0.05	0.3
17	B3/5	Kempston Hill	2M	MM	-	231.4	0.6	-19.8	-0.22	0.1	A	A	±0.3	U	t					0.2
49	L1/6	Burnmoor (E to C)	3C	CC	-	292.3	6.2	17.8	-0.14	0.1	A	A	±0.4	S	c(x2)	291.7	6.1	17.4	-0.54	0.2
52	W9/7	Parc-y-Meirw	A	A	-	301.4	-0.4	17.8	-0.14	0.1	A	A	±0.2	U						0.1

Table 4.1 (contd.)

23	G9/13	Kell Burn	"A"	A	-	129.8	1.8	-19.7	-0.12	0.1	C	Z								
46	G3/3	Laggangarn	2M	SO	-	124.8	0.0	-19.6	-0.02	0.1	C	Z								
30	A2/12	Duncracaig	2A	A	-	320.7	3.1	28.2	0.03	0.1	A	A	±1.5	E	a	322.7	3.0	28.9	0.73	0.6
32	A2/19	Achnabreck	2M	MM	-	159.5	2.1	-29.9	0.05	0.1	A	A	±0.2	U	f					0.1
18	P3/1	Corogle Burn	A	A	-	198.1	1.9	-29.9	0.05	0.1	B	B	±1.2	S	f	198.1	1.9	-29.9	0.05	0.2
15	B7/3	Dulnainbridge	3M	MM?	-	230.9	0.9	-19.5	0.08	0.1	A	Z								
12	H3/6	Barpa nan Feannag	T	SO	-	160.7	0.7	-29.8	0.15	0.2	C	Z								
48	L1/1	Castle Rigg	C	MM	-	157.1	2.8	-29.8	0.15	0.1	A	A	±0.2	S	i	157.0	2.2	-30.4	-0.45	0.2
53	S5/2	The Sanctuary	Q	CHH	IF	320.0	0.3	28.4	0.23	0.1	B	B	±1.0	E	ol	320.4	0.2	28.5	0.33	0.5
37	A4/1	Escart	A	A	IF	206.5	0.9	-29.7	0.25	0.1	A	A	±1.0	U	ts					0.3
6	N1/13	Guidibest	C	CO	-	196.1	1.0	-29.7	0.25	0.2	A	B	±0.8	E	oc	195.2	1.2	-29.7	0.25	0.2
26	M2/14	Loch Buie	C, 4M	MM	-	65.7	6.2	18.2	0.26	0.1	A	A	±0.2	U						0.2
20	P1/8	Comrie	2M	MM	-	296.8	5.3	18.2	0.26	0.1	A	C	±0.9	S	isg	300.8	5.3	20.9	2.96	0.5
7	H1/15	Dursainean	M	SS	-	227.9	2.0	-19.3	0.28	0.1	C	C	±0.5	U	q					0.3
47	G4/2	Blair Hill	2M	MM	-	48	6.7	28.5	0.33	0.2	A	C	±0.7	E	s	48.8	6.4	27.7	-0.47	0.3
8	H1/14	Lower Bayble	M	SS	-	24.8	0.4	28.5	0.33	0.1	C	C	±0.3	U	q					0.2
49	L1/6	Burnmoor (C to D)	3C	CC	-	63.6	4.3	18.3	0.36	0.1	A	A	±1.0	E	c(x2)	64.6	4.2	17.8	-0.14	0.6
13	H3/3	South Clettraval	T,M	SS	-	126.5	-0.1	-19.2	0.38	0.1	B	B	±0.1	S		126.2	-0.1	-19.1	0.48	0.1
27	M2/8	Bunessan	M	F	IF	330.7	0.2	28.6	0.43	0.1	B	B	±1.3	S	m	329.3	0.1	28.1	-0.07	0.4
16	B2/4	Esslie South	T,RC	CO?	-	306.2	0.2	18.4	0.46	0.2	C	C	±1.5	S	ioc	305.6	0.6	18.6	0.66	0.7
21	P1/1	Muthill	2M	MM	-	57.3	1.8	18.7	0.76	0.1	A	A	±3.0	U	sy					1.5
10	H1/2	Callanish II	C	SS	-	129.5	0.3	-18.8	0.78	0.1	A	A	±0.2	P		128.9	0.4	-19.4	0.18	0.1
	A3/4	Tayvallich				34.1	2.1	29.3	1.13	0.1										
	S1/5	Treswigger				317.2	1.9	29.3	1.13	0.1										
	A2/12	Duncracaig				151.9	1.1	-28.8	1.15	0.2										
	A4/4	Ballochroy				44.2	6.2	29.4	1.23	0.1										
	P1/10	Fowlis Wester				30.9	0.6	29.4	1.23	0.1										
	A3/4	Tayvallich				32.8	1.9	29.5	1.33	0.1										
	N1/8	Battle Moss				343.0	0.0	29.6	1.43	0.1										

Site 46: Laggangarn. The site consists of two cross-incised menhirs which are probably the remains of an early Christian settlement, but may be the Christianised remains of two prehistoric stone circles. See appendix 4. 2 for details.

Site 55: Altarnun. This is a stone circle, restored in 1889 (J. R. Inst. Cornwall Editor's note 1889) away from which run two lines of stones marking the parish boundary between Altarnun and North Hill. These rows, one of which forms the alleged indication, are thus probably modern, although some stones in them may have formed a stone row (common on Dartmoor but anomalous here on Bodmin Moor) which was subsequently utilised and added to (Page 1906: 397).

Four alleged indications involve chamber tombs or cairns; some knowingly, and others owing to misidentifications. Such monuments, however, should not without extreme caution be proposed as the components of astronomical sightlines: they have an obvious primary function as burial places, and are often, as on North Uist, numerous. The study of their individual orientations, with a view to investigating possible secondary astronomical significance (such as the solar orientation of Newgrange [Patrick 1974]) is obviously one of great interest. But to propose an odd few sightlines which make use of chamber tombs, either to be stood upon or viewed from a distance, seems dubious practice indeed. Another criticism of tombs and cairns, which strictly lies in the domain of the following section but will be mentioned here, is that they provide very imprecise backsights and indicators. The sites involved are:

Site 7: Dursainean, & Site 8: Lower Bayble. Both indications involve an inter-site line using as foresight Thom's site H1/13 (not H1/15, as stated in the master source list). H1/13 is in fact a ruined passage-grave (see appendix 4. 2).

Site 12: Barpa nan Feannag. This is a ruined chambered cairn (see appendix 4. 2).

Site 16: Esslie South. This is a denuded ring-cairn, with central cist, surrounded by a recumbent stone circle (Smith 1880: 303).

It is felt that in any independent investigation starting from scratch, these lines (designated class C) would more satisfactorily be omitted altogether.

The authenticity of a number of further sites is in some doubt, though not as seriously so. They are as follows:

Site 7: Dursainean. The site itself is a triangular block of gneiss, as wide (1. 8 m) as it is high, listed as a standing stone by the Ordnance Survey (OSAR: NB 53 SW no. 7) but on no other lists known to the author; as a prehistoric menhir it is unconvincing.

Site 11: Airigh nam Bidearan ("Callanish V"). The site consists of five small upright slabs about 0. 5 m high, three of which are close (within 10 m) and in line (Mackenzie 1904; Cooke et al., 1977: 118). This arrangement may form the remaining part of a field or enclosure wall (OSAR: NB 22 NW no. 1).

Site 18: Corogle Burn (Glen Prosen). This site does not seem to be mentioned on any ancient monument lists. One end of the alignment consists of two prostrate slabs close together which may well be fallen menhirs, but at the other end, some 20 m distant, are two small irregular blocks about 1 m high which are unconvincing as prehistoric structures.

Site 27: Bunessan. This is one of a series of standing stones traditionally identified as a pilgrim route marker (see appendix 4. 2) and is not certainly prehistoric.

One further inter-site line involves two chamber tombs, but its interpretation as an astronomical indicator is perhaps less dicey than in the cases listed above, since at each site there is a tall menhir separate from the chamber tomb which could conceivably have functioned separately as part of a precise sightline. Even this interpretation, however, is inherently somewhat doubtful; the practice of erecting tall menhirs near to chamber tombs or cairns, perhaps simply as markers, is not uncommon. The line in question is the following:

Site 13: South Clettraval. For details of this site see appendix 4. 2: the foresight is Unival chamber tomb (Scott 1947; Ruggles & Norris 1980), referred to by Thom as "Leacach an Tigh Cloiche".

Finally, one further site is in a class apart from the others in the sample, and should also be singled out as a rather dubious contender for an astronomical indicator:

Site 53: The Sanctuary. The site consists of a number of roughly concentric rings of timber post holes and stone holes, excavated by Cunnington (1931). It is probable, though not certain, that the post holes held the timber supports of a sequence of huts (Piggott 1939; Musson 1971: 368-372); however the outer circle (ring A) was undoubtedly free standing. Cunnington's stone holes N1 and N2 lie just outside ring A to the NW, forming a radial alignment with stone 3 of the circle, and it seems likely that they were associated with the two radial lines of stone holes further round to the west which are thought to represent the beginning of the avenue to Avebury. Evidently in the absence of further excavated evidence the idea of N1 and N2 being lunar indicators to be viewed from the centre, which appears to be the intended indication, should be treated with some caution. See also appendix 4. 2.

Apart from the line from Dursainean, which was already assigned to class C, these sites of somewhat doubtful status have been designated class B. The lines of classes A, B and C have been represented in fig. 4. 1b by heavy, light and no shading, respectively. It will immediately be seen that sites of dubious archaeological status contribute significantly to the double-peaked structure.

Twenty-four sites are involved in the sightlines which remain as class A. All but two are to be considered as possible places of lunar observation, the others being used as distant foresights. It is thus of some interest to examine their coherence. They consist of the following:

5 simple alignments of three or more menhirs or stones;
1 double alignment;
8 standing stone pairs;
3 single menhirs;
5 stone circles, with up to 6 outliers within 500 m;
1 circle with radial alignments (Callanish I); and
1 site consisting of 5 nearby circles.

Abbreviated descriptions of the 22 putative observing sites are given in table 4.1; the foresights (viewed from sites 10 & 36) are both stone pairs. The sites are mainly distributed throughout western Scotland and the Western Isles, with a concentration in Argyll, Kintyre and the Inner Isles.

It is by no means clear that any significance should be attached to the concentration of Level 2 lunar sites in and around Argyll, since this is an area of high megalithic site density anyway and is one most thoroughly studied by Thom; similarly the overall distribution of interesting sites seems merely to reflect those regions most fully investigated (MSB: table 12.1). What is more, no typological trends readily emerge: the sample seems merely to reflect the relative abundances of types of megalithic site found in the areas in question. Thus no independent backing is forthcoming for the hypothesis at Level 2 on the basis of typological or spatial coherence of the sites involved—although of course this does not rule out the possibility of trends being revealed by more apt sampling methods.

4.3.4 Intrinsic status of the data

After archaeological reassessment we are left with 25 lines of class A (reasonable), 5 of class B (somewhat dubious) and 8 of class C (very dubious). We now turn to reassessing the lines in their own terms, that is as potential astronomical indicators. This will lead us to dismiss some of them altogether, and to relegate others to more dubious status.

A convenient approach is to focus upon the two essential components of a lunar (or other astronomical) sightline: the structure(s) that form the indication on the ground, and the horizon behind which the moon is actually seen to rise or set. There are 6 lines which we are forced to dismiss from further consideration for various reasons. At one site, the claimed horizon can not in fact be seen from the alleged indication:

> Site 9: Callanish I. The southern horizon, claimed by Thom to be indicated by the avenue looking south (MSB: 124; MLO: 68-69), is in fact obscured by a local outcrop (Cooke *et al*. 1977: 120). Thom did not seem to realise this because his horizon profile was calculated from Somerville's (1912) plan and the 1-inch Ordnance Survey map; although this is odd because he mentions the "large natural rock to the south" as a good vantage point from which to observe events to the north (MSB: 123-124). Hawkins (1965) was also unaware of the outcrop blocking the distant horizon, but corrected the mistake later (1973: 240). This was one of six lines picked out by Thom (MSB: 119-121) as having particularly good accuracy.

At two sites indications corresponding to the description in the master source list could not be found:

Site 15: Dulnainbridge. The type of indication given in the master source list is "alignment to stone". In fact, there are three menhirs at this site, about 100 m apart and about 15° off line, the end ones not being intervisible. Thom's site description (MSB: table 12.1) is "alignment etc." which is unenlightening. Viewing from the central to the SW stone gives at least roughly the quoted azimuth, but the horizon is hidden by trees; if this is the intended indication, one would have expected three similar possible indications, equally plausible on the ground, to have been included in the master source list.

Site 46: Laggangarn (2 lines). The two lunar indications are contended to be marked by outlying stones, but neither was located. Each horizon is about 1 km distant, with low shrub-covered ground intervening. If the stones are hidden amongst this vegetation, they will be well below the skyline and thus most unconvincing as potential indicators.

Site 12 (Barpa nan Feannag) was not visited, and no antiquity can be found on any map corresponding to the "stone on horizon" listed as the foresight. As this was a class C site anyway, it has been excluded; likewise site 23 (Kell Burn), which, as mentioned above, has now been largely destroyed and was doubtful anyway.

General considerations lead us to relegate 2 further lines to the status of very dubious (class C). In both cases, there exists evidence that what now appears to be a plausible indication was once merely an undistinctive part of a more complex structure:

Site 20: Comrie. Documentary evidence exists (see appendix 4.2) that the two stones here were once part of a stone circle or 'four-poster'. Hence it is most unlikely that the orientation of the line joining them was of any particular significance in the original structure. This is another of the six lines picked out as of particularly good accuracy.

Site 47: Blair Hill (The Thieves). The two uprights here stand within a low ring of earth and rubble. A third large stone lies prostrate between them. Thom has incorporated all three into a geometrical construction (MSB: fig. 6.12) such that the assumed axis of the ring is parallel to the line joining the stones. However the three stones actually appear merely to be the remains of a complete inner ring of stones, an impression confirmed by Atkinson (appendix 4.2). In this case the line joining the two remaining uprights is of no special significance. (The Ordnance Survey, on the other hand [OSAR: NX 47 SW no. 2] record that the enclosure is modern, but give no references.)

A further line is considered to be very dubious because the outlier forming the indication is very close to the circle centre from which it is to be viewed. The site was already allocated to class C on archaeological grounds anyway. It is:

Site 16: Esslie South. The type of indication here is quoted as "site to stone". No distant stone was located at the stated azimuth; only an anomalous menhir a mere 3 m outside the outer circle, which can be seen in Thom's diagram (Thom, 1961: fig. 5). If this is indeed the intended foresight, not only does it fail to conform to Thom's own distance criterion (MSB: 94) but it is also of doubtful archaeological status (see appendix 4.2).

Finally, the following two sites have been relegated to the status of somewhat dubious:

Site 6: Guidibest (Latheronwheel). This line is marked as "circle to outlier", but the only candidate for the outlier appears to be a possible menhir stump about 50 m from the circle centre, one of a group of stones of uncertain status noted by the Ordnance Survey (OSAR: ND 13 NE no. 3). If this identification is correct, the outlier does not conform to Thom's own distance criterion (MSB: 94).

Site 51: Boroughbridge (Devil's Arrows). The middle stone of this line of three massive menhirs was originally one of a pair 6-8 ft apart (see appendix 4.2), and it is not clear whether the fourth stone was actually in the line.

The upshot of these general considerations is that only 19 of the original 38 lines are still considered to be of reasonable status, 7 being in class B, 6 in class C and 6 having been dismissed altogether. The resulting effect on the declination curvigram is shown in fig. 4.1c. It transpires that when class A lines are considered alone, virtually all trace of a double-peaked structure disappears.

4.3.5 Correctness and objectivity of the data

So far we have not questioned the verity of the indicated declinations quoted by Thom, and this is the purpose of the present section. Our aim here is not merely to seek gross disagreements between his measurements and ours (which, after all, only serve to demand further fieldwork to settle matters) but rather to identify and attempt to eliminate possible overall sources of subjective bias that would influence the results as a whole. There are two.

Consider the present deteriorated state of a putative astronomical indicator. Define the "accuracy" of the indication as the range of azimuths that might have been meant to have been indicated, taking into account the possible original positions of the structures on the ground. If stones have tilted or fallen, this accuracy may be no fair reflection of the original precision. But it is all we have to go on. The accuracy is the sole thing we can measure; the precision can only be deduced, as a generality about the sample, from the statistics. If any individual indication is simply assumed to have been originally precise, and an indicated declination quoted on this basis, then bias may enter in deciding exactly where it "presumably" originally pointed.

The second possible source of bias concerns the confusion of two distinct hypotheses. Hypothesis A, which is the one genuinely being tested at Level 2, is that structures on the ground themselves indicated with sufficient precision that part of the horizon where the moon rises or sets at a significant time. Hypothesis B, which is in fact tested at Level 3, is that a natural feature on the horizon forms a very precise foresight, and that structures on the ground do no more than mark the observing position and point out which horizon feature is to be used. If we are gathering together data to test A, then we must simply measure the accuracy of each putative indicator as defined above. To test B, on the other hand, we must look for horizon features within, and possibly slightly outside, the azimuth range that might have been indicated. All equally tenable potential foresights, whatever their declinations, should be included in the analysis. If, however, we assess each site individually, deciding on the basis of the measured declinations whether any horizon features could have functioned as precise foresights, and then proceed on this basis to classify the line as either rougher or more precise making use of foresights (cf. MLO: 11-12), bias certainly enters.

Seven of the remaining 32 lines are given in the master source list as "indicated foresight" and the quoted declination should be replaced by the full range that might have been indicated by the structure on the ground. The declination of the foresight should only be used in a separate Level 3 analysis. Curiously, only 3 of the 7 have been carried forward to Level 3 by Thom. In each of these 3 cases, though, a horizon diagram is drawn in MLO, enabling the data he has presented at the two levels (i. e. in MSB and MLO) to be easily compared (see table 4.2). In moving from MSB to MLO, we find the following: at Escart (MLO: 60) the foresight assumed to be indicated appears to have shifted from the top of Sheirdrim Hill to a knob some way to the right, and the calculated azimuths appear to have shifted by about $0^{\circ}.3$ in any case; at Stillaig (MLO: 66) there has been a definite change in the measured azimuths and altitudes, as the point given in MSB is now in mid-air some $0^{\circ}.4$ to the left of the top of Cruach Breacain; and at Knockrome (MLO: 65; fig. 4.3b here) the assumed foresight has shifted from the top of Crackaig Hill to a point some way down the slope to the right (although the declination is unaltered). Of the four indicated foresights not carried through to MLO, at one (the Sanctuary) the indication is inaccurate and the horizon featureless apart from some barrows; at another (Balinoe) the indication is highly inaccurate and there are many equally tenable potential foresights in the possibly indicated horizon range; and a third site (Craighouse) consists of a supposed alignment all but one menhir of which is fallen, and the nature of the foresight could not be ascertained as the site is now deep inside a Forestry Commission plantation. The last indicated foresight, at Bunessan, is the top of a flattish hill, not of itself very precise, indicated by the flat side of a menhir, an indicator of poor accuracy. A nearby boulder noted by Thom as part of the indication (MLO: 67) is in fact behind the menhir and is certainly part of an old field wall. Three other lines of the 32 (Castle Rigg, Comrie and Esslie South) are not listed as indicated foresights but the declinations given appear to be those of assumed foresights.

These observations are indicative of a considerable degree of subjectivity present in Thom's data, and also that the accuracy of measurement (at least in MSB) may not always match up to the quoted declination values and standard deviations used. Whether these factors, however, necessarily tend to bias the sample as a whole can only satisfactorily be ascertained by independent reassessment and remeasurement. Such work has been carried out by the author wherever possible and the results are tabulated alongside Thom's in table 4.1.

Table 4.2 Comparison of foresights given in both MSB and MLO

Site name	MSB				MLO			
	Az	Alt	Dec	Diff	Az	Alt	Dec	Diff
Knockrome	203.4	1.0	-30.4	-0.45	204.00	0.92	-30.33	-0.38
Stillaig	325.5	0.8	27.9	-0.27	326.00	0.75	28.00	-0.17
Escart	206.5	0.9	-29.7	+0.25	207.28	0.82	-29.56	+0.39

An independent estimate of the accuracy, as defined above, of the indicating structure on the ground has been obtained for each of the 32 lines (table 4.1, column 14). The following cases are worthy of special comment:

Site 8: Lower Bayble. The foresight is in fact a ruined passage-grave, as noted in section 4.3.3 above, about 1.5 km distant and subtending about $0^o.6$ in azimuth; yet Thom picks out this line as one of six having particularly good accuracy, "better than $\pm 0^o.1$". He must presumably have been sighting onto one of the remaining kerbstones. (While the azimuth of the centre could be found to $\pm 0^o.1$, its original height is uncertain.)

Site 21: Muthill. The central stone in the alignment of three noted by Thom appears not to be original (see appendix 4.2).

Site 24: Balinoe. The two sides of the menhir point in directions some 10 degrees apart. It might have been more constructive to consider two separate horizon ranges, each corresponding to the indication of one flat face and only about $2\frac{1}{2}^o$ wide; but this would not have been consistent with Thom's interpretation of Knockrome, where the assumed foresight comes between the two flat face indications, and neither would it have included the assumed foresight here.

Site 30: Duncracaig. The NW menhir in this alignment is displaced to the SW of the line formed by the other three; Thom's azimuth value appears to assume that the line joining the two end menhirs defines the indication, but the line of three seems an equally, if not more, likely choice.

Site 36: Stillaig. The flat faces of the single menhir (site A10/5) are in fact oriented some 10° to the left of the two distant menhirs (site A10/6).

Site 49: Burnmoor (3 lines). This site consists of five stone rings (Burl 1976: 93-97), the three relevant here (designated C, D and E by Thom) being within 150 m of one another. All three alleged indications involve sighting from ring centre to ring centre. If such lines were used as reasonably precise indicators, the centres would have had to have been marked say by wooden posts. The centres of circles C and D contain small cairns, and circle E contains 5 cairns in two of which were found "burnt bones" last century (Williams 1856); the only reasonably consistent hypothesis would seem to be that the cairns post-dated the circles.

Site 50: Kirksanton. Thom notes an alignment of three standing stones forming the indication, but there seems to be no trace of a third, and no evidence for further stones apart from an adjacent stone circle (destroyed). (See appendix 4.2.)

The relevant horizons at 4 of the 32 lines could not be directly measured owing to trees; resurveys at a further 9 were rendered impossible by crops or persistent bad weather. Measured values at 6 sites may be somewhat in error (see section 4.2), leaving 13 lines which have been reliably resurveyed. The only gross inconsistency between Thom's results and ours occurred at site 20 (Comrie), where there is a discrepancy of 4° in azimuth, yet the estimated accuracy is $\pm 1^{\circ}$. The spread of declinations corresponding to the estimated range of possibly indicated azimuths depends on the value of the azimuth itself, the slope of the horizon at the point concerned and the latitude of the site. Semi-widths of the declination ranges are listed in column 21 of table 4.1, and for convenience the standard deviation for the purposes of graphical display have been taken equal to this value—a procedure quite adequate for the present purpose. This corresponds to a 70% probability that the intended declination falls within the estimated declination range.

The resulting graph is shown in fig. 4.1d, which can be directly compared with previous curvigrams. In all cases the standard deviation used is that quoted in column 21 of table 4.1; where sites have been reliably resurveyed (marked "S" or "P" in column 15) our values are taken for the mean declination, and in other cases Thom's values are taken. It will be seen that no significant trace remains of the double-peaked structure, even when lines of dubious status are included. In other words, the claims of Level 2 are not borne out by Thom's data sample after reassessment, even setting aside the various archaeological and intrinsic reservations aired in sections 4.3.3 and 4.3.4.

4.3.6 Conclusions and discussion

The conclusions of section 4.3 as a whole may be summarised as follows. The data sample at Level 2 consists effectively of 38 lines, and the crucial evidence is the double-peaked structure of the graph in fig. 4.1b. After taking into account archaeological evidence and general considerations about the proposed structures as astronomical indicators, 6 of these lines had to be dismissed altogether or excluded from further consideration, 6 more were considered very dubious, and a further 7 were considered somewhat dubious although not altogether unreasonable. This left 19 lines felt to be plausible enough. Closer examination of the data then showed palpable evidence of subjectivity, and when first-hand reassessments and remeasurements of individual lines were taken into account, the crucial double-peaked structure vanished without significant trace. This was true even including the lines felt to be dubious (fig. 4.1d).

These conclusions do not, of course, disprove the existence of reasonably precise astronomical indicators, but they do seem emphatically to deny that the data provided by Thom gives reasonable support for the hypothesis at Level 2. What, though, of Level 1—the idea of lunar indications precise only to half a degree or so at best? This possibility is not challenged by fig. 4.1d, as a general build-up of probability around the mean lunar extremes will probably still be evident, even when the adjacent lines of fig. 4.1a (each one of which would also have to be suitably reassessed and modified if necessary) are reinstated. However, as noted in section 4.3.1, before we can have any positive proof we must await the results of independent surveys of complete and coherent samples of sites.

One comment may be made about Level 1, though, merely on the basis of the present results. The 38 lines considered here represent an effectively arbitrary cross-section of the 262 surveyed lines in the master source list. Thus, by a process of extrapolation, we would expect something like half of

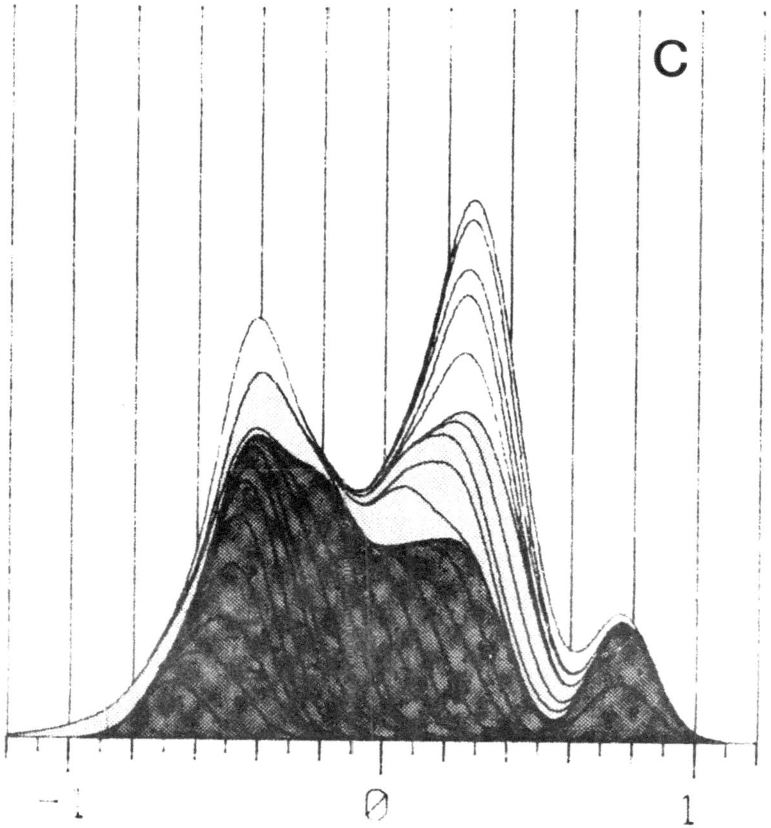

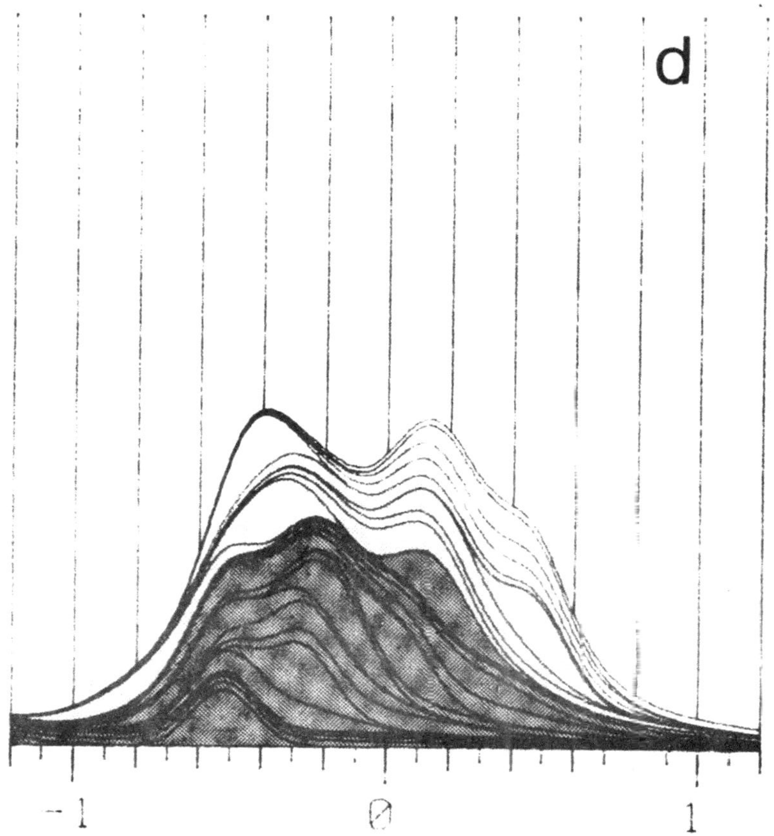

Fig. 4.1c Remaining indications after the general reconsiderations of section 4.3.4. Status classes are marked by differential shading.

Fig. 4.1d Revised indications after the detailed reconsiderations of section 4.3.5. Status classes are marked by differential shading.

the latter to be inherently unreasonable, dubious or uncheckable, and a significant proportion of the remainder to be subject to overestimation of accuracy or to subjective selection. As the 262 lines represent the entire basis for postulating astronomical indications in general (as opposed merely to deliberate orientations of an even rougher nature), a thorough reassessment at Level 1 is clearly overdue.

4.4 A CRITIQUE OF THE PRECISE LUNAR FORESIGHTS

4.4.1 The data set at Level 3

Only at this level do we encounter Thom's most widely acclaimed general hypothesis: the idea that the megalith builders used distant natural foresights, such as horizon notches, to obtain extreme precision (down to a minute of arc or so) in observations of the sun and moon. Artificial structures on the ground, such as alignments of menhirs, are relegated under this hypothesis to the secondary purposes of (i) marking with sufficient precision the desired observing position and (ii) singling out the particular horizon feature which is to be used. An item of data designed to test this hypothesis, or "line", will now consist of a putative foresight together with a claimed observing position and method of indication.

Thom (MLO: table 7.1—hereinafter the "Level 3 source list") has presented a selection of 40 lines at 23 sites. Azimuths are quoted to the nearest minute of arc, and geocentric lunar declinations (that is, the declinations that would have been measured from the centre of the earth, obtained by applying a correction to the measured altitude corresponding to a fixed horizontal parallax of 57' are calculated and quoted to 0'.1. The results are displayed in a curvigram (MLO: fig. 7.1) which differs from figs. 4.1a-4.1d in that the two halves are folded together, so that the absolute value of the difference of the measured declination from the mean lunar extreme is plotted. (In other words, declination differences of -25' and +25' would both be plotted as 25'.)

The relevant information from the Level 3 source list is rearranged and presented in table 4.3, and the information plotted in fig. 4.2a. A standard deviation of 0'.75 has been used, which appears to match Thom's humps, in all but two cases where a declination was only quoted to 1' by Thom and a standard deviation of 1'.5 was used. The mean extreme declinations used to calculate the relevant differences are ±29° 2'.3 and ±18° 44'.6.

The humps of fig. 4.2a can be seen to fall into three groupings with a high peak around 24'. These effects are readily explicable in terms of an additional small perturbation of the moon on top of its 19-year cycle, of period 6 months and amplitude 9'.4. The moon's apparent semidiameter is 15'.9, so the position of either of the moon's limbs at the maximum, mean and minimum of the additional wobble will give declination differences from the mean extremes of (respectively) 25'.3, 15'.9 and 6'.5. The correspondence with the observed peaks is manifest.

The data at Level 3 is of a fundamentally different nature from that at Level 2, and for this reason different techniques of reassessment are more convenient. In particular, the primary concern with natural horizon features rather than archaeological structures makes it more expedient to root out problems associated with the data selection before proceeding to archaeological considerations.

4.4.2 Suitability of the data

A clear diagram is drawn in MLO of all but one of the claimed foresights, greatly facilitating reassessment and discussion, but what have been assumed to be the observing positions and methods of indication are not always clear from the text. These latter factors turn out to be critical.

A major potential source of subjective bias, as at Level 2, is the confusion of distinct hypotheses. It is valuable, then, to identify at the outset exactly what is the hypothesis being tested, and to check that each item of data is actually relevant for testing it. It can be split into three essential parts as follows:

(I) Certain distant horizon features, such as notches, hill slopes and peaks, were used as astronomical foresights;
(II) For each foresight there was a well-defined observing position (perhaps more than one); and
(III) Permanent structures were erected so as to identify from the observing position the particular horizon feature(s) to be used.

This is what is claimed at the much-discussed "classic" Thom sites such as Ballochroy (Thom 1954; MSB: 142; MLO: 36-37), Kintraw (MLO: 37-40) and Temple Wood (MLO: 45-51). An item of data to test this hypothesis must consist of

(i) A distant horizon feature (putative foresight);
(ii) An alleged observing position; and
(iii) A putative means of identifying the foresight from the observing position.

A typical item of data, then, might consist of a prominent notch together with a stone circle from whose centre an outlier is positioned roughly beneath the notch; or else a hillslope indicated by sighting along a line of menhirs or the flat side of a single menhir from behind.

The putative foresight for each line in the Level 3 source list is clearly shown in a profile diagram somewhere in MLO, in all cases except the following one:

> Site 18: Corogle Burn. The horizon here is only about 1 km away, flat and featureless. Thus the line should not have been brought forward from Level 2.

This line, then, does not possess attribute (i) and must be dismissed out of hand.

The assumed observing position is implicit in each of Thom's profile diagrams, defining the position of the altitude/azimuth grid with respect to the profile. The sensitivity of the deduced declination of the foresight to changes

Table 4.3 Table of precise lunar foresights

Key to Column 8 (Status of line)

Z Dismissed out of hand (section 4.4.2)
Y No (genuine) indication exists, so dismissed (section 4.4.2)

Key to Column 9 (Status of re-survey)

S Surveyed
E Surveyed, but measurement errors large
U Unsurveyed

Table 4.3

1 Site Ref. no.	2 Thom's site ref.	3 Site name	4 Page ref. in MLO	5 Horizon feature	6 Declination quoted by Thom	7 Difference from mean extreme	8 Status of line	9 Status of re-survey
					° '	'		
39	A4/17	Tarbert, Gigha	62	A_1	+28 34.6	-27.7	Y	
14	H3/11	Unival	70	A_2	-29 29.5	-27.2		S
45	G3/2	Haggstone Moor	72	A_2	+18 17.8	-26.8	Y	
41	A4/9	Beinn an Tuirc	57	A_1	+18 18.0	-26.6	Y	
52	W9/7	Parc-y-Meirw	73	A	+18 19.7	-24.9		U
9	H1/1	Callanish I	69	A_2	-29 27.1	-24.8	Z	
25	M1/3	Quinish	67	A	-29 26.9	-24.6	Y	
5	N1/1	Mid Clyth	93	A_2	-29 26.6	-24.3	Y	
42	A4/2	High Park	60	A_1	-29 26.4	-24.1	Y	
11	H1/5	Airigh nam Bidearan	69	A_3	-29 26.3	-24.0	Z	
33	A6/4	Knockrome	65	A	-29 25.9	-23.6		S
11	H1/5	Airigh nam Bidearan	69	A_4	-29 20.5	-18.2	Z	
2	N1/17	Dirlot	96	A_1	-29 19.1	-16.8	Y	
22	P4/1	Lundin Links	55	A	-18 59.7	-15.1	Y	
9	H1/1	Callanish I	69	A_1	-29 11.6	-9.3	Z	
36	A10/5	Stillaig	66	A	+28 52.4	-8.1		E
5	N1/1	Mid Clyth	93	A_1	-29 8.7	-6.4	Y	
29	A2/8	Temple Wood	46	A_1	+28 56.5	-5.8		E
14	H3/11	Unival	70	A_1	-29 4.6	-2.3		S
29	A2/8	Temple Wood	46	A_1	+29 2.5	+0.2		E
28	A2/5	Kintraw	39	A_2	-18 40.6	+4.0	Y	
11	H1/5	Airigh nam Bidearan	69	A_2	-28 58.1	+4.2	Z	
18	P3/1	Corogle Burn	-	-	-28 55	+7.3	Z	
43	A4/14	Campbeltown	62	A_2	-28 53.8	+8.5	Y	
40	A4/5	Beacharr	61	A	+29 12.0	+9.7	Y	
30	A2/12	Duncracaig	52	A_1	+29 13.6	+11.3		E
2	N1/17	Dirlot	96	A_2	-28 48.5	+13.8	Y	
29	A2/8	Temple Wood	46	A_2	-28 48	+14.3		U
31	A2/13	Dunadd	64	A	+29 17.0	+14.7	Y	
19	P1/10	Fowlis Wester	54	A	+29 18.8	+16.5		S
42	A4/2	High Park	60	A_2	-28 45.7	+16.6	Y	
29	A2/8	Temple Wood	46	A_1	+29 19.1	+16.8		E
43	A4/14	Campbeltown	62	A_1	-28 44.0	+18.3	Y	
37	A4/1	Escart	60	A	-28 39.3	+23.0		U
39	A4/17	Tarbert, Gigha	62	A_2	+29 26.3	+24.0	Y	
28	A2/5	Kintraw	39	A_1	-18 20.3	+24.3	Y	
41	A4/9	Beinn an Tuirc	57	A_2	-28 37.3	+25.0	Y	
45	G3/2	Haggstone Moor	72	A_1	+19 9.9	+25.3	Y	
30	A2/12	Duncracaig	52	A_2	+29 28.5	+26.2		E
11	H1/5	Airigh nam Bidearan	69	A_1	-28 35.4	+26.9	Y	

in observing position will depend upon its distance. Although the assumed observing position is not usually explicitly stated in the text, it can be deduced from other details about the site: for example, if the indicator is a stone alignment, then any position a short distance back along the line will serve equally well, and the uncertainty in the declination of the foresight will in most cases be negligibly small. In 5 cases, however, the alleged foresight is not actually visible from behind the indicating alignment:

Site 9: Callanish I (2 lines). The southern profile (MLO: 69) is in fact obscured by a local outcrop. See the remarks on this site in section 4.3.4.

Site 11: Airigh nam Bidearan ("CallanishV") (3 lines). Most of the southern profile here (MLO: 69) is also obscured by local ground (Cooke et al. 1977: 121). Only the leftmost of the four horizon points, A_1, is in fact visible.

Each of these profiles was calculated by Thom from Ordnance Survey maps rather than directly surveyed, so that the intervention of local ground was unsuspected. In each case the relevant horizon can be brought into view by moving some metres to the right or left, but the foresights are now unmarked from the observing position. The idea that a distant horizon might have been indicated at a point from which it could not quite be seen is most unconvincing. Thus we also dismiss these 5 lines out of hand.

Fourteen of the remaining 34 putative foresights are not, it transpires, actually indicated at all. They are as follows:

Site 2: Dirlot (2 lines). The site consists of rows of tiny, mostly sunken stones. The direction of the two distant horizon features (MLO: 96) is unrelated to any structure on the ground.

Site 5: Mid Clyth (2 lines). The former existence of a suitable indicator (the outer edge of an extension to the stone sector) is only inferred, on dubious evidence (see apendix 4.2).

Site 22: Lundin Links. The profile here (Bass Rock) is where the moon would rise when it set on Cormie Hill, which is claimed to be indicated. However no direct indication of Bass Rock exists at all (MLO: 55-56).

Site 45: Haggstone Moor (2 lines). Various individual stones scattered within an area about 2 km square, are proposed as marking positions from which the Mull of Kintyre, certainly the most conspicuous stretch of horizon to be seen, is to be viewed. The backsights consist of a menhir called Long Tom (OSAR: NX 07 SE no. 3), a fallen menhir not found on any ancient monuments lists but of genuine appearance (it is situated amidst a 10-year old Forestry Commission plantation and will soon be near impossible to find) and a stone "S" of very dubious archaeological status (see appendix 4.2). Viewed from Long Tom a cairn is situated beneath the tip of the Mull of Kintyre, but a notch further up the Mull is considered to be the foresight from here (MLO: 72). No other local indications are proposed at all.

The remaining 7 unindicated foresights are all supposed to be viewed from single menhirs: Tarbert, Gigha (2 lines); Beacharr, Kintyre; High Park, Kintyre (2 lines) and Campbeltown, Kintyre (2 lines). Each of these menhirs has recognisable longer faces, in some cases reasonably flat; but in no instance is any of these faces oriented anywhere near one of the proposed lunar foresights.

The claimed indications for 4 further putative foresights do not in fact seem to be gunuine:

Site 28: Kintraw (2 lines). A diagram of a proposed lunar profile at Kintraw is given, without comment, amidst discussion of the possible solar significance of the site (MLO: 39). In Thom's site plan (MLO: 38) this profile appears to be indicated by the orientation of a "fallen stone", S_2. This stone, however, is unmarked by Simpson (1966: 55) and, if it is a fallen menhir, its prostrate orientation can presumably be of no more than chance significance. [1]

Site 41: Beinn an Tuirc (2 lines). This inaccessible site in the middle of Kintyre consists of a single standing menhir, "MacKay's Cross" (RCAHMS, 1971: no. 130—Arnicle). When Thom visited the site it was covered with peat and moss, and he located several more suspected stones by prodding (MLO: 56-57). Some of these formed the alleged lunar indications. Since his visit the site has been prepared for planting by the Forestry Commission. Their drainage ruts show this to be an area of unusually thin peat cover: many of Thom's stones can now be identified as natural boulders or bedrock, and others appear to have been boulders which were moved in the furrowing. Certainly there is no convincing evidence for any further artificial structures here.

Finally, the alleged indications for three more putative foresights are in fact some degrees away from their assumed targets:

Site 11: Airigh nam Bidearan. The azimuth of the remaining foresight here, A_1, is in fact some 7° to the left of the orientation of the stone alignment (see MLO: 69).

Site 25: Quinish. This site is the remains of a documented five-stone group, of which RCAHMS (1980) record one standing, one fallen, and the possible stump of a third. Their visit was in 1972. Prodding and the removal of a thin layer of turf in 1976 revealed a total of three fallen stones, a plan of which is presented in fig. 4.4. If the four stones once formed an alignment, its orientation was almost certainly at least 10° to the left of the lunar profile drawn by Thom (MLO: 67).

Site 31: Dunadd. The original orientation of the assumed backsight, a large fallen menhir, is unknown. The azimuth from here of a standing stone between the farm approach road and the river is some 3° to the right of the lunar profile (MLO: 63-64).

The overall consequence is that a total of 21 of the remaining 34 items of data given in the Level 3 source list do not actually represent indicated horizon features, and so are not relevant to the hypothesis being tested. This leaves only 13 of the original 40 lines as truly suitable for testing it. In fig.

4. 2b the humps corresponding to the 13 lines are shaded and the 21 unindicated foresights are shown in white. It can be seen that amongst the shaded humps there is still some evidence of grouping around 17' and 25', but not nearly so markedly as before. It was mostly unindicated foresights that accounted for the high peak at 24'.

The following objection might be made at this stage: that we have been so restrictive as to exclude potentially significant evidence. After all, Thom's hypothesis was nowhere as rigidly stated as at the beginning of this section: surely at some sites indicators of the foresights might not have been thought necessary, or might have consisted of non-durable markers such as wooden posts, or of menhirs that have since disappeared. Thus can we be confident that the peaks in the white humps of fig. 4. 2b do not express real trends in the data?

The essential point about a hypothesis as rigid as the one we have stated is that it limits the admissible data at each site to a small range of possibilities, and hence gives the strongest hope of isolating real trends and providing some sort of statistical proof. A looser or more flexible hypothesis will admit so much possible data at each site that either the investigator must keep making his own choices as to which of it to disregard (which is tantamount to imposing his own subjective interpretation) or else he must include so much data, most of which cannot help but be irrelevant, that any genuine trends will be submerged in a welter of random "noise".

A specific example, that of indicated and non-indicated foresights, will clarify this point. Consider first the hypothesis that each foresight was clearly indicated. When we visit a site—be it a flat-sided slab, an alignment, a circle with an outlier or whatever—the hypothesis restricts us to looking in a few well-defined directions for horizon features that might have been deliberately indicated. All such candidates must then be measured and included in any subsequent analysis; the rigidity of the hypothesis ensures that there will not be overwhelmingly many of them. (They might be weighted in terms of "likelihood" or "prominence" in some predefined sense, but that is another matter.) Consider now, on the other hand, the hypothesis that natural foresights were not always indicated, or that the indicators have disappeared, so that the remaining site merely records the observing position. In order to test it objectively we should need to measure and include all likely foresights around the horizon at each site: at a typical site in hilly country there will be hundreds. No more than a handful could possibly have been used simultaneously, even by the most clever site placement, and by the very nature of the hypothesis the investigator cannot know which. To include them all in any analysis would be a total waste of time, since any genuine trends would be washed out by the random declinations of the majority of the horizon features; yet to include only those with astronomically "interesting" declinations would be biasing the data (in other words choosing a theory to fit the site) and would prove nothing.

Suppose we now specifically consider possible lunar foresights. At any site, several possible foresights will typically be found in one or another of the eight stretches of horizon near to a mean lunar extreme declination. A well-

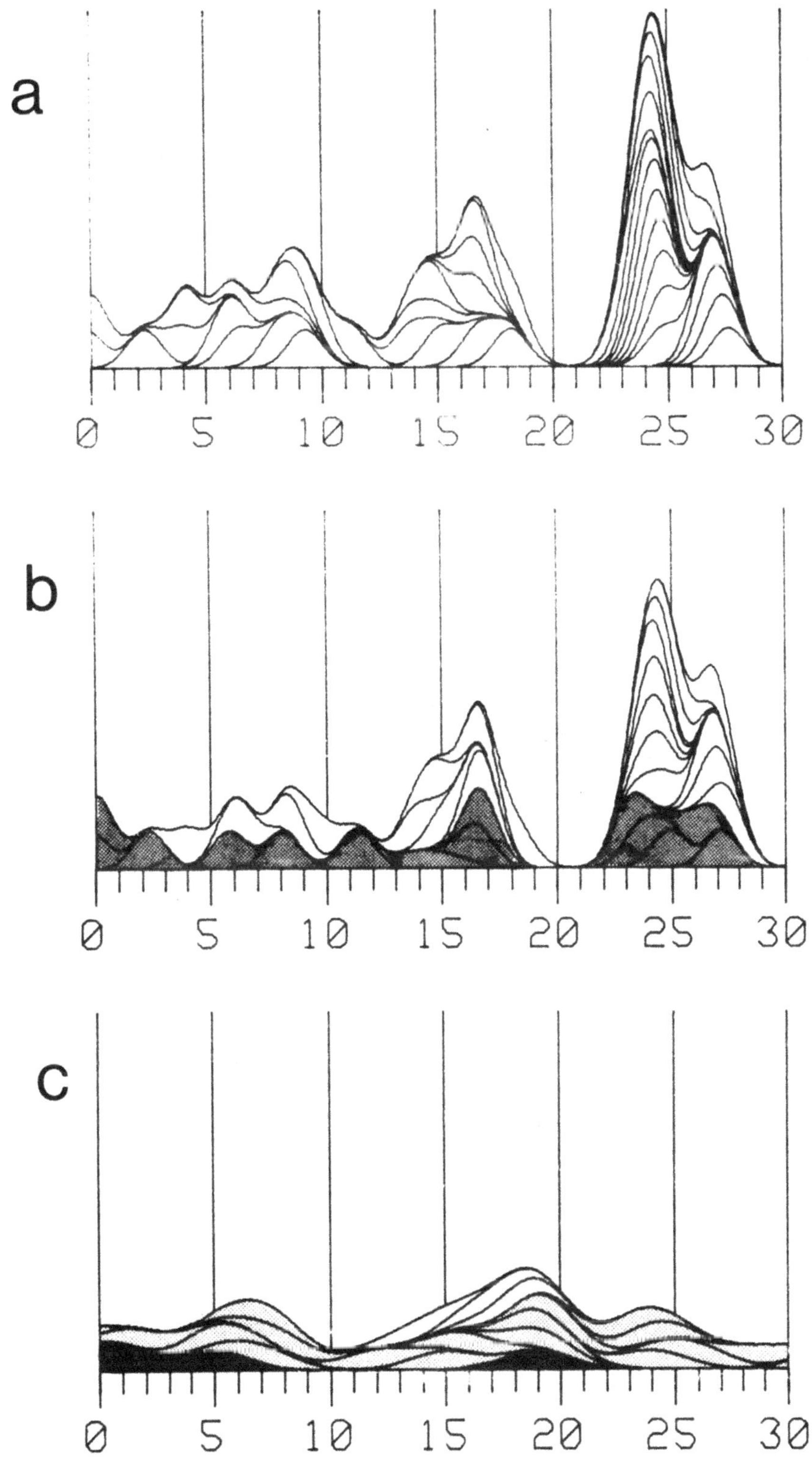

Fig. 4.2a Indications listed in the Level 3 source list. See section 4.4.1.

Fig. 4.2b Remaining indications after the suitability study of section 4.4.2. Only the shaded humps correspond to indicated foresights; others are unindicated.

Fig. 4.2c An analysis of all possibly-indicated horizon dips and notches at the remaining sites. Classes of horizon feature (see section 4.4.4) are marked by differential shading.

placed site might make use of foresights in two of these directions simultaneously, and in lucky cases three (Thom even proposes four at Brogar, but see section 4.5.1). In other fortunate cases, more than one foresight might be usable in a single range, as is proposed by Thom at several sites like Airigh nam Bidearan (MLO: 69); but then there will almost certainly be other hilltops, notches or whatever in the range as well (as Airigh nam Bidearan illustrates). Whatever happens, it is clear that no more than a small proportion of the possible lunar foresights at any typical site could possibly have been used, and if the hypothesis is that no indication is present, the investigator can have no clue as to which. In consequence, under this looser hypothesis significant peaks could only emerge above the background noise in a diagram like fig. 4.2b after an analysis of a very large number, perhaps hundreds, of sites: the majority of data would necessarily be entirely random. High peaks with relatively little background noise could only arise by the most exceptional chance—unless, that is, the investigator preferentially selected those potential foresights which best fitted his theory. The presence of such peaks amongst the white humps of fig. 4.2b, in other words, seems to be tantamount to proof not of any genuine trends in the data but of the investigator's own bias in his selection of it.

Examples of the sort of selection going on can in fact be seen amongst the profile diagrams in MLO corresponding to the 21 lines in question. At Campbeltown (MLO: 62), for example, one hilltop and one shallow dip have been included as potential foresights, but two notches and another dip, both equally tenable and both lying in the extreme moon's path, have been missed out. At Dirlot (MLO: 96) one notch and one hillslope have been included, but another notch to the right and a similarly-inclined hillslope to the left have been ignored. At Haggstone Moor (MLO: 72) the tip of the Mull of Kintyre has been considered from the fallen menhir and a notch further up the Mull from Long Tom, but neither of the opposite combinations enters the analysis. Virtually every site manifests some degree of selection.

It is crucial to try to separate investigator-imposed interpretations from genuine trends, in the quest for some sort of statistical "proof". In the case of the Level 3 source list, we are forced to focus our attention upon those remaining 13 potential foresights which are actually indicated.

4.4.3 Archaeological and intrinsic status of the data

This is a convenient point at which to step briefly aside and review the status of the remaining data. We consider two of the indicating structures to be somewhat dubious on archaeological grounds:

Site 14: Unival. The proposed indication is the flat face of the tall menhir a (Ruggles & Norris 1980) adjacent to the ruins of Unival chamber tomb (Scott 1947); thus its interpretation in terms of a separate function as a lunar indicator is somewhat dubious.

Site 19: Fowlis Wester. The indication here involves standing atop or behind a small cairn (Young 1943: 175-176) and sighting along the right-hand side of a menhir (Young's "A") little more than 10 m away. As noted in section 4.3.3, the use of cairns in this manner is inherently improbable and should be treated with caution. See also appendix 4.2 (first part).

Several intrinsic doubts could also be aired about the interpretation of some of the sites. For example, some of the horizons are rather too near to facilitate observations of the extreme precision claimed, for factors such as vegetation growth could alter the observed declination appreciably: the most notable example is notch A_1 at Temple Wood (see Patrick [1979] for a critical discussion). Then again the proposed lunar indications do not always seem to be the most obvious at the site: the most notable example is Knockrome, where there are two menhirs in line with a third 1 km away at Ardfernal. The long alignment provides a much more precise and convincing indication than the one proposed, but it does not turn out to have any conceivable solar or lunar significance. One further line deserving discussion is the following.

> Site 52: Parc-y-Meirw. The proposed foresight here is a staggering 91 miles (146 km) away in Ireland, and as such could only be visible under the most exceptional atmospheric conditions, even if the climate in the 3rd millennium BC was generally more favourable than at present. Added to this, Thom's foresight is not near the highest point of Mount Leinster but low on an adjacent hill: there are large uncertainties in the refraction correction at such distance and low altitude, rendering impossible the reliable accurate determination of any indicated declination. McCreery (1979) has argued that the proposed foresight could quite probably never have been seen at all.

Despite these doubts, it would still be desirable to account for those groupings amongst the shaded peaks of fig. 4.2b either in terms of some underlying real effect or in terms of subjectivity in the selection of data. This is the aim of the following section.

4.4.4 Correctness and objectivity of the data

The various errors involved in measuring a site, such as uncertainties in the refraction correction and in the precise intended observing position, do not normally justify quoting azimuths, altitudes and declinations to better than 1 arc minute. Patrick (1979) has elaborated on this with particular reference to Temple Wood. The direct comparison of reliable resurveys by the author (S) with Thom's measurements or calculations (T) of three putative foresights, listed in table 4.4, reinforces this point; so too do three direct comparisons between measurements by Thom and Patrick listed in the same table. Small discrepancies such as those for notch A_1 at Unival, although this particular profile was only calculated (rather than surveyed) by Thom, are also manifest in many other resurveys by the author. Two of the three T-S comparisons show gross discrepancies in azimuth: 15' for A_2 at Unival and 9' at Fowlis Wester. The author's azimuth at the latter site, however, is in agreement with revised values published by Thom & Thom (1978b: table 3). All this suggests that a rather greater standard deviation, say 1'.5, should be used throughout the declination curvigram.

Table 4.4 An analysis of possibly-indicated dips and notches

Key to Column 6 (Class of feature)

1 The unique, or else undoubtedly the most prominent, feature in the range of horizon possibly indicated

2 One of two or more features in that range

3 A feature in a range of horizon adjacent to that possibly indicated

4 A feature outside even the adjacent horizon ranges

Key to Column 7 (Source of data)

T Thom in MLO

P Patrick (1979)

S Reliable resurvey by the author

E Estimate obtained by comparing a resurvey by the author in which measurement errors were high with measurements of other features quoted by Thom or Patrick

U Unresurveyed; estimate obtained from the profile diagram drawn by Thom in MLO

Table 4.4

1 Site ref. no.	2 Thom's site ref.	3 Site name	4 Horizon feature	5 Reference	6 Class of feature	7 Source of data	8 Azimuth		9 Altitude	10 Geocentric declination		11 Difference from mean extreme
							°	′	′	°	′	′
14	H3/11	Unival	A_1	MLO:70	1	T	155	35	-11	-29	5	-3
						S	155	33	-13	-29	7	-5
			A_3	fig. 4.3a	2	S	154	50	-13	-28	55	+7
			A_2	MLO:70	3	T	157	16	-11	-29	29	-27
						S	157	31	-13	-29	36	-34
			A_4	fig. 4.3a	3	S	154	15	-13	-28	47	+15
19	P1/10	Fowlis Wester	A	MLO:54	1	T	30	55	34	+29	19	+17
						S	30	46	32	+29	21	+19
29	A2/8	Temple Wood	A_1 from S_1	fig. 4.3c	3	T	316	59	278	+28	56	-6
						P	317	0	279	+28	57	-5
			A_1 from Q	fig. 4.3c	1	T	317	13	278	+29	2	0
						P	317	14	278	+29	3	+1
			A_1 from S_5	fig. 4.3c	4	T	317	53	277	+29	19	+17
						P	317	56	277	+29	20	+18
			Dip from S_1	fig. 4.3c	3	E	316	18	290	+28	50	-12
			Dip from Q	fig. 4.3c	2	E	316	32	290	+29	4	+2
			B	MLO:46	4	T	203	46	52	-29	19	-17
			Right of B	MLO:46	3	U	204	36	42	-29	18	-16
			Left of A_2	MLO:46	3	U	207	7	36	-28	43	+19
			A_2	MLO:46	4	T	207	56	18	-28	48	+14
30	A2/12	Duncracaig	Between A_1 & A_2	MLO:52	2	E	321	56	174	+29	21	+19
			B	MLO:52	2	E	322	50	165	+29	34	+32
33	A6/4	Knockrome	Right of A	fig. 4.3b	2	S	204	8	52	-29	26	-24
			Further right	fig. 4.3b	2	S	204	40	44	-29	23	-26
36	A10/5	Stillaig	None									
37	A4/1	Escart	Dip near A	MLO:60	3	U	207	5	50	-28	42	+20
52	W9/7	Parc-y-Meirw	Left of A	MLO:73	2	U	300	41	-21	+18	16	-29
			Right of A	MLO:73	2	U	300	55	-23	+18	21	-24
			Croaghaun left	MLO:73	2	U	301	24	-23	+18	37	-8
			Croaghaun cen.	MLO:73	2	U	301	48	-22	+18	52	+7
			Croaghaun right	MLO:73	2	U	302	19	-23	+19	8	+23

In order to assess how objectively each putative foresight has been chosen by Thom, we must ask the following for each of the 13 lines. Do the horizon features identified in the Level 3 source list represent the only ones, or at least the best ones in some sense, that could have been meant to be indicated by the alleged structure on the ground? For if equally tenable ones also fall within the range of horizon which, as far as we can tell, was possibly indicated, it is essential that they also be included in the analysis so as to eliminate subjective bias.

Actually performing such an assessment presents a string of problems. How, for example, do we define the "best" feature in a stretch of horizon? Should "best" mean "most prominent"? If so, how do we quantify prominence? Then again, do we take into account just notches, or many other features such as dips, hill tops, knobs, hill slopes of an appropriate inclination, glitches, boulders on the skyline and so forth? All we can say for certain is that in order actually to prove anything, as opposed merely to finding a particular theory to fit each site, the investigator has to settle upon a particular schema and stick to it. Has Thom done this?

There are several indications amongst the 13 lines that this has not been the case. Types of foresight vary: we have notches at Unival, Fowlis Wester and Temple Wood; sides of hill slopes at Duncracaig and Knockrome; the shoulder of a hill at Stillaig; knobs at Parc-y-Meirw and Escart. At Escart the indication (taking the alignment orientation quoted by Thom) is a little to the right of the proposed foresight; at Stillaig to the left. At Escart the chosen knob gives a declination difference of +23', contributing to the peak in fig. 4.2b at 25'; the adjacent dip, however, gives +20' and would have fallen between two peaks. At Parc-y-Meirw the chosen knob gives -25', right on a peak; two nearby notches would have given -29' and +7'. The right-hand notch A_2 at Unival (MLO: 70) is outside the indicated horizon (fig. 4.3a). The Temple Wood notch A_1, albeit arguably the most prominent feature within about 20°, is only actually indicated by the circle as seen from group Q (see Thom's site plan, MLO: 46). Other possible indications give azimuths scattered over more than 10° of horizon (fig. 4.3c), and the newly-discovered second circle (appendix 4.2) adds to the possibilities.

Only three features (all notches) can fairly be said to be uniquely indicated; they are A_1 at Unival, A at Fowlis Wester, and A_1 at Temple Wood as seen from group Q. It is ironic that the first two indications were both noted in the previous section as being of somewhat dubious archaeological status. The third, involving sighting towards rather than away from the centre of a stone circle, is of a type rarely considered elsewhere by Thom.

A schema has been constructed in table 4.4, by way of example for direct comparison with Thom's interpretation of the 13 lines. Only notches and dips have been considered as possible foresights. Declinations have been calculated using original surveys and resurveys where available, and by estimating values from Thom's profile diagrams in other cases. Each feature falling within the range of horizon possibly indicated has been included, provided it yields a declination within 35' of a mean lunar extreme; and has been classified from 1 to 3 according to the specificity of the indication. For all the details see table 4.4. We have also included three features used by Thom but

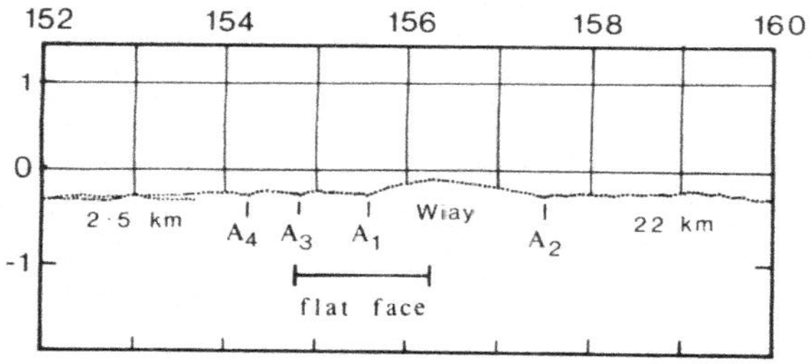

3a - Unival :
Horizon to SSE

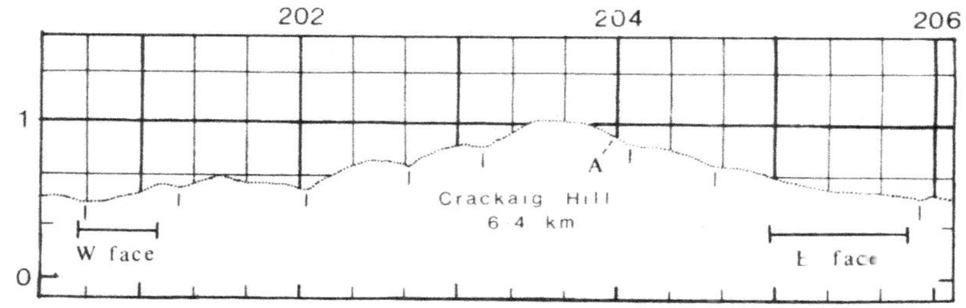

3b - Knockrome : Horizon to SSW

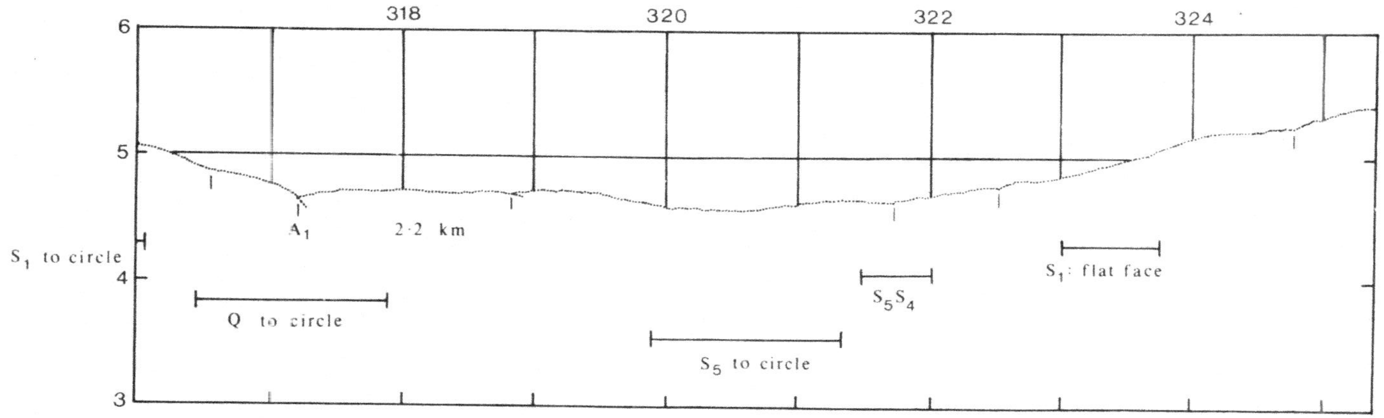

3c - Temple Wood : Horizon to NW

Fig. 4.3 Horizon profiles at three sites.

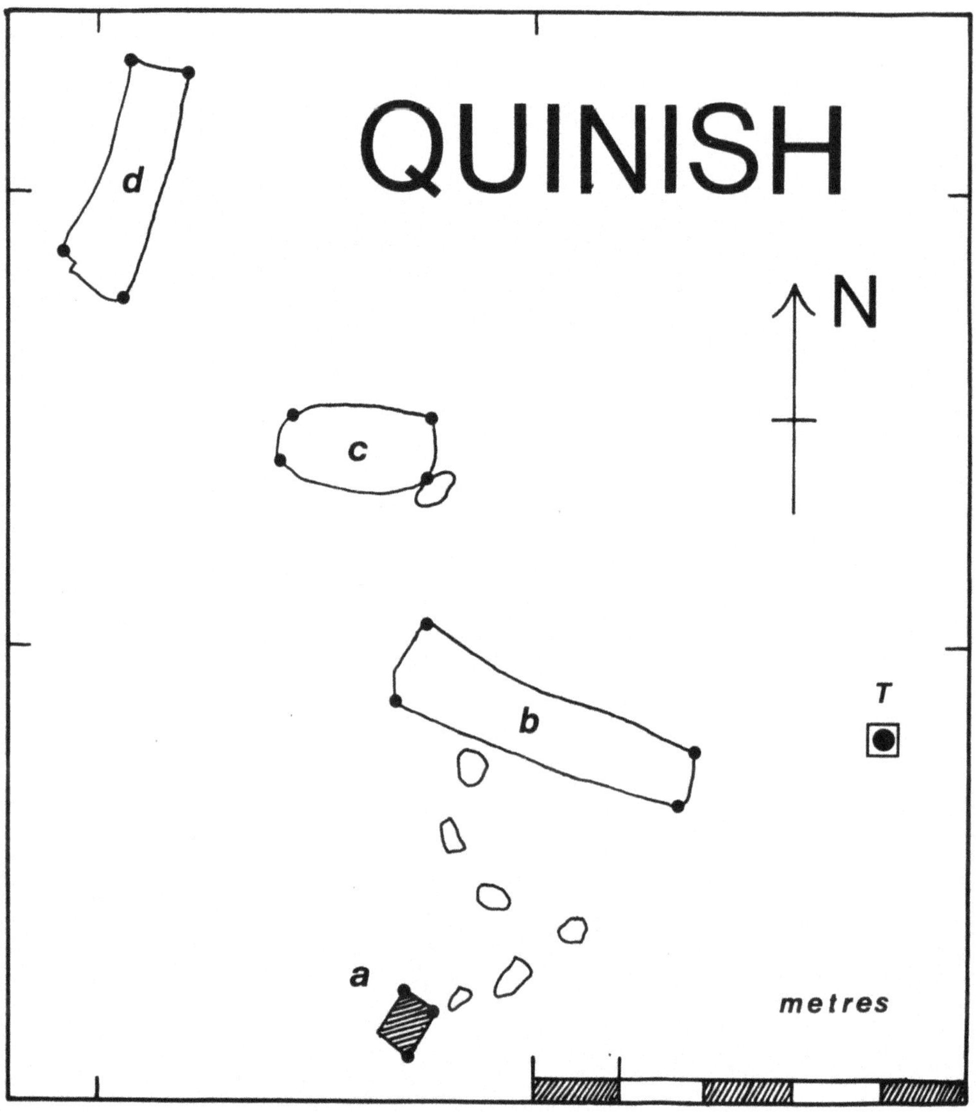

Fig. 4.4 Siteplan of site 25: Quinish, Mull.

which are well outside the indicated horizon range; these we have designated class 4. The resulting declinations have been plotted in fig. 4.2c, using a standard deviation of 1'.5 throughout. Class 1 features have been heavily shaded, classes 2 and 3 are mixed together and lightly shaded, and class 4 are in white. Any trace of significant peaks has completely disappeared.

It is important to stress that this analysis does not actually prove or disprove anything; it is merely an example of the sort of method that must be employed if genuine effects, emerging in a sample as a whole, are to be provenly separated from investigator-imposed ones. What the analysis does clearly demonstrate is that, the choices of putative foresight being as wide as they are, and the remaining sample being as small as it is, the hypothesis of the use of horizon features as precise lunar foresights is entirely unproven on the basis of the evidence presented in MLO. Much more data, data which can be seen to be free of subjective bias, is necessary before an analysis of the type performed by Thom (MLO: 77-82) could be deemed valid.

4.5 FURTHER COMMENTS ON LUNAR INDICATIONS

4.5.1 Level 4

What can be considered yet a further level of evidence in favour of megalithic lunar observatories has appeared in later work by Thom & Thom (1978b). This analysis of 25 putative observing lines at 13 sites carefully takes into account the particular astronomical and atmospheric corrections for each foresight at the time of year and day when it would necessarily have been used, rather than assuming mean parallax and refraction corrections as at Level 3. The results appear fully to confirm the Thoms' earlier conclusions; a curvigram (Thom & Thom 1978b: fig. 4) shows significant peaks at the relevant declination differences.

Some of the lines in this new list have been carried over from Level 3, with recalculated declinations. At Knockrome and Fowlis Wester the measured azimuths and altitudes have been corrected or changed. All but two unindicated foresights have been excluded, but lines at Kintraw and Haggstone Moor have slipped through. At Lundin Links a foresight on the (indicated) Cormie Hill profile has been substituted for one on the unindicated Bass Rock profile. Four new sites have been introduced, which account for 13 of the 25 lines. At each of them there are some doubts, either archaeological or intrinsic:

> Site 35: Ballinaby. The proposed indication here (MRBB: 169-170) is the flat side of the tall menhir. However the author estimated that various flat parts of the longer faces could indicate anywhere between the Thoms' col and about 6° to the right. There is also a second, smaller menhir some 200 m NNE of the tall one, and MacKie (1975: 141) notes the stump of a third, farther away in the same direction. Pennant (1772: 224) records "three upright stones, of a stupendous size". The third stone was not known of at the time of the author's visits, but the line of two is roughly at right angles to Thom's indication, and without solar or lunar significance.

Site 38: Dunskeig. This site is not mentioned on any ancient monuments lists. It is on a ridge about 150 m SE of Dun Skeig itself. One stone is 1.0 m high and reasonably convincing, but the second is no more than a small rounded boulder sticking about 0.4 m out of the ground, of itself a dubious candidate for a prehistoric structure. See also appendix 4.2.

Site 44: Knockstapple. This slab is wide but the faces are highly irregular. An original indication could have been anywhere between azimuths of 316° and 332° as estimated by the author. While this includes notch B (MLO: 63), it also includes numerous other potential foresights.

The fourth site is the Ring of Brogar, which is discussed in great detail by the Thoms (MRBB: ch. 10). They claim it is a lunar observatory making use of four foresights in distinct directions. A curvigram of Brogar sightlines alone (MRBB: 135) shows noticeable groupings around significant declination differences. The most salient general points to be made are these:

(i) It is the mounds around the Ring that are extensively interpreted as backsights and indicators. The use of mounds and cairns is archaeologically dubious (see section 4.3.3); they form imprecise reference points unless supplemented by lost markers (thereby giving the investigator more room for manoeuvre in imposing an interpretation upon the site); and why use mounds when local flagstone is abundantly available? These points have been elaborated by Atkinson (1975: 42-43).

(ii) As at Temple Wood, foresights are not always claimed to have been observed from where they were claimed to have been indicated. Of the 9 lines tabulated in MRBB only 6 are indicated from the observing position. This again gives the investigator more room for manoeuvre.

(iii) Two of the proposed foresights, Kame of Corrigal and Ravie Hill, are in themselves unconvincing, consisting of barely visible changes in hill slopes. Not that this means they were not used as foresights; it simply raises the doubt that the presence of more than two distinct foresights at the same site may reflect the investigator's ability to infer foresights in promising directions rather than necessarily demonstrating the builders' ability to hit upon the perfect location.

(iv) The calculations are so accurate that an approximate date of use has been deduced by the Thoms from the minute change in the obliquity of the ecliptic. It comes out around 1600 BC. Renfrew (1979: ch. 5) failed to obtain any meaningful radiocarbon date from the bottom of the ditch at Brogar, but Ritchie (1975) has obtained two dates around 2300 bc (3000 BC) from the similar rock-cut dutch at nearby Stenness. This perhaps suggests a date of construction for both monuments considerably earlier than that obtained by the Thoms.

All these points need to be taken into account in an adequate reassessment of Brogar.

An important point that enters at Level 4 is the possibility of circularity in the interpretation and subsequent reduction of putative foresights. The systematic variation in the obliquity of the ecliptic, variations in lunar parallax depending upon the particular lunar extreme being observed, and periodic variations over a 200-year cycle owing to the ellipticity of the moon's orbit (MRBB: ch. 2; Morrison, 1980), not to mention random variations in refraction owing to the daily changes in the weather (MLO: ch. 3), taken together impose an effective blur of some minutes of arc upon the precise declination deemed to be significant. At Level 3, indeed, these uncertainties should theoretically blur out to a great extent the peaks in the declination curvigram even if the moon's 9' wobble <u>was</u> observed—unless, that is, observations made at 19-year intervals were averaged over 200 years before any permanent backsight was erected. (This point has not, as far as the author is aware, been made in print and backed up by rigorous arguments. It endorses the assertion that the peaks in fig. 4.2a are entirely investigator-imposed. (In fact this point is acknowledged in sections 1.28 and 1.29 of paper 1 in this volume —eds.) At Level 4, such uncertainties have been taken into account as far as possible by the Thoms. However it is not clear whether the original choice of function for any particular foresight is influenced by initial assumptions such as the approximate date of use. Suppose, for example, that the Thoms at Brogar had thought from the outset in terms of a date around 3000 BC, when the obliquity of the ecliptic was numerically some 10' greater than around 1500 BC. Would they have been able to deduce different functions for each foresight in terms of the lunar wobbles, maybe using different observing positions, and then, after an analysis such as in MRBB, come out with a date around 3000 BC? If so, it would strongly suggest that the remarkable proximity of each of the Thoms' sightlines to their expected targets, noted by the Thoms themselves as confirmatory evidence, is simply a product of the wide range of interpretative choices available.

It is hoped to provide a full and detailed discussion of these points, illustrated by independent survey work, in a forthcoming publication. There is no point in going further with Level 4 here, as the 25 lines have been rationalised and increased to 42 in a paper to appear alongside this one (paper 1 of this volume), and it is towards these data, with full recourse to astronomical and statistical jargon, that such a reassessment will most valuably be directed.

<u>4.5.2 Extrapolation</u>

There are severe practical difficulties in observing and recording at all precisely the mean lunar extreme declinations and the small additional wobble. The moon's most northerly and southerly declination limits are reached monthly (as opposed to annually for the sun), and as the moon only rises or sets roughly once a day it will not usually do so when it is actually at the declination limit. Watched at every rising or setting over the relevant two or three days, it may never be seen nearer this limit than about 12'. The monthly limits themselves vary over a 19-year cycle, and it is their maxima and minima which form the extreme declinations referred to throughout the paper so far. The precise determination of one of these extremes (also known as "standstills" or "extrema") involves deducing each month what the declination limit actually was, for month after month; recording and comparing these observations; and marking the maxima and minima. It is <u>not</u> a simple matter, as with the solar solstice, of recording the rising or setting on the appropriate night. For a fuller non-technical discussion see Heggie (1972).

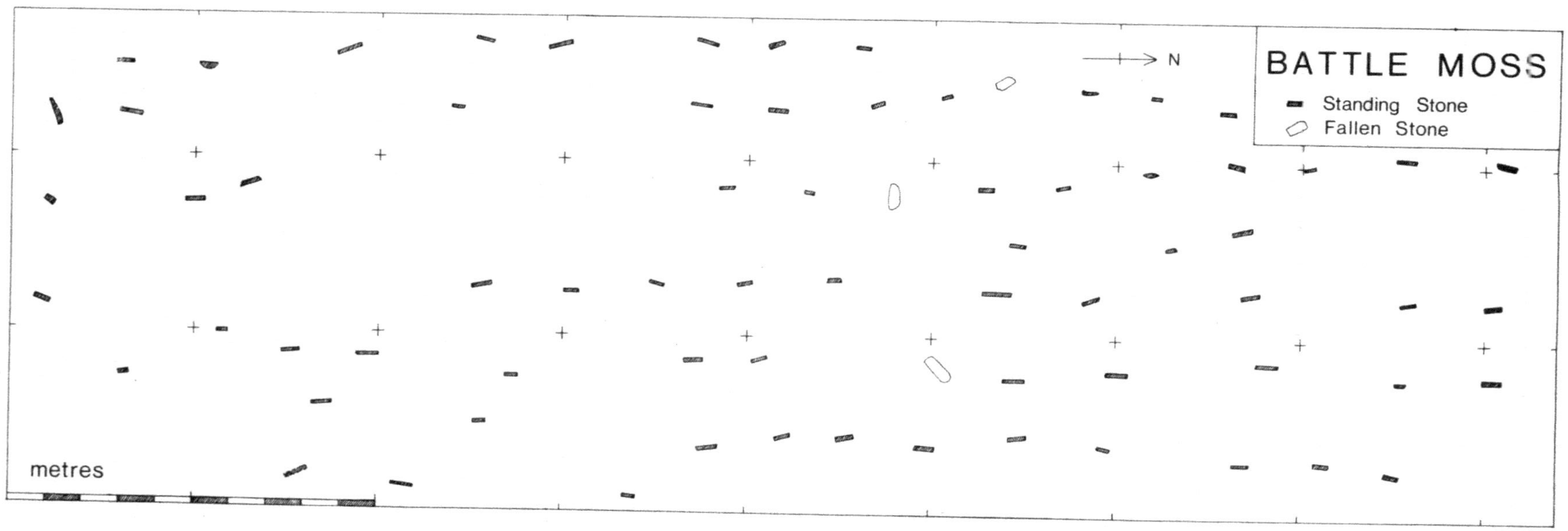

Fig. 4.5 Siteplan of site 3: Battle Moss (Loch of Yarrows), Caithness.

The most crucial problem is that of determining the declination limit each month. Related to the apparent motion of the moon is the so-called "declination deficiency" (MLO: 23-24) for which Thom calculated a fixed value. Given this constant, there exists for any particular sightline an "extrapolation length" (dependent upon the distance of the foresight from the observer and other factors such as the latitude). By relating observations on two nights nearest the monthly limit to this extrapolation length, the observing point for the theoretical limit can be measured and marked on the ground (MLO: ch. 8). Thom claims (MLO: 104-105) that this extrapolation distance was recorded in the structures at certain sites. He further claims (MLO: ch. 9) that several of the Caithness stone rows were used as extrapolating grids or sectors for calculating geometrically a further small correction.

The sites at which Thom has claimed the extrapolation distance is marked (MLO: table 9.1) are as follows.

Site 2: Dirlot. The site consists of rows of tiny stones, sunken and overgrown; not all of the ones marked by Thom (MLO: 96) could be located. It is not clear whether Thom's sector construction is a "best fit" to the site plan in some sense, and if so what is the likely error in the deduced sector radius. The foresight to which its radius would correspond as extrapolation distance is deduced to have been present at Achkeepster, at a position where the Ordnance Survey mark a site of three standing stones. Thom suggests that the extrapolation length for the distant notch (see Level 3) is marked by the distance from the site to a corner in some rows of stone slabs (MLO: 98), but these are almost certainly a modern stone fence (see appendix 4.2).

Site 3: Battle Moss (Loch of Yarrows). This site consists of the remains of some rows of small stones, planned by Thom (MLO: 98). A resurvey by the author is reproduced in fig. 4.5 for comparison. Thom gives the radius of the rows as between 750 and 830 ft (230 and 250 m), but the resurvey suggests that parallel rows are equally consistent with the data. A best-fit analysis has not been attempted. The corresponding foresight is a cairn on the horizon.

Site 4: Camster. This site is almost completely sunken in peat. What is visible seems to correspond with Thom's plan (MLO: 100), but to quote a sector radius on the basis of so few extant stones seems questionable. Further, the proposed foresight is to be viewed from one of the Grey Cairns nearby rather than from the site itself.

Site 5: Mid Clyth. Thom's analysis of the main and SW sectors yield radii not irreconcilably greater than the extrapolation length for an unindicated notch to the NNE. The foresights to the SSE (see Level 3) are so distant—about 50 miles (80 km)—that extrapolation would not be remotely feasible: the extrapolation length would be around 1000 metres.

Site 18: Corogle Burn. The extrapolation distance here is claimed to be marked by the length of the alignment. See sections 4.3.3 and 4.4.2. The two blocks at one end of the alignment are archaeologically unconvincing, and no potential precise foresight actually exists.

Site 19: Fowlis Wester. The distance from the cairn to the menhir M (MLO: 54) gives the extrapolation length for a second horizon at azimuth 320°. The axis of the ring W turns out to be a spurious indication for this, however, as Young's excavation (Young 1943) shows the original stone holes in different positions: in MRBB (174) Thom has changed the indication to a line between two other stones, B and L; however Young's stonehole for L shows that this orientation was probably originally some 2° out too (appendix 4.2, second part). Furthermore there is no actual foresight; an artificial one has to be postulated (MLO: 54). The extrapolation length for the indicated notch (see Level 3) is alleged to be marked by the distance to a boulder N, but this is of doubtful status.

Site 23: Kell Burn. See section 4.3.3. This is a highly dubious site, now destroyed.

Site 29: Temple Wood. The distance from the central stone S_1 to the circle is almost exactly the correct extrapolation length for the Bellanoch Hill notch to the SSW. Thom (MLO: 49) also notes a hard patch which may be a ploughed-out cairn but which, again from S_1, gives the extrapolation length for the nearer notch A_1.

Site 36: Stillaig. The distance between the two sites is itself near to the required extrapolation distance for the line.

Only at the latter two sites do combinations of indicated foresights and convincingly marked lengths on the ground coincide. The circle to S_1 distance for the SSW sightline at Temple Wood has the practical advantage of being roughly at right angles to the sightline, that is roughly parallel to the direction in which extrapolation would take place. The set-up at Stillaig, on the other hand, where the distance is stored parallel to the sightline, is more unconvincing. Indeed, taking the sample overall, the proposed structures storing the extrapolation distances seem to be entirely randomly oriented with respect to the alleged sightlines.

Taken as a whole, the sample is more suggestive of structures (and geometrical constructions fitted to structures) being chosen to fit available sightlines, and of sightlines being chosen to fit available structures, than it is of convincing simultaneous occurrences of both.

In fact, the idea of permanent markers of the extrapolation distance turns out to be unlikely on theoretical grounds. The reason is that the declination deficiency is not in fact a constant after all, but varies owing to an effect Thom neglected to take into account. Heggie (1972) estimated the variation as ±20%; McCreery (1980) claims, by calculating sample values from the Astronomical Ephemeris, that the widest variation is slightly more than a factor of two. This means that the extrapolation distance for any given sightline would vary, and would have to be calculated afresh every 19th year when the sightline was to be used. The use of any fixed value for the extrapolation length would result in errors of up to about 9', the amplitude of the wobble itself. The same doubts apply to the use of permanent calculating grids of fixed dimensions.

These considerations all taken together lead firmly to the conclusion that extrapolation methods, if they were employed, were not recorded permanently.

This leads us back to the practical difficulties of determining the lunar extremes. Recalling that observations need to be made month after month, that two consecutive nightly observations either side of the monthly limit are needed in order to be of any use, and noting that a third is needed for the determination of the current extrapolation length, we can estimate in how many months during, say, a 6-month period either side of the extreme the determination will be successful. (Outside this period it will be of little or no use anyway.) There are two main problems: the fact that the moon rises or sets in daylight approximately half of the time; and the incidence of bad weather. To this must be compounded the fact that both monthly limits on either side of the extreme itself must be successfully determined for a particular sightline to be successfully set up. Failure will result in a delay of 19 years. Both Heggie (1972) and McCreery (1980) have discussed these matters more fully; Moir (1979) and Ellegård (1981) discuss related problems. Heggie concludes that

> "only at about one in eight [extremes] will it be possible to set up a sightline. Since there are at most two such periods every 18 years, a new sightline can be erected only once every 70 years on average. .... in any case several sightlines must be erected before it is possible to construct one corresponding to the mean distance of the moon. By the time sufficient observations had accumulated, the slow change in [the obliquity of the ecliptic] would have become sufficient to interfere seriously with the project. It is considerations of this kind which might compel us to search for some alternative explanations of the apparent high [precision] of some of the sites."

The conclusions of section 4.4 about Level 3, together with the doubts expressed in section 4.5.1 about Level 4, are strongly suggestive that an alternative explanation very largely lies in unwitting subjective bias in the selection of data. This bias has given rise to conclusions which do not, in fact, reflect trends which are inherent in the data.

4.5.3 Stonehenge: an object lesson

Quite apart from any other theories and speculations about the monument, Thom & Thom (MRBB: 151-162) have suggested that the site of Stonehenge is the centre of an observatory from which artificial foresights served to mark some of the solar and lunar extremes. The lunar foresights are not claimed necessarily to have been precise enough to record the 9' wobble, but they do all lie within the 30' span of the moon's disc rising or setting at one of its eight extreme azimuths. Atkinson, however, has commented in appendix 4.2 on the status of the alleged foresights, and it is clear that there can be no reasonable doubt, on archaeological grounds alone, that the claimed sightlines are totally fortuitous. (The Thoms have also suggested the top of Chain Hill as the site of a further lunar foresight, but found no surviving candidate for the foresight itself.)

The question immediately arises: how was it then possible for the Thoms to find no fewer than four apparent sightlines, presumably by luck? The answer must lie in the density of structures in the vicinity of Stonehenge which the Thoms would have been prepared to consider as foresights had they occurred in the right directions. Obviously, without any idea of this background density, no adequate idea can be gained about the reliability of any astronomical hypothesis in the absence of independent evidence. The Thoms state (MRBB: 151) that "the hypothesis that the site of the monument is a lunar observatory invites us to examine eight sight-lines to find out if any of these could have carried distant foresights on the apparent horizon". Surely this remark pinpoints most clearly the flaw in approach which is at the heart of unconscious selective bias. It is bound to result in a restrictive data set which is biased to suit the hypothesis of the investigator. The pursuit of objective site data must surely be an essential part of any future methodology.

4.6 A SUMMARY OF THE CONCLUSIONS

The evidence in favour of megalithic lunar indicators and observatories is cumulative. The first three levels in the structure were identified in section 4.1. Level 1 suggests the existence of solar and lunar indications precise to half a degree or so. It was pointed out in section 4.3.1 that an adequate reassessment at this level would involve independent resurveys of complete samples of sites in certain areas, work which is in progress.

Level 2 asserts the existence of lunar indications precise enough to pick out one or other of the lunar limbs at one of the extreme declinations. Thirty-eight lines were identified in section 4.3.2 as providing the evidence. After detailed reconsideration of the archaeological status of the sites, and of the intrinsic status of the lines—that is, their inherent likelihood as potential astronomical indicators—it was found that only 19 of the 38 were of reasonable status, 7 being considered somewhat dubious, 6 very dubious, and a further 6 being dismissed altogether for various reasons. All but the latter 6 lines were nonetheless included in a reassessment of the correctness of the data and the objectivity of its selection. After this was completed, no overall evidence remained for the preferential observation of the lunar limbs.

Level 3 asserts the use of distant natural foresights, indicated by structures on the ground, to obtain extreme precision and hence measure the 9' wobble of the moon. The data set consists of 40 lines. It was shown in section 4.4.2, however, that only 13 of the 40 actually represent indicated horizon features: a further 21 features are not indicated at all (or else the supposed indication is not genuine or is some degrees off line), 5 cannot in fact be seen from the alleged indication because of the intervention of local ground, and 1 is non-existent. It was argued that the presence of apparently significant trends in the 21 lines is tantamount to proof of bias in the selection of data. Some of the remaining lines were considered archaeologically or intrinsically suspect (section 4.4.3) and only 3 were found to represent foresights uniquely indicated. In section 4.4.4 an attempt was made to remove subjectivity from the core data. It was found that no evidence whatsoever remained for observations of the 9' wobble.

Some comments were made in section 4.5.1 about a fourth level of evidence which is represented in a recent publication by Thom & Thom and more fully in paper 1 of this volume. Some archaeological and intrinsic dubieties were listed, and the possibility was raised of circular argument in calculations accurate to a minute of arc or better which purport to establish a reasonably accurate date of use. It is hoped to present a thorough critique of Level 4 in the near future.

The related problem of extrapolation was tackled in section 4.5.2. Observing the detailed lunar motions is no straightforward matter, and some means of extrapolating the moon's theoretical position from a series of actual observations is an essential prerequisite to setting up markers as precise as Thom claims at Levels 3 and 4. Thom has proposed that fixed lengths for use in the extrapolation procedures were stored at some of the sites; however a reassessment of the evidence shows that either the alleged foresight or the associated extrapolation length claimed to be marked is doubtful in all but 2 cases. Quite apart from this, astronomical arguments show that the "fixed" lengths are not in fact fixed at all, so that permanent measures on the ground would have been of little or no use anyway.

Recent investigations (see section 4.5.2) have highlighted the enormous, probably insurmountable practical difficulties involved in observing and marking the moon's motions to high precision. The investigations of this paper have shown that apparent trends in the crucial data at Levels 2 and 3, and quite possibly at Level 4 (although a definite verdict must await thorough reassessment) can quite adequately be explained away by subjective bias in its selection. Taken together, these factors lead us overwhelmingly to the conclusion that lunar motions were <u>not</u> in fact observed and recorded to high precision by the megalith builders. The extent to which astronomical orientations and indications of a rougher nature were incorporated into megalithic structures is a question which must await the outcome of reassessments at Level 1.

4.7 EPILOGUE

Whatever subsequent work may prove or disprove, Professor Thom's contribution, in megalithic astronomy as in megalithic mensuration and geometry, is of incomparable importance in a pioneering sense. It has not been the aim of this paper either to vindicate Thom's work or to bring it into discredit: merely to attempt to step nearer the truth. A Popperian approach dictates that we identify and attack only the strongest evidence, taken as a whole. The critique has been harsh, but it has attempted to be fair. Our conclusions strongly suggest that future work in archaeoastronomy in Britain should concentrate on rougher astronomical orientations or indications, that is at a far lower level of sophistication on the part of the megalith builders than envisaged by Thom, taking the weighty burden of proof more carefully into account.

It is probably worth emphasising that the conclusions of this paper have no bearing upon the validity of Thom's inferences in the fields of megalithic mensuration and geometry. There both the data and the sequence of argument are formally quite independent of any astronomical considerations.

NOTE

1. Dr. Heggie has pointed out to me that a proposed indication at Kintraw <u>has</u> been given by Thom (1969: 8-9). It is the line from the tall menhir through the centre of the small cairn beside it. This means that the two proposed foresights at Kintraw should have been considered, together with any other equally well-indicated horizon features on the profile, alongside those reassessed in sections 4.4.3 and 4.4.4. However it is clear that the conclusions of the latter section will be unaffected. The given indication is regarded by the author as somewhat dubious on archaeological grounds, for reasons similar to those at Fowlis Wester (section 4.4.3).

ACKNOWLEDGEMENTS

I am most grateful to Douglas Heggie, Richard Atkinson, Guy Morgan and Roger Few for reading the manuscript in advance of publication and for suggesting amendments. I am also grateful to my colleagues Guy Morgan, Roger Few, John Cooke and Stephen Burch for allowing me to include a few relevant results from our joint work (sites surveyed in 1978 or before—see appendix 4.1) in advance of its full publication. My sincere thanks are due to the following, who assisted me during the site work of 1979: Clare Huxley, Mari Williams, Mary Wilson, Ray Norris, Cilla Norris, Roger Few, Oliver Strimpel, Sheila Brookes, Phil Appleton, Chris Jennings, Colin Martin and Gordon Moir. Professor Atkinson's comments on many of the sites (appendix 4.2) have been invaluable. I am especially indebted to the staff of the following institutions: the Ordnance Survey, Southampton, for letting me have access to essential triangulation station information at only a nominal charge; the Royal Commission on Ancient and Historical Monuments (Scotland), Edinburgh, for giving me access to inventory information in advance of publication; the Bodleian Map Room, Oxford, for their co-operation and patience; and the University Computing Centre, Cardiff, for their endless help with computing problems. Finally, the financial support of the Science Research Council is gratefully acknowledged.

REFERENCES

Atkinson, R. J. C., 1975. Megalithic astronomy—a prehistorian's comments. J. Hist. Ast., 6, 42-52.

Burl, A., 1976. The stone circles of the British Isles. London: Yale University Press.

Coles, F. R., 1900. Report on stone circles in Kincardineshire (north) and part of Aberdeenshire. Proc. Soc. Ant. Scotl., 34, 139-198.

Coles, F. R., 1911. Report on stone circles in Perthshire. Proc. Soc. Ant. Scotl., 45, 46-116.

Cooke, J. A., Few, R. W., Morgan, J. G. & Ruggles, C. L. N., 1977. Indicated declinations at the Callanish megalithic sites. J. Hist. Ast., 8, 113-133.

Crawford, O. G. S. & Keiller, A., 1928. Wessex from the air. Oxford.

Cunnington, M. E., 1931. The "Sanctuary" on Overton Hill, near Avebury. Wilts. Arch. Mag., 45, 300-335.

Ellegård, A., 1981. Stone age science in Britain? Curr. Anthrop., in press.

Gale, R. (ed.), 1709. "Antonini Iter Britannarium", by Augustus Antoninus. London.

Hawkins, G. S., 1965. Callanish, a Scottish Stonehenge. Science, 147, 127-130.

Hawkins, G. S., 1973. Beyond Stonehenge. London: Arrow Books.

Heggie, D. C., 1972. Megalithic lunar observatories: an astronomer's view. Antiquity, 46, 43-48.

Henshall, A. S., 1972. The chambered tombs of Scotland, Vol. 2. Edinburgh.

Hutchinson, W., 1794. The history of the county of Cumberland. Carlisle.

J. R. Inst. Cornwall Editor's note, 1889. The re-erection of the Nine Stones. J. R. Inst. Cornwall, 9, 496.

Leadman, A. D. H., 1890. The Devil's Arrows, near Boroughbridge, Yorkshire. The Reliquary, 4 (new series), 1-4.

McCreery, T., 1979. Megalithic lunar observatories—a critique, Part I. Kronos, 5(1), 47-63.

McCreery, T., 1980. Megalithic lunar observatories—a critique, Part II. Kronos, 5(2), 6-26.

Mackenzie, W. M., 1904. Notes on certain structures of archaic type in the Island of Lewis. Proc. Soc. Ant. Scotl., 38, 173-204.

MacKie, E. W., 1975. Scotland: an archaeological guide. London: Faber.

M'Laughlin, T., 1865. Notice on monoliths in the Island of Mull. Proc. Soc. Ant. Scotl., 5, 46-52.

Moir, G., 1979. Hoyle on Stonehenge. Antiquity, 53, 124-128.

Morrison, L. V., 1980. On the analysis of megalithic lunar sightlines in Scotland. Archaeoastronomy (supplement to J. Hist. Ast.), 2, 65-77.

Musson, C. R., 1971. A study of the possible building forms at Durrington Walls, Woodhenge and The Sanctuary. Appendix VII in G. J. Wainwright & I. H. Longworth (eds.), Durrington Walls, Excavations 1966-1968. London: Society of Antiquaries.

Page, W. (ed.), 1906. The Victoria history of the counties of England: Cornwall, Vol. 1. London.

Patrick, J., 1974. Midwinter sunrise at Newgrange. Nature, 249, 517-519.

Patrick, J., 1979. A reassessment of the lunar observatory hypothesis for the Kilmartin stones. Archaeoastronomy (supplement to J. Hist. Ast.), 1, 78-85.

Pennant, T., 1772. "A tour in Scotland" and "Voyage to the Hebrides", Vol. 1.

Piggott, S., 1939. Timber circles: a re-examination. Arch. J., 96, 193-222.

R. C. A. H. M. S., 1924. Inventories of Ancient Monuments, Vol. 8: East Lothian. Edinburgh: H. M. S. O.

R. C. A. H. M. S., 1928. Inventories of Ancient Monuments, Vol. 9: Outer Hebrides, Skye and the Small Isles. Edinburgh: H. M. S. O.

R. C. A. H. M. S., 1971. Argyll: an inventory of the Ancient Monuments, Vol. 1: Kintyre. Edinburgh: H. M. S.O.

R. C. A. H. M. S., 1980. Argyll: an inventory of the Ancient Monuments, Vol. 3: Mull, Tiree, Coll and northern Argyll. Edinburgh: H.M. S. O.

Renfrew, C., 1979. Investigations in Orkney. London: Society of Antiquaries.

Ritchie, J. N. G., 1975. The stones of Stenness, Orkney. Proc. Soc. Ant. Scotl., 107, 1-60.

Ruggles, C. L. N. & Norris, R. P., 1980. Megalithic science and some Scottish site plans: Part II. Antiquity, 54, 40-43.

Scott, W. L., 1935. The chambered cairn of Clettraval, North Uist. Proc. Soc. Ant. Scotl., 69, 480-536.

Scott, W. L., 1947. The chamber tomb of Unival, North Uist. Proc. Soc. Ant. Scotl., 82, 1-49.

Simpson, D. D. A., 1966. Excavations at Kintraw, Argyll. Proc. Soc. Ant. Scotl., 99, 54-59.

Smith, L. T. (ed.), 1907. "The itinerary of John Leland in or about the years 1535-1543", by John Leland.

Smith, R. A., 1880. Notes on stone circles in Durris, Kincardineshire, and its neighbourhood. Proc. Soc. Ant. Scotl., 14, 294-309.

Somerville, B., 1912. Astronomical indications in the megalithic monument at Callanish. J. Brit. Ast. Assoc., 23, 83-96.

Stukeley, W., 1776. Itinerarium Curiosum, 2nd. edition.

Thom, A., 1954. The solar observatories of Megalithic Man. J. Brit. Ast. Assoc., 64, 396-404.

Thom, A., 1955. A statistical examination of the megalithic sites in Britain. J. R. Stat. Soc., A118, 275-295.

Thom, A., 1961. The egg-shaped standing stone rings of Britain. Archives Internationales d'Histoire des Sciences, 14, 291-300.

Thom, A., 1966. Megalithic astronomy: indications in standing stones. Vistas in Ast., 7, 1-57.

Thom, A., 1967a ("MSB"). Megalithic sites in Britain. Oxford: Oxford University Press.

Thom, A., 1967b. Contribution to "Hoyle on Stonehenge: some comments". Antiquity, 41, 91-98.

Thom, A., 1969. The lunar observatories of Megalithic Man. Vistas in Ast., 11, 1-29.

Thom, A., 1971 ("MLO"). Megalithic lunar observatories. Oxford: Oxford University Press.

Thom, A. & Thom, A. S., 1978a ("MRBB"). Megalithic remains in Britain and Brittany. Oxford: Oxford University Press.

Thom, A. & Thom, A. S., 1978b. A reconsideration of the lunar sites in Britain. J. Hist. Ast., 9, 170-179.

Williams, B., 1856. Letter on ancient monuments in the county of Cumberland and its borders. Proc. Soc. Ant. Lond., 3, 224-228.

Wilson, G., 1872. Notice of sculptured stones at Laggangarn, Mull of Sunnoness, Airrelich and Cassendeoch, Wigtownshire. Proc. Soc. Ant. Scotl., 10, 56-61.

Young, A., 1943. Report on standing stones and other remains near Fowlis Wester, Perthshire. Proc. Soc. Ant. Scotl., 77, 174-182.

APPENDICES

APPENDIX 4.1

REFERENCE SITE LIST

Column Headings

1 Site reference number used in this paper
2 Site reference in site lists to be published by the author (* = provisional)
3 Site reference used by Thom
4 Site name
5 Location
6 All-figure National Grid Reference
7 Description of the site
8 Number of lunar indications (Level 2) at the site reassessed in section 4.3
9 Number of sightlines to distant natural foresights (Level 3) reassessed in section 4.4
10 Number of sightlines to distant foresights (Level 4) listed in Thom & Thom (1978b)- see section 4.5.1
11 Whether extrapolation length(s) are alleged to be stored at the site— see section 4.5.2
12 Latest year visited by the author
13 Year(s) surveyed
14 Whether commented upon by RJCA in appendix 4.2
15 Alternative name for the site

Key to Column 7 (Site description)

The following description codes are adapted from those used by Thom, but the descriptions themselves have been compiled independently by the author.

C Stone Circle or Ring
RC Recumbent Stone Circle
T Chamber tomb, cairn or mound
M Menhir
BM Broken menhir (i.e. earthfast stump)
FM Fallen menhir
R Stone rows
P.. Possible..
X.. Several..
../.. ..surrounding..
A(..) Alignment consisting of.. (NB. Two menhirs are only considered an alignment if they are both slabs oriented along the line of centres or there exists documentary evidence of a third which has disappeared)
Q See text (section 4.3.3)

Reference site list

1	2	3	4	5	6	7	8	9	10	11	12	13	14	15
1	-	-	Brogar	Orkney	3294 10132	C,M,XT	-	-	9		79	79		
2	-	N1/17	Dirlot	Caithness	3123 9486	R	-	2	-	y	79	79	y	
3	-	N1/7	Battle Moss	Caithness	3313 9440	R	-	-	-	y	79	79		Loch of Yarrows
4	-	N1/14	Camster	Caithness	3260 9438	R	-	-	-	y	79	-		
5	-	N1/1	Mid Clyth	Caithness	3296 9384	R	-	2	-	y	79	79	y	Hill o' Many Stanes
6	-	N1/13	Guidibest	Caithness	3180 9350	C,PBM,3PFM	1	-	-		79	79		Latheronwheel
7	LH29	H1/15	Dursainean	Lewis	1529 9334	PM	1	-	-		79	-	y	
8	LH31	H1/14	Lower Bayble	Lewis	1516 9318	FM	1	-	-		79	-	y	Clach Stein
9	LH16	H1/1	Callanish I	Lewis	1213 9330	C/T,5A(XM)	1	2	-		80	79,75		
10	LH19	H1/2	Callanish II	Lewis	1221 9326	C	1	-	-		80	79,75		Cnoc Ceann a'Gharaidh
11	LH24	H1/5	Airigh nam Bidearan	Lewis	1234 9299	A(3PM), 2PM	1	4	-		80	75		Callanish V
12	UI17	H3/6	Barpa nan Feannag	North Uist	0857 8720	T	1	-	-		-	-	y	
13	UI22	H3/3	South Clettraval	North Uist	0750 8713	T,M	1	-	-		77	77	y	
14	UI30	H3/11	Unival	North Uist	0800 8669	T,M	-	2	-		79	79,77		Leacach an Tigh Cloiche
15	-	B7/3	Dulnainbridge	Inverness	3011 8246	3M	1	-	-		79	-		
16	-	B2/4	Esslie South	Kincardine	3717 7916	RC/T	1	-	-		79	79	y	
17	-	B3/5	Kempston Hill	Kincardine	3876 7894	2M	1	-	-		79	-		
18	-	P3/1	Corogle Burn	Forfar	3349 7601	A(2FM,2PM)	1	1	-	y	79	79		Glen Prosen
19	-	P1/10	Fowlis Wester	Perth	2924 7250	M,T,PM,C	-	1	1	y	79	79	y	
20	-	P1/8	Comrie	Perth	2755 7225	2M	1	-	-		79	79	y	
21	-	P1/1	Muthill	Perth	2824 7159	2M,M	1	-	-	y	79	-	y	
22	-	P4/1	Lundin Links	Fife	3405 7027	3M	-	1	1		79	-		
23	-	G9/13	Kell Burn	East Lothian	3643 6642	PM	1	-	-	y	79	-		
24	CT9	M4/2	Balinoe	Tiree	1974 7426	M	1	-	-		79	79		Balemartin
25	ML2	M1/3	Quinish	Mull	1414 7552	A(M,3FM)	-	1	-		79	79,76		
26	ML28	M2/14	Loch Buie	Mull	1618 7251	C,4M	1	-	-		76	76		
27	ML30	M2/8	Bunessan	Mull	1398 7223	M	1	-	-		79	76	y	
28	AR5	A2/5	Kintraw	Argyll	1830 7050	M,T	-	2	1		79	-		
29	AR13	A2/8	Temple Wood	Argyll	1827 6979	C,6M	2	4	3	y	79	79	y	
30	AR15	A2/12	Duncracaig	Argyll	1833 6964	A(4M),A(2M)	1	2	2		79	79		Ballymeanach
31	AR27	A2/13	Dunadd	Argyll	1840 6933	FM,M	-	1	-		79	-		
32	AR30	A2/19	Achnabreck	Argyll	1856 6899	M,FM	1	-	-		79	-		

33	JU4*	A6/4	Knockrome	Jura	1548	6715	2M	1	1	1		79	79		
34	JU6*	A6/6	Craighouse	Jura	1513	6665	A(M,3FM)	1	-	-		79	-		Carragh a'Ghlinne
35	IS11*	A7/5	Ballinaby	Islay	1221	6673	A(2M)	-	-	2		79	79,78		
36	-	A10/5	Stillaig	Cowal	1935	6678	M	1	1	1	y	79	79		
37	KT5	A4/1	Escart	Kintyre	1846	6667	A(5M)	1	1	1		79	-		
38	KT8	-	Dunskeig	Kintyre	1762	6571	M,PM	-	-	1		79	79	y	
39	KT12	A4/17	Tarbert	Gigha	1655	6523	M	-	2	-		79	-		
40	KT15	A4/5	Beacharr	Kintyre	1692	6433	M	-	1	-		79	79		
41	KT23	A4/9	Beinn an Tuirc	Kintyre	1734	6351	M	-	2	-		79	79		Crois Mhic-Aoida
42	KT29	A4/2	High Park	Kintyre	1695	6258	M	-	2	-		79	79		
43	KT36	A4/14	Campbeltown	Kintyre	1724	6213	M	-	2	-		79	79		
44	KT41	A4/19	Knockstapple	Kintyre	1703	6125	M	-	-	1		79	79		
45	-	G3/2	Haggstone Moor	Wigtown	207	572	2M,FM,PM,T	-	2	1		79	79	y	Taxing Stone; Long Tom
46	-	G3/3	Laggangarn	Wigtown	2222	5718	2M	2	-	-		79	-	y	
47	-	G4/2	Blair Hill	Kirkcudbright	2404	5716	A(2M,FM)	1	-	-		79	79	y	The Thieves
48	-	L1/1	Castle Rigg	Cumbria	3292	5236	C	1	-	-		79	79		
49	-	L1/6	Burnmoor	Cumbria	3174	5024	3C; 2C	3	-	-		79	79		
50	-	L1/11	Kirksanton	Cumbria	3136	4811	2M	1	-	-		79	79	y	Giants' Graves
51	-	L6/1	Boroughbridge	North Yorks.	4391	4665	A(3M)	1	-	-		79	-	y	Devil's Arrows
52	-	W9/7	Parc-y-Meirw	Pembroke	1999	2359	A(4M,FM)	1	1	-		79	-		
53	-	S5/2	The Sanctuary	Wiltshire	4118	1680	Q	1	-	-		80	79	y	
54	-	-	Stonehenge	Wiltshire	4123	1422		-	-	-		80	-	y	
55	-	S1/2	Altarnun	Cornwall	2236	0781	C,2A(XM)	1	-	-		79	-		Nine Stones

APPENDIX 4.2

COMMENTS ON THE ARCHAEOLOGICAL STATUS OF SOME OF THE SITES

by R. J. C. Atkinson

The following comments have been selected by Dr. Ruggles from a somewhat longer list, as being relevant to sites named in his preceding text. With one exception they are based on personal knowledge, gained either from visits made between 1949 and 1958, when I was on the staff of the University of Edinburgh, or during a tour of some 1,600 miles undertaken in September 1979 for the specific purpose of inspecting sites with which I was not previously familiar, or for which my memory and my earlier notes were inadequate.

Site 2: Dirlot. There seems no reason to suppose that the zig-zag row of stones on edge to the E. of the grave-yard is anything but the remains of a relatively modern stone fence. Fences made of contiguous slabs of the local flagstone survive commonly in this area.

Site 5: Mid Clyth. There is no surviving indicator of the required azimuth at Mid Clyth itself. It has been inferred that there was formerly an eastern annexe to the main rows, symmetrical with that to the west; but the surviving evidence on the ground gives little support to this hypothesis. It would be surprising if such an eastern annexe had been almost totally destroyed, whilst much still remains of the adjacent eastern part of the main rows.

Site 7: Dursainean & Site 8: Lower Bayble (Clach Stein). Both alignments use Thom's site H1/13 as a foresight. In 1957 I noted this as the much-ruined remains of a passage-grave of Hebridean type, of squarish outline. This identification, under the name Garrabost, is confirmed by Henshall (1972: 464-5). Thom's 'circle' is the remains of the squarish setting of upright kerbstones. The site is noted by RCAHMS (1928: no. 54), as Cnoc nan Dursainean, Garrabost.

Site 12: Barpa nan Feannag. This is a ruined chambered cairn. It is described by Henshall (1972: 503-4).

Site 13: South Clettraval. This is the Hebridean passage-grave excavated by Sir Lindsay Scott (Scott 1935).

Site 16: Esslie South. Along the upper part of the sloping margin of this site on the NNW side there is a length of dry walling, approximately straight, which looks like the remains of a small rectangular building of medieval or post-medieval date, constructed of material robbed from the ring-cairn. In view of this apparent disturbance, the outlying stone in an anomalous position on the margin of the outer circle, which seems to be Thom's foresight, must be regarded as of dubious status. For a plan made in 1889, see Coles (1900: 161).

Site 19: Fowlis Wester. The east ring here is undoubtedly the kerb of a small cairn, originally surrounded by a free-standing ring of ten small stones. The original height of this cairn cannot have been more than about 1.5 m and was therefore just below the eye-height of an observer standing behind it. For an observer standing on top of the cairn the stone M would be well below the altitude of the NNE horizon and would not, at night, provide a good local foresight. Mrs. Young's excavation (Young 1943) showed that the centre of the stonehole of Thom's stone L lies about 1.2 m NE of the alignment from B given in his figure (MRBB: fig. 13.5), so that the azimuth of BL could originally have been 2° greater than the 321° stated by Thom, and would not have marked the horizon notch in question.

Site 20: Comrie. This site has been described, planned and illustrated under the name of Tullybannocher, a farm which lies about 700 m SW of the stones (Coles 1911: 50-51). This account says:

> "On the Ordnance Map four stones are shown on this spot. At what date destruction began is uncertain, but Mr. MacIntyre, tenant of Kindrochet, informed me that up to seventeen years ago three stones remained, and that one of them, which had fallen, was then removed as an obstruction to the plough by the farmer; therefore there are at present only two."

The map referred to was presumably the Six-Inch sheet then current. The inference is that the two existing stones are the survivors of a stone circle or of a 'four-poster', and that the alignment which they mark had no original significance.

Site 21: Muthill. The low stone which now lies between two much taller companions looks very much like one ploughed up in the vicinity and dumped here to get it out of the way. It is not mentioned or planned in the account of the site by Coles (1911: 72). This states that the current Ordnance Map marked the sites as 'remains of a stone circle'; but there appears to be no other evidence for a circle here, despite the situation on flat ground close to a river, which would be typical in this part of the country. The single monolith to the NW is noted on the previous page of the same article.

Site 27: Bunessan. The rev. T. M'Laughlin (1865: 49) records a tradition in Mull that a series of standing stones, each visible from the next, marked the pilgrimage route to Iona, off the west coast of the Ross of Mull. Given the impossibility of dating isolated menhirs, even by excavation in most cases, this is a tenable hypothesis. The stone here in question is one of his series, but he gives no specific account of it, and merely marks it on a simplified map (op. cit.: 48). If his hypothesis were correct, the stone would date from the Early Christian period or later.

Site 29: Temple Wood. The recent excavations have revealed remains of a second circle to the north of that already known, and further details of the latter. There is no evidence that any of the standing stones to the east of the road has been moved in modern times.

Site 38: Dunskeig. These two small stones, previously unrecorded, have undoubtedly been deliberately erected; but as they are less than 7 m apart, and the taller one to the SE is now leaning markedly from the vertical, they provide only a very rough indication of direction. They could both be surviving grounders of a former field-wall, of which the rest has been robbed to build the present wall flanking the adjacent trackway.

Site 45: Haggstone Moor. The small upright stone "S" stands in the line of a ruined dyke, or field-wall, and may well be no more than a normal grounder. It is quite inconspicuous from a distance of a few hundred metres.

Site 46: Laggangarn. There seems to be no good reason to doubt the attribution of this pair of standing stones to the Early Christian period; but it is possible that they are the survivors of a prehistoric stone circle, subsequently christianised by the incision of crosses. The Rev. George Wilson (1872: 56-58) gives the dimensions of seven stones which then survived (but unfortunately no plan), and mentions seven more, re-used locally as lintels or gate-posts, which he believed came from this site. He says: "An examination of the groundplan suggests the idea that there have been two circles of stones; but I have heard no tradition to that effect."

Site 47: Blair Hill (The Thieves). The surviving stones appear to be the remains of a circular or elliptical ring, probably of ten uprights spaced about 2 m apart, centre to centre. The form of the surrounding bank, even if it is of much later date, confirms the probability of such an arrangement.

Site 50: Kirksanton (Giants' Graves). The earliest reference to these stones (Hutchinson 1794: 529) says:

> "At a place called Kirksanton, is a small tumulus, on the summit of which are two huge stones pitched endwise, eight or nine feet in height, and about fifteen feet asunder; near adjoining to this monument, several large stones stood lately, placed in a rude manner."

The adjacent stones seem to have been those of a circle, since destroyed. There is now no trace of a tumulus on the site of the two standing stones; but they lie on the spine of a slight natural ridge, which would be an appropriate site for a barrow. On the other hand, a barrow containing a pair of stones of this size would be anomalous, here or anywhere else. The third stone mentioned by Thom cannot be identified, and the farmer asserts that no such stone has been moved or destroyed during the last twenty years.

Site 51: Boroughbridge (Devil's Arrows). There appears to be no evidence for the re-erection of any of the surviving three stones here. There was formerly a fourth one "which stood 7 or 8 feet from the central arrow, and Dr. Gale relates that it was 21 feet in height" (Leadman 1890: 2). The reference is to Roger Gale's account in his edition of his father's book (Gale 1709). Stukeley (1776: 74) says "Three now stand; one was taken away, as all report, to make a bridge over the bec a little eastward." John Leland, c. 1538, recorded four stones, the two middle ones being 6-8 feet apart (Smith 1907).

Site 53: The Sanctuary. The stoneholes numbered N1 and N2 cannot be related stratigraphically to any of the phases of construction on this site, although it is reasonable (but not necessary) to assume that they are broadly contemporary with the two stone circles rather than with the putatively earlier timber buildings. There seems little doubt that they actually held stones, because John Aubrey's sketch-plan of the mid-seventeenth century shows fallen stones in corresponding positions.

Site 54: Stonehenge. Of the four sites suggested by Thom as solar or lunar foresights three are questionable and one is demonstrably modern. Gibbet Knoll, when I visited it in the 1940s, was then still visibly the remains of a large square earthwork, identified by the late B. H. St. J. O'Neil as a Civil War gun-battery commanding from the east the old road from Devizes to Salisbury. Since then it has been obliterated by cultivation, apart from the west side, the surviving profile of which is far too sharp for it to be of Neolithic or Early Bronze Age date. Peter's Mound has been shown by excavation in 1977 to be of modern date. The small oval mound in the centre of Figsbury Rings appears to lie over modern ridge-and-furrow cultivation, and was the site in the 1920s of a large elm tree, as can be seen in an air-photo (Crawford & Keiller 1928: plate IX). The mound M at Hanging Langford is an integral part of a complex of earthworks dated by excavation to the late pre-Roman Iron Age and to the Romano-British period. Nothing on the ground suggests that it ante-dates the rest of the complex; but it has not been tested specifically by excavation.

5. A REASSESSMENT OF THE SOLSTITIAL OBSERVATORIES AT KINTRAW AND BALLOCHROY

Jon Patrick

5.1 INTRODUCTION

Recent studies by Thom (1971, hereinafter "MLO") and MacKie (1977a; 1977b) have advanced the theory that prehistoric observers were making an accurate study of the Sun's motion in the region of the Argyll peninsula, Scotland, at about 1750 BC $\pm$ 100 years. This date is determined (Thom 1971) from 6 supposedly accurate sightlines from 4 sites on the Kintyre peninsula. Three of these sightlines are taken at Ballochroy and another from Kintraw. This paper discusses important evidence that has been overlooked in the assessment of these sites.

5.2 KINTRAW

The site at Kintraw, 19 km north of Lochgilphead, consists of a large and a small cairn and a single tall menhir. Thom considers that the site for the large cairn was deliberately chosen for observation of the setting of the winter solstitial sun. The last flash of sun-light would be lined up against an horizon 'notch' formed by the mountains known as the Paps of Jura, some 43 km to the southwest. Successive observations could result in a fairly accurate determination of the day of the solstice; however because of an intervening ridge, about 2 km from the site, the notch is not actually visible from the cairn. To overcome this difficulty, Thom claims that the observers made their first sightings from a 'platform' about 130 m northeast of the cairn on the side of a very steep hill. MacKie (1974) has attempted to find support for Thom's theories by the excavation of this possible observing platform.

Attention has centred on the positive evidence supporting Thom's theories and some critical negative factors have been ignored. The astronomical theory requires that the successive observations on the platform be carried out along a line running to the southeast (or left as you view the Paps). In fact movement in this direction causes the notch to disappear from view behind the nearby ridge and this alone seems to discredit the platform theory.

Part of the 'platform' was excavated by MacKie (1974; 1977a: 84-92) and no evidence of human activity was found. However, a petrofabric analysis of a stone layer under the topsoil convinced MacKie that the layer was man-made and therefore he considered that Thom's theory was verified. MacKie has pointed out that the 'observation platform' is part of a terrace that runs for twenty to thirty metres to the southeast and gradually becomes broader and rises in that direction. The contours on the map shown in fig. 5.1 show this fact clearly. His argument that the terrace layer is man-made would be

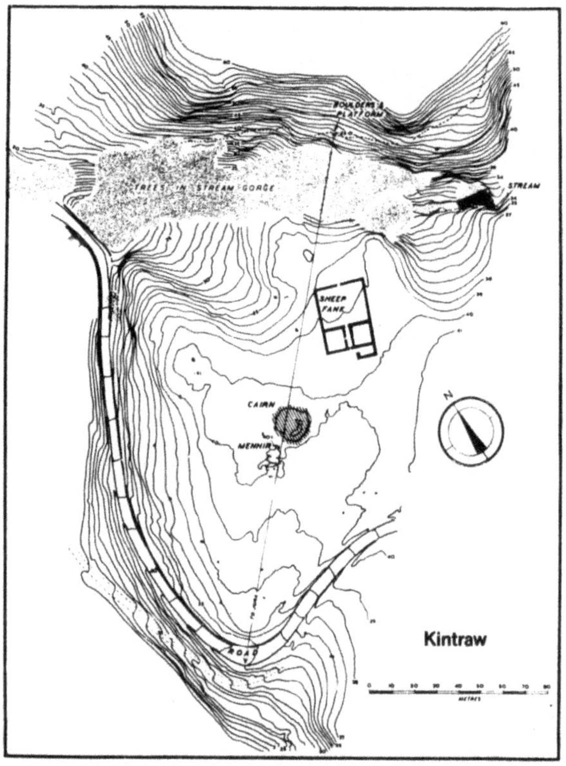

Fig. 5.1 Photogrammetric contour plan of the Kintraw site showing the cairns, menhir, stream gorge and platform. (Courtesy, E. MacKie.)

credible if the layer of stones were shown not to extend under this terrace any further than the limit of visibility of the notch.

Fig. 5.2a is a sketch drawing of MacKie's purporting to illustrate the view of the notch from the platform. Fig. 5.2b is a photograph taken by the author at eye height (1.75 m) whilst standing on the platform. It is apparent that the notch is still slightly obstructed by the near topography. Indeed the author could not satisfy himself that it was possible to see the base of the notch using a theodolite with 30 power magnification. A photograph of the notch recently published by MacKie (1977b: 106) has not resolved this conflict, and appears to have been taken from a position northwest of the author's photograph.

A visit to the higher reaches of the hillside about 30 m in height above the 'platform' casts further doubt on the validity of the platform. The steep slope flattens out to a much more gentle crest where it is possible to walk or even run freely for about 80 metres along the proposed line of observation without the risk of falling into the gorge, unlike the perilous situation on the 'platform'. More significantly there is a very scenic panoramic view of Loch Craignish and its islands with a totally unobscured view of the Paps of Jura. Near the most easterly end of the crest is a large stone about 1 metre long, 0.5 metre wide and 0.5 metre high. It is standing on its long edge in a 'recumbent' position. It is very roughly pointing to the Paps of Jura and it is certainly the only stone anywhere on this site that can reasonably claim this distinction. However it is unsteady, and there is no convincing evidence that this stone has been deliberately erected.

The azimuth and altitude to the Paps of Jura from this stone are 244° 01'.5 and 00° 23'.7 approximately, which gives a declination (assuming the upper limb of the sun was observed) of -23° 57'.1. This must be regarded as at variance with the declination of -23° 53'.9 determined at the platform given the claims for extreme accuracy. If the stone on the hillcrest was used for the observation it seems unusual that the site was transferred to the lower plateau where visibility was such a problem. The fact that the Hillcrest stone is at a high point of troublesome access is not acceptable as an objection, for the nearest alignment of standing stones about $1\frac{1}{2}$ km to the south-east at A'Bheinn is on much higher ground and well removed from the easy access of valley floors.

The large cairn at Kintraw was excavated by Simpson in the two seasons of 1959 and 1960. Under the cairn were revealed two small standing stones resembling a portal and a recumbent stone (see Simpson 1966 for plan). These two features are the only definite archaeologically attested alignments and they should be preferred to any dubious long distance alignment that happens to fit a theory. A similar alignment is also in the Culcharon cairn (Peltenburg 1972). The azimuth of the cairn alignment is 132° 54' (reliable to about 20') and the horizon altitudes are +3° 47' and +12° 52' to the south and north respectively; this gives corresponding declinations of -19° 02' and +33° 36'. The southern direction of the alignment is near to one of the minor declination nodal points of the moon (-18° 48' c. 2000 BC) but also points at Sirius, the brightest star in the sky, around 1850 BC. At about 1750 BC Sirius and the moon had the same declination and there seems no basis for deciding that one

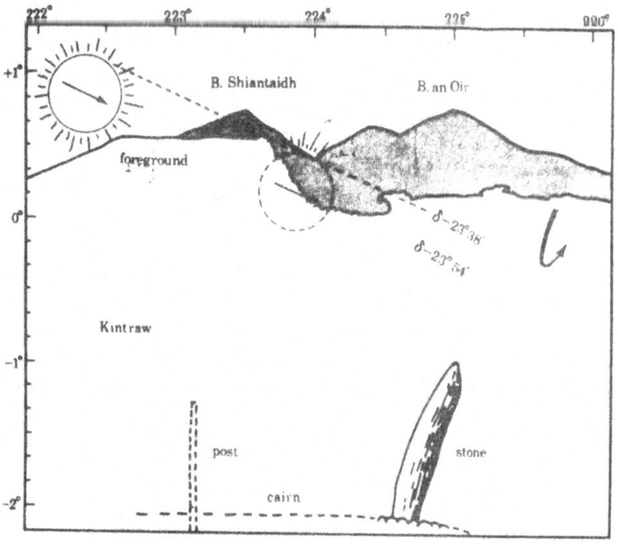

Fig. 5.2a The view of the Paps of Jura from the Kintraw platform as reported by MacKie (1974).

Fig. 5.2b The view of the Paps of Jura as photographed from the two boulders on the Kintraw platform with the camera at eye height.

body was observed in preference to the other. The alignment to the north could indicate Capella circa 1600 BC. The plethora of real alignments shows that we must have some indication of the date of a monument from archaeological evidence, otherwise one can almost always find a convenient celestial body to fit an alignment.

If the long-distance alignment to the Paps of Jura was determined under such troublesome conditions then one could expect that any features built into the cairn would incorporate the results of the observers' effort. The lack of any solar alignments adds to the evidence that the site was not used in the manner assigned to it by Thom.

5.3 BALLOCHROY

Another site used by Thom is Ballochroy, about 130 km south of Kintraw and consisting of 3 standing stones and a cist in an alignment that points towards Cara Island 11 km to the southwest. The arrangement of stones is used by Thom (MLO:42) in conjunction with Kintraw and other sites, as one of his accurate lines for computing the mean date of construction of his solstitial observatories. This site defines the winter solstitial sunset in a way similar to Kintraw. However, the visibility of Cara Island from Ballochroy must mitigate against its use as an accurate line. Fig. 5.3, Thom's sketch of Cara Island as viewed from Ballochroy, shows the island to be fairly visible whereas in fact, it is barely discernible above the headland, while the neighbouring island of Gigha (not shown in Thom's sketch) is readily visible. MacKie (1974) acknowledges that the cist at the end of the alignment was probably covered by a cairn, though later (1977a) he chooses to dismiss this fact. Opportunely Burl (1980) has now uncovered documentary evidence that this was the case. Moreover any sort of tree vegetation, even when still quite young, would obscure the view of the island altogether.

An additional claim for Ballochroy is that the face of one stone points directly to sunset at the summer solstice. The horizon in this direction is about 40 km distant, and so this alignment was also included as an accurate determination of the sun's declination in Thom's computations to establish a precise date of construction for the monuments (MLO: ch. 4).

The sketch plan of MacKie (1974) provides a profile of nine degrees of the horizon in this north west direction. As in most parts of Scotland there is a variety of notches and hills potentially suitable for highly accurate observations.

On each of two stones one face indicates approximately features that coincide with particular summer solstitial declinations. However both Thom and MacKie ignore the third and tallest stone. It appears to be just as valid a marker yet it seems to have been overlooked because it doesn't indicate any suitable declination.

The occurrence of both winter and summer solstitial alignments at any one site should not be regarded as unusual. The latitude of the British Isles is such that solstitial alignments generally fall close to azimuths of 45°, 135°, 225° and 315°. Indeed scrutiny of Thom's solstice alignments show that most azimuths lie close to these values with no divergence greater than 10°

Fig. 5.3 The Ballochroy alignment seen looking to the southwest with Cara Island on the horizon.

the largest being generally caused by high horizon altitudes. This means that at a site like Ballochroy where the thin sections of the stones are set at right angles to an alignment indicating the winter solstitial line, the other faces will automatically indicate approximately the summer solstice. If the design of the site was explained by a non-astronomical theory, then only one of the alignments would be accepted as deliberate orientation.

5.4 GENERAL POINTS

It is claimed by MacKie (1977a) that the exact coincidence of the declinations determined at Kintraw and Ballochroy is "almost sufficient on its own to demonstrate the existence of accurate solar observation points in prehistoric times". He considers this claim further justified as Thom has stated Kintraw and Ballochroy are the two best solar observatory sites. It seems that Thom's claim is based on the close compatability of the results from the two sites, so MacKie's justification of the hypothesis on the basis of the mutual consistency of the sites is plainly a circular argument. As well, MacKie's (1977a) claim that "the two sites are moreover exceptionally clear and convincing" is disputed by the evidence presented herein. However the most significant deficiency in both Thom's and MacKie's work has been their failure to demonstrate that agreement of their data is not consistent with accidental occurrence. This is not an unreasonable possibility given the correlations of summer and winter solstitial alignments, the density of Later Neolithic and Bronze Age monuments in the district and the likely number of available positions throughout Argyll where their results would be achievable, and the freedom of definition that they permit themselves in selecting an 'alignment'.

However these criticisms should not obscure the fact that this alignment of stones points to the winter solstice. If many stone alignments are shown to be directed towards the sun at either of the solstices, then it is fair to claim that there was a general recognition of the basic solar cycles by the monument builders. To extend this recognition into a theory of a full scale scientific investigation by the builders exceeds the limited evidence available and depends almost entirely on a series of subjective assumptions.

The problem of obtaining dates for the erection of standing stones is a difficult task. Even with the aid of radiocarbon dates the absolute dating for the erection of a structure may not be possible, let alone accurate. If accurate astronomical alignments can be shown to exist at an ancient site, independently of their conformity to a current hypothesis under examination, then they can certainly be used for dating. However there are many difficulties of procedure involved in dating astronomical alignments. In MLO (ch.4) six supposedly accurate sightlines from 4 sites on the Kintyre peninsula are used to compute the date of 1750 BC ± 100 years for their mean erection date. Three of these lines come from Ballochroy and one from Kintraw. This figure itself is meaningless when one considers that if each is considered accurate, the dates range from 2200 BC to 1400 BC and this does not take account of the greater date range if the variability of refraction is considered. While this time span roughly coincides with the Later Neolithic and Early Bronze Age period, during which many cairns and some standing stones were erected, the archaeological background cannot provide the precision necessary to confirm the astronomical dating. Indeed it is possible that the erection of standing stones could date back at least to 3300 BC on the basis of the excavations from Duntreath (MacKie 1974) and Newgrange (O'Kelly 1971).

These uncertainties illustrate the central danger of using astronomical alignments to date monuments. Initially one has to select an approximate date of construction and then choose the celestial body that best fits the alignment at the assumed date. Thom has taken 6 alignments on the Argyll peninsula that best conform to his theory and analysed them to arrive at a mean date of construction. The circularity of the process is manifest; the fixing of a mean date for this group of structures whose archaeological associations are unknown is meaningless. The possible use of sites as astronomical alignments could be tested more effectively by a different method. Each particular type of monument should be shown to be oriented on a particular celestial phenomenon more frequently than would be expected if the orientations were random. While one may claim that an accurate alignment is deliberate because of its close correlation to an astronomical theory, it is begging the question to accept this as a definitive measure of the sun's declination and so calculate the date of construction of the monument.

The case for highly accurate astronomical observations depends on a series of assumptions:

(a) predicting a date for the site to within a few hundred years;

(b) the selection of one face of a stone rather than any other for use as an indicator of the alignment;

(c) the choice of a particular part of the sun or moon involved in the observations;

(d) the assumption that one particular 'notch' or hilltop was used as a distant foresight;

(e) the accurate reporting of the physical characteristics of each monument and the proposed sightlines.

It is reasonable to ask any investigator to corroborate the date of the site by other means than the fact that the assumed alignment agrees with the hypothesis under analysis. Assumption (b) could be eliminated by a comprehensive study that showed that the stone faces were significantly more frequently oriented on astronomical phenomena than randomly so. Similar studies may be attempted for assumptions (c) and (d), but it is felt by the author that the inaccuracies of the primary data may well lead to inconclusive results. The fulfilment of any such studies requires a clear definition of four factors:

1. the types of indicators that constitute alignments;

2. the accuracy to which an indicator must point to a foresight;

3. the definition of all acceptable foresights both in terms of their shape and distance from the alignment;

4. the accuracy to which the declination indicated by the foresight is to conform to the declination of a celestial body. This may also involve a definition of the timespan of the observatories.

The descriptions of proposed sightlines can be more usefully verified by a photograph than by sketches. It is impossible to eliminate the subjective element completely, yet without some reasonably systematic method of recording these alignments, many unnecessarily complicated and unreliable results are bound to occur.

ACKNOWLEDGEMENTS

Mr. Evan Hadingham assisted me greatly with the field work, in the definition of the many problems associated with this work and in editing this paper. Dr. Douglas Heggie was very helpful in discussions on the technical aspects of the work and Mr. Aubrey Burl gave advice on the archaeological problems. The field work was carried out on an expedition directed by Dr. Gerald Hawkins to whom I am most grateful. The expedition was financially supported by Educational Expeditions International, Cambridge, Mass., U.S.A. Figs. 5.1 and 5.2a were provided with the kind permission of Dr. Euan MacKie and likewise Professor Thom provided fig. 5.3.

REFERENCES

Burl, A., 1980. Science or symbolism: problems of archaeo-astronomy. *Antiquity*, *54*, 191-200.

MacKie, E. W., 1974. Archaeological tests on supposed astronomical sites in Scotland. *Phil. Trans. R. Soc. Lond.*, *A276*, 169-194.

MacKie, E. W., 1977a. *Science and society in prehistoric Britain*. London: Paul Elek.

MacKie, E. W., 1977b. *The megalith builders*. Oxford: Phaidon.

O'Kelly, C., 1971. *An illustrated guide to Newgrange* (2nd edn.). Wexford: John English.

Peltenburg, E. J., 1972. Excavation of Culcharron cairn, Benderloch, Argyll. *Proc. Soc. Ant. Scotl.*, *104*, 63-70.

Simpson, D. D. A., 1966. Excavations at Kintraw, Argyll. *Proc. Soc. Ant. Scotl.*, *99*, 54-59.

Thom, A., 1971 ("MLO"). *Megalithic lunar observatories*. Oxford: Oxford University Press.

6. SOME ARCHAEOLOGICAL AND ASTRONOMICAL OBJECTIONS TO SCIENTIFIC ASTRONOMY IN BRITISH PREHISTORY

Gordon Moir

6.1 Introduction

This paper attempts to assess the evidence for megalithic astronomy presented by Prof. Alexander Thom in his book "Megalithic Sites in Britain" (Thom 1967, hereinafter "MSB"). In particular, the evidence will be considered in relation to those astronomical practices postulated by Thom—no attempt will be made to assess the likelihood of some form of astronomy being current in Neolithic or Early Bronze Age Britain. My view is that ceremonial or ritual astronomy was almost certainly part of British megalithic cultures. Here, the major concern is to investigate whether what can be termed scientific astronomy was practised.

The writer's comments are based on recent visits to and observations at many of the sites considered by Thom to be astronomically important. The evidence provided by these sites is not convincing. Many of the sites are ill-adapted for scientific use. Many contain elements that the proposed astronomy does not explain. An almost bewildering variety of archaeological sites are invoked by Thom in support of his hypotheses. Horizon features which are claimed as "indicated" are not so indicated. Those which are indicated contain more than one possible feature. Foresights are too small to be seen properly with the unaided eye. The archaeological literature relevant to many of the sites seems to have been ignored by Thom. In places this has led him into serious errors of identification.

This paper deals only briefly with the evidence for astronomical lines associated with stars and with the moon. Criticism is concentrated on the solar sites, in particular on those sites which Thom claims provide the best evidence for a megalithic calendar. First, it is helpful to define and discuss some terms.

6.2 Levels of astronomical knowledge

Ritual astronomy is a body of traditional astronomical knowledge used to fix feast days, dates and times of ceremonies and mystical or religious rites. Agricultural and stock control (and possibly transhumance) activities are regulated by the feasts or ceremonies. Great astronomical accuracy is not required to maintain the economy and well-being of the society. In practice, other constraints generated within the culture may interact with the astronomy to finalise the exact day or time of a particular event. No writing and only rudimentary numeracy is required. An example is the astronomical practices of the Hopi Indians of the south-west U.S.A. (McCluskey 1977).

Scientific astronomy presupposes literacy and well-developed concepts of numeracy—to record, process and store observations. Systematic observations are carried out to determine the periods and cycles of the sun, moon and planets. The periods discovered are used to regulate a calendar, possibly issue eclipse warnings, predict the reappearances and retrograde times of the planets, provide a system for horoscopy, or omen interpretation, etc. As an example we can take Mayan astronomy. In its highest developments, when astronomical facts are determined more and more accurately and successful mathematical theories are devised to test against the facts, we have theoretical or mathematical astronomy, as in late-Babylonian astronomy.

It should be noted that Aaboe (1974) reserves the term scientific astronomy for only the most highly developed theoretical or mathematical astronomy. He then distinguishes between two levels of prescientific astronomy. His less advanced level corresponds to what is here termed ritual astronomy, his more advanced level to scientific astronomy in its present usage.

The following example may help to clarify the distinction that is to be made between ritual and scientific astronomy. Consider a standing stone, like the Heel Stone, that indicates the direction of the midsummer rising sun. It cannot define the solstice exactly. But by counting and recording the days from one midsummer festival to the next over a long enough period of time, it may be used to determine the length of the tropical year more and more accurately—a scientific use. On the other hand, it may be used to fix the day of the midsummer festival each year, quite independent of the preceding years. Providing that cultural mechanisms exist to ensure that everyone agrees on the same day for the festival (e.g. by accepting the local festival fixer's word), it matters little whether 365 or 369 days have elapsed since the previous midsummer festival. Indeed, a society unable to count beyond the score or so necessary for animal husbandry would find it impossible to tell how many days had elapsed. Moreover, the notion of carefully measuring such a long period of time could very well appear bizarre or pointless to such a society. The existence of a solstitial indicator does not of itself imply an interest in scientific astronomy.

6.3 Sighting lines and megalithic astronomy

A sighting line generally consists of a man-made backsight—the place where the observer stands, and a man-made or natural (usually horizon) foresight—showing the direction in which he is to look. If the foresight and backsight are well separated (which usually means the foresight is a natural horizon feature) then the direction is precisely defined. Such an arrangement will be termed a precise alignment. Normally this can be used to detect small changes in declination. If the backsight and foresight are close together, so that no great accuracy is possible, we have an orientation. Orientations are sufficient to enable a culture to practice ritual astronomy. However, as in the case of the Hopi with sighting lines up to 80 miles long, precise alignments by no means necessarily imply scientific astronomy. The uses to which the Hopi put their precise alignments could as readily be accomplished with orientations.

It is clear that Thom views megalithic astronomy essentially as a scientific astronomy using precise alignments. Though he records many orientations, his chief interest is in the precise alignments and the evidence for scientific astronomy that can be deduced from them. This includes the determination of the day when a solstice occurs, the division of the year into a 16-month calendar, and observations of the moon at declination extrema.

There is no evidence for writing in Britain before the 1st century BC. Apart from Thom's own work on the "Megalithic Yard" and on "Megalithic Geometry", there is little evidence for the use of numbers in Britain in the period of interest. From a detailed study of the numbers of stones in stone circles, Burl (1976a) finds only a rudimentary numeracy, with different number bases for counting in different parts of the country. The only possible archaeological artefact showing the least interest by its manufacturers in counting or measuring is the hazel "measuring rod" from Borum Eshøj, a Danish Early Bronze Age site (Glob 1974:38). Without writing and reasonably well-developed numerical concepts (as opposed to mere counting) it is difficult to see how astronomical data can be recorded, processed and stored, and thus how any form of scientific astronomy developed. It is here that the analogy proposed by MacKie (1977:ch. 10; paper 3, section 3.3.3, this volume) between the astronomer-priests of the Maya and "astronomer-priests" of the British Neolithic, breaks down. The Maya could write; they had a number system based on place value and could record large numbers and operate with them.

There is plenty of evidence for orientation from megalithic Britain. Burl (paper 7 of this volume) discusses this evidence for various classes of monuments. In particular, his explanation of the Aberdeenshire recumbent stone circles (1979:34; 1980) is a good example of ritual astronomy using orientations.

6.4 Maes Howe

Evidence that the megalithic cultures of Britain had an interest in ritual astronomy can be found at the chambered tomb of Maes Howe. The entrance passage faces south-west, in the direction of midwinter sunset. I am assured by the custodian of the monument, Mrs. Mathers, that the setting sun's rays on the shortest days of the year shine along the passage, illuminating the rear wall of the central chamber, below the entrance to the rear cell. Spence (1893) seems to have been the first to comment on a midwinter sunset phenomenon at Maes Howe, taking as the foresight the Barnhouse standing stone. However crouching on the floor, next to the blocking stone for the rear cell, in the position indicated by Mrs. Mathers, it seemed to me that the view along the passage shows that a distant notch on the horizon very probably controls the phenomenon. This notch is formed by the gentle slope of the western ridge of Mid Hill and the steep side of Ward Hill on Hoy. The Barnhouse stone is well below the horizon, and was not visible to my (by then, dark-adapted) eye. If the notch does define the path of the sun's rays, then we have here a precise alignment. Also it may well be possible to determine accurately the day of the solstice by marking the highest position reached by the sun's rays on the chamber wall, or some such procedure.

This does not necessarily mean that it was so used by its builders. To interpret Maes Howe as an astronomical observatory is to ignore the two side cells; the height, and architecture, of the central chamber; the massivity of the mound; and the similarity with other structures undoubtedly intended for funery practices. The structure is not well-adapted for scientific astronomy. On the other hand, if Maes Howe is accepted as a tomb, or place where rituals occurred connected with burial or the bones of the dead, then the siting and orientation may be understood as linking the celestial event with the dead or with the rituals. The astronomy is then one part, an integral part, of the ritual. To postulate that the monument was designed and used both as a tomb and independently as an observatory is to multiply hypotheses unnecessarily.

This is not the place to analyse the Maes Howe midwinter sunset phenomenon further. Apart from the need for accurate measurements at the site to be made, the other tombs in the Maes Howe group (Henshall 1963:ch.4) need to be considered and the possibility that the orientation is fortuitous eliminated. However it is worth remarking on the similarity between the Maes Howe phenomenon and New Grange. At New Grange (O'Kelly 1971; Patrick 1974) an architectural feature of the monument—the roof box—permits the light of the midwinter rising sun to illuminate the end recess of the cruciform chamber. Welfare and Fairley (1980:93) draw attention to the fact that the blocking stone at Maes Howe is not tall enough to fit to the top of the entrance and suggest that this was to allow a beam of light to enter the chamber even when the door was blocked, as at New Grange. (This further argues against its use as an observatory.)

The analogy with New Grange would be more nearly completed if the sun's rays at Maes Howe penetrated into the rear cell, which at present they reportedly do not. If small changes since it was built—growth of peat on the horizon, changes in the obliquity of the ecliptic, tilting of the Orkney Islands with respect to the earth's surface—were taken into account, it is quite possible that that the light did originally enter the rear cell. The carbon-14 dates for the two monuments are perhaps separated by 500 years. This argues a remarkable continuity of ritual connected with the midwinter sun over what were very probably distinct megalithic cultures.

6.5 The astronomical lines in MSB table 8.1

I wish now to turn to a critical examination of the evidence that Thom brings forward in support of his hypotheses on megalithic astronomy. The bulk of this evidence is to be found in MSB table 8.1. The data of this table do not constitute a statistical demonstration of the existence of any sort of megalithic astronomy, as Heggie (1977) has pointed out; see also the discussion on the need for a statistical approach by Ruggles (paper 4 in this volume). However the sheer number of proposed astronomical lines in table 8.1 demands our attention. There are 262 lines at 145 sites. Almost all these lines imply declinations that Thom is able to explain in astronomical terms—as associated with bright stars, or with the sun at significant points in a 16-month calendar, or with the moon at one of its 4 limiting positions. A breakdown of the 262 lines is given in table 6.1.

An immediate point to be made about the 262 lines from MSB table 8.1 is the great variety of site they are found at. As well as stone circles (of

Table 6.1 Analysis of lines in MSB table 8.1

Line		Number	Percentage
Sun	35 solstitial 83 calendrical	118	45.1
Moon		44	16.8
Star		63	24.0
No celestial explanation		37	14.1
Total		262	100.0

many different types) and single standing stones, there occur cairns, kerb-cairns and cairns of more specialised type, different forms of chambered cairn, stone rows (of small stones), stone alignments (consisting of megaliths), pairs of standing stones, sites not obviously of megalithic period, modern sites (i.e. fakes or re-erected), and some sites which appear to be natural features (e.g. H6/5 Berneray). These naturally give rise to a large variety of types of sighting line. Examples will be given in the detailed discussion of the accurate calendrical lines. Again, as Heggie (1977) and Ruggles (this volume) stress, any statistical examination of megalithic astronomy must limit itself to well-defined categories of sites. This table 8.1 most certainly does not do. Because of the variety of sites and sighting lines there is great variability in the standard of evidence provided by table 8.1.

6.6 Stellar lines: Ballimore and Tayvallich

Two examples of particularly bad sites are to be found among the 63 star lines. The site A10/3 Ballimore (National Grid reference NR 933817) is described by Thom (MSB: table 12.1) as a "kist" with passage, and a menhir. From table 8.1 the astronomical line is to be viewed along the passage over the stone, at an azimuth of 228°, and corresponds to the star Rigel (β Orionis) setting around 1950 BC. Ballimore is the Clyde-group long cairn Auchnaha (Henshall 1972: 327, ARG 14). Quoting from her description:

> "... the cairn is aligned NE to SW, facing uphill. The cairn has been greatly denuded. .... The cairn was fairly complete until 1794, when it was said to be about 120 ft. long, 'of considerable breadth and amazing depth' (O.S.A.). About 70 ft. beyond the apparent SW end is an upright stone 2 ft. 10 in. high, set at an angle to, and slightly N of, the main axis of the cairn".

The "passage" is in fact a three-compartment chamber, about 27 ft (8 m) long. Entrance to it is gained from the concave facade at the NE end of the cairn, which is about 9 ft. (3 m) deep. Deducting, then, about 30 or 40 ft (9-12 m) from the length of the cairn given in 1794, leaves at least 80 ft. (24 m) of stonework in the SW end of the long cairn. Bearing in mind the general agreement of archaeologists that long cairns were not built after 3200 BC—i.e. all built by the beginning of what Colin Burgess (1980) calls the Meldon Bridge period—we can see that Thom's Neolithic astronomer, as well as planning to observe through at least 24 m of stones, also intended waiting at least 1200 years before making his observations.

The other site is A3/4 Tayvallich (NR 729863) (Thom 1966:fig. 7). It is described as "rings and an alignment" (Thom 1966: table 1). The two orientations at the site are associated with Castor (α Geminorum). One of the stones of the alignment is a fine standing stone. The other two stones of the alignment appeared quite natural, similar low stones occurring elsewhere in the vicinity. The standing stone itself is set across the line of the alignment. The rings are very low, grass covered structures, and could be of any period. The area around the menhir has plough-rigs, and it might be reasonable to interpret the two rings as structures associated with the cultivation—almost certainly medieval or later. The horizon shown by Thom (1966: fig. 7) is not an accurate representation of what is seen at the site. The statement accompanying the site plan that "from this site the distant skyline is nowhere else visible but along the line of stones" is misleading. One could move 15 or 20 m either side of the alignment without obstructing the view in any way. In short, the site is quite unconvincing.

The two sites just described have been chosen as particularly bad examples, but my experience has been that many of the sites in MSB table 8.1 leave much to be desired. However they are both stellar sites, and Thom & Thom (1978: 179) now recognise that the new carbon-14 dates and recalibration mean that:

> ".... the large range of possible dates makes it impossible to be certain of the association of a given alignment with a definite star."

So the 63 star lines in table 8.1 must be deleted, at least until a certain date can be assigned to a given site from non-astronomical considerations.

6.7 Lunar lines

In a recent paper reviewing lunar lines (Moir 1980a) it is noted that only 10 out of the 44 lunar lines in MSB table 8.1 are considered worthy by Thom for inclusion in his book on lunar observatories (Thom 1971, hereinafter "MLO"). (Note that in Moir 1980a a misprint makes this read "10 out of 34".) From this it is concluded that Thom himself regards few of the 44 lunar lines as reliable. McCreery (1979; 1980a) subjects the whole idea of lunar observatories to searching criticism, and deals in detail with 7 out of the 10 lunar lines referred to above. He describes many inconsistencies, and difficulties, in their use. Accordingly, as examples of scientific astronomy, the 44 lunar lines will be removed from table 8.1. Whether this is valid or not the reader may judge from Ruggles' complete resurvey and analysis of lunar lines to be found in paper 4 of this volume.

6.8 Solar lines

There remain the 118 solar lines. These are more than half of the lines for which an astronomical explanation has been proposed. There is nothing inherently unfeasible about the idea of solar orientations—there is very little doubt that both major phases at Stonehenge were orientated on midsummer sunrise; Maes Howe and New Grange offer further examples. There is no fundamental objection to the idea of solar horizon observations. There are many ethnographic parallels (see, e.g., Maistrov 1974; paper 8 of this volume). However, just because specialists within a culture make observations of the sun on a distant mountainous horizon it is unwise to conclude that some form of scientific astronomy is being practised.

In assessing the evidence for scientific astronomy in Thom's work on solar lines we need to see if there are any British field monuments that are constructed so as to allow precise horizon observations to be made and that this appears to be their main function. We also need to decide if these observations could be used to determine the length of the tropical year or to provide an accurate calendar. It would further strengthen the credibility of any hypothesis in scientific astronomy if the relevant monuments were all of the same type, or of a restricted number of types. (And if all the monuments of a particular type were thus explained, even greater strength must be credited to the astronomical hypothesis.) However, the solar sites contain examples of every one of the wide variety of categories of sites that Thom includes in MSB.

6.9 Solstitial lines—Ballochroy and Kintraw

Of the 118 solar lines, 35 are solstitial. Of these 35, three—namely one at A2/5 Kintraw and two at A4/4 Ballochroy—have received great attention in the literature. This is because they are good examples of precise alignments. In fact very few of the 35 lines are precise; the great majority are orientations. Kintraw and Ballochroy are atypical examples of solstitial sites.

The alignment at Ballochroy has been criticised by Burl (1979:67; 1980) who shows that a now-vanished cairn obstructed the line to the midwinter sunset, which is over the western end of Cara island. This makes it unusable for scientific (though not for ritual or symbolic) purposes. But the ground in the middle-distance obstructs the view of part of the western end of Cara island anyway, when viewed from the stones. The sketch-plans Thom shows of this (MSB: fig. 12.2) are not accurate: the shape of Cara island is wrong, and not all of the intervening sea can be seen, because of the ground in the middle-distance.[1] The alignment does not accurately indicate the west end of Cara.

Ballochroy is an unusual alignment in that the three stones are at right angles to the alignment. The other solstitial line here is indicated by the flat face of the centre stone to Corra Bheinn on Jura, giving midsummer sunset. If the Cara island line is rejected, because of the intervening cairn, then as a scientific site the other two stones are redundant—they are not needed for the line to the NW. It seems a little forced to interpret the site as containing a ritual line to the midwinter sunset, linking that phenomenon with the dead (in the cairn), and also containing a scientific line, to the midsummer sunset. See also the discussion by Cardona (1979a).

Kintraw has been discussed by several authors in detail (Burl 1976b: 199; Cardona 1979b; MacKie 1974; MLO:37; Hadingham 1975: 111). The cairns at the site have been excavated (Simpson 1966). In a recent paper (Moir 1980b) I drew attention to the differences between Simpson's plan and Thom's. It now appears that Simpson's plan is in error as regards the diameter of the smaller cairn, and hence Thom's plan (MSB: fig. 12.1) must be accepted as accurate. However Simpson's excavation does remind us that the site was a centre for ritual activity, and consequently any astronomy associated with it can be most easily understood in that context.

Since the publication of MSB, MacKie's excavation of the "observing platform" at Kintraw has extended the discussion. In considering this stone platform, it will be recalled that MacKie (1974: 183) found a

> "complete lack of what one might call the normal archaeological signs of human activity—no potsherds or artefacts of any kind were found during the excavations, nor any fragments of charcoal such as might have come from fires. Neither were there any signs of post-holes, kerb-stones or any other obvious structural remains".

However a petrofabric analysis applied to the stones of the platform by a soil scientist (Bibby 1974) seemed to provide positive evidence that the stone platform was man-made. McCreery (1980b) has considered Bibby's work in detail and throws great doubt upon it. McCreery concludes that there is no reason to suppose the stone platform is anything but natural. Pending a reply from Bibby or further work at the Kintraw platform there is no secure evidence for scientific astronomy at Kintraw. (MacKie replies in paper 3—eds.) In any event, as several commentators have pointed out, the mere fact of the platform being man-made is not sufficiently strong in itself to support the hypothesis that scientific astronomy, as opposed to ritual astronomy, was practised at Kintraw.

The determination of the exact day a solstice occurs is not an easy matter. This is because the change in the sun's declination—and hence horizon position—on either side of the solstice is very small. MacKie's diagrams and figures (MacKie 1974: 172; 1977: 75) make this clear. In practice, the sun will not attain its declination maximum at the moment of observation (i.e. at sunrise or sunset) and this further increases the difficulties. In order to make such small changes in declination possibly detectable, a far distant foresight is needed—as at Ballochroy (line to Corra Bheinn 19 miles (30 km)) or Kintraw (line 27 miles (43 km) long). However this almost inevitably means a low horizon altitude—less than 1°. Fluctuations in refraction about the mean value for a given altitude are more pronounced at low altitudes. These fluctations can cause greater changes in the sun's apparent position than those due to the small changes in declination. Thus however precise an alignment, it cannot be used to determine with any great accuracy the day of the solstice. Similar remarks apply to alleged accurate moon observations (Moir 1979). In a non-numerate society it is difficult to see the point in determining the exact day of the solstice, unless it is to satisfy some ritual criterion.

A good idea of the effects that refraction anomalies can cause near the horizon can be obtained by examining the excellent photographs of the distortions in the sun's shape shown by O'Connell (1958). Variations in refraction are caused by many factors, some of which have been studied by Prof. Thom himself (1958). In this paper he does not give his actual raw data, only averages. I suspect the measurements were very scattered, though he was measuring the total refraction, not just the astronomical refraction. Thom (MSB: 153) draws attention to the problems posed by refraction. He concludes that perhaps Megalithic Man was attempting at Ballochroy to study the anomalies in the sun's extreme position caused by variable refraction. Is this not simply the imposition of a modern scientific outlook on to an essentially primitive people? It neglects the possibility that other people, other cultures, can have fundamentally different world-views from our own.

It may be of interest here to consider some data on a near-solstice observations given by McCluskey (1977). For the Hopi, the day of the culminating dance of the Soyal festival occurs at midwinter. The horizon observations to determine the commencement of the festival are made 11 days earlier. The notch in which the setting sun is to be observed in 79 miles (127 km) distant. Despite the fact that the sun's declination changes by about 4 arc minutes each day when it is 11 days before the solstice—and so is perceptible—the dates for actual Soyal festivals given by McCluskey range over 6 days. Whether this variability is due to refraction anomalies, indifference of the

Hopi to accurate observations, changes in observer from year to year, consideration of other factors by the Hopi in fixing their festival days, or whatever, is not said. With the clear skies of Arizona, observations made at 6000 ft (1800 m) (and thus above the chief sources of refraction anomalies) and a sighting-line 127 km long, we have here all the pre-requisites for accurate observations. Yet the Hopi appear content to work with an error of up to 3 days. But as McCluskey concludes, the Hopi's "observational solar calendar appears to have been more than adequate for the practical needs of agriculture".

6.10 Calendrical lines

Having considered, then, the solstitial lines in MSB table 8.1 we are left with the 83 calendrical lines. Thom's megalithic calendar (MSB:ch. 9) divides the year into 16 "months". (In fact, of the 83 lines ascribed to the sun which are not solstitial only 71 correspond to the division of the year into 16 parts. The other 12 give the proposed intermediate calendar dates (MSB:116) that further subdivide the year into 32 parts.) Thom's proposed calendar is not a farming calendar which subdivides the year according to the needs of agriculture, but a systematic calendar that subdivides the year into (nearly) equal periods. Though such a division may suit a civil bureaucracy, astronomers and accountants (rents became due on quarter days), it fits less well with the husbandry of cattle and corn.

As with the solstitial lines, many of the 71 calendrical lines are orientations. Thom selects 21 of the calendar lines for especial consideration in MSB table 9.2. This table contains "all those lines where the declination is considered to be known to ± 0.1 degrees". In other words, it contains precise alignments. These 21 lines will be analysed so as (i) to show that many are dubious or unusable for scientific astronomy, and (ii) to illustrate the wide variety of archaeological sites incorporated by Thom in his work. There are 3 other lines in table 9.2. These are from the site N1/15 Watten, but as they are not included in table 8.1 (because the sighting lines are not indicated at the site) they will not be considered further. The 21 lines are found at 19 different sites. One site has 3 lines (Learable Hill), the rest one each. In analysing table 9.2 it has been assumed that the 3 lines given as G8/8, A3/18 and M4/2 are misprints for, respectively, G8/9, H3/18 and H4/2. For each site Thom's code and name, any alternative names, the grid coordinates, the declination and any references will be given, followed by a discussion of the site.

A2/5 Kintraw; Danish King's Grave; NM 830050; dec = 21°.9. (MSB: fig. 12.1; MLO: fig. 4.3; Burl 1976b: 198; Cardona 1979b; Simpson 1966; Moir 1980b.) The backsight is Simpson's cairn B, a kerb-cairn (Ritchie & Maclaren 1972). The direction is indicated by the small cist attached to the inside of the kerb of the cairn on the NW side. The foresight is either a hill (corresponding to a declination of 21°.9) or the notch next to it (22°.0). It is not clear why Thom prefers the hill here rather than the notch. Apart from the rather imprecise directional indication given by the small cist, presumably it was covered over by the cairn material and could not have been seen. Excavations at a group of kerb-cairns at Claggan (Ritchie & Thornber 1974) have provided 4 radiocarbon dates, ranging from 1058 bc to 462 bc.

If these are representative of this type of monument, it would appear that kerb-cairns were being built in the later part of the Bronze Age, at least 500 years later than the proposed solstitial activity at Kintraw (see above).

A2/8 Temple Wood; NR 827979; dec = -8°.2. (Thom 1966: fig. 5; Patrick 1979.) The backsight is the centre of the stone circle, the foresight is stone S6 (see Thom 1966: fig. 5). The horizon altitude is 7°.1 for this line. The stone is far below the horizon and cannot be said to indicate unambiguously any position on the skyline. Though the orientation could be interpreted in a ritual sense, it is quite unsuitable for precise, scientific observation. Thom considers the stone S6 as an outlier to the stone circle. However it is closer to, and more obviously part of, the group of standing stones about 300 m to the SE, the Nether Largie (or Kilmartin) stones. Two of these stones form an alignment with S6. But in neither direction does this alignment have an obvious astronomical explanation. In MLO Thom takes the view that S6 is part of the Nether Largie group, and assigns to it a special role in the complex lunar observations he proposes for the site (MLO:50), evidently abandoning its interpretation as a calendrical outlier to the stone circle. The recent excavations by Scott at Temple Wood stone circle (as yet unpublished) have shown that the site has several distinct phases, spread over a long period of time. Clearly it would be most unwise to separate the circle from the standing stones, but much archaeological work is still needed before we can be sure which phase (if any) at the stone circle is associated with the erection of the standing stones. Indeed, the elucidation of an area like the Kilmartin valley, with its extensive complex of prehistoric remains, is an enormous challenge to archaeologists. The creation of esoteric astronomical theories to "explain" part of such a complex does little to help them.

A6/5 Tarbert, Jura; NR 609822; dec = -8°.1. (Rideout 1932.) The line is defined by 2 standing stones. The foresight stone stands in the churchyard at Tarbert and has a cross incised upon it, but it seemed to me to be megalithic. However, Graham Ritchie (pers. comm.) suggests that it is probably early Christian. The backsight stone cannot be said to indicate the foresight unambiguously, but the line is acceptable. It is defined by 2 monuments of the same type (i.e. standing stones) and these are the only 2 in the vicinity. The only reservations one has is to enquire what declination is implied by the reverse direction, and whether the churchyard stone is truly megalithic.

A8/1 Mid Sannox; NS 014456; dec = -16°.3. (Thom 1966:44.) (NOT VISITED.) The backsight is a large standing stone "roughly orientated" on the foresight, Goatfell, the highest peak on Arran.

G4/14 Caulside; NX 529571; dec = 16°.8. (MSB: fig. 6.3; Thom 1966: fig. 19; RCAHMS 1914: no. 16.) (NOT VISITED.) The line joins two of the stones on opposite sides of the stone circle, the foresight being a peak, Fell of Laghead (NX 614617). This is not the highest hill in the area. Several such diameters may be drawn connecting opposite pairs of stones—Thom shows 5. If more of the stones in the eastern half of the circle remained, no doubt more such diameters could be drawn.

G6/2 Auldgirth; NX 918852; dec = $8^{o}.9$. (NOT VISITED) Thom (MSB:137) refers to this site as a fake. Indeed it is; built by one Capt. Riddel. Quoting from Balfour-Browne (1948):

> "Another of his (Riddel's) imitations was a Bronze Age stone circle which he erected on a knoll a short way up the river. It is now a most realistic affair, correct in lay-out and so weathered as to take in anyone. It should be a warning to all antiquaries".

Having recognised the spuriousness of the site, one wonders why Thom included it in his table 8.1.

G8/9 Eleven Shearers; NT 790184; dec = $0^{o}.5$. (MSB: fig. 12.9; RCAHMS 1956: no. 409.) About 18 low stones form a fairly straight line. This is the backsight. About 2 km distant the top of Craik Moor provides the foresight. Thom marks this "cairn (modern?)". The piles of stone on the top of Craik Moor are in fact the remains of a hillfort rampart (RCAHMS 1956: no. 650) and thus of a very different period. In the opinion of the Royal Commissioners, the Eleven Shearers

> "... are nothing more than the grounders of an ancient field-dyke.... and probably erected at some period during the occupation of the neighbouring fort " (i.e. Hownam Rings).

It seems very likely that the site is some form of boundary—the stones do not form an exact straight line; beyond the 18 referred to above, a further 7 or 8 stones veer away from the line; the stones are no more than knee-high. The site is at 1000 ft, 300 m, above sea-level and this would suggest that if it is a cultivation boundary, it is probably earlier than the Iron Age period proposed by the Royal Commissioners. To further counter the hypothesis that the site is astronomical, it may be observed that one end of the row is not visible from the other. (I am indebted to Douglas Heggie for pointing this out to me.) Also there seems a high degree of redundancy in the provision of 18 stones where 2 (or one flat-sided slab) would do.

The next 4 sites involve the use of the island of Boreray, 50 miles (80 km) out into the Atlantic, as a foresight for the setting sun. The backsights are stones on the Outer Hebrides: Clach Mhic Leoid, Clach an't Sagairt, The Rueval stone and An Carra. These sites were considered in detail in Moir (1978). From MSB fig. 11.5 it can be seen that Thom envisages the foresight, Boreray, being used in two different ways. For 2 of the stones, An Carra and The Rueval stone, the upper limb of the sun disappears behind Boreray—i.e. the island is silhouetted against the sun's disc. For the other 2 the lower limb sets tangentially to the right-hand side of Boreray.

H2/2 Clach Mhic Leoid; NG 040973; dec = $0^{o}.0$. (RCAHMS 1928: no. 135.) This flat-sided stone slab is over 10 ft (3 m) high. The azimuth of Boreray from the stone is 271°. The stone is described (MSB:129) as being orientated at 280°, but elsewhere (Thom 1966:12) its orientation is given as $283^{o}.7$. Thus the stone has an error of at least 9° in its orientation, corresponding to an error of 5° in declination. The sun will take about 13 days to change its declination by 5° at the equinoxes. It cannot be denied that the sun sets more or less down the side of Boreray at the equinoxes when viewed from

Clach Mhic Leoid. However the flat face of the stone does not indicate the foresight accurately. Moreover, the stone is too tall to be looked over except at some distance from it, at an unmarked spot. Further, since Boreray is so distant, and low on the horizon, it would seem to be advantageous to position a backsight for it as high above sea-level as possible. The erectors of Clach Mhic Leoid have chosen to put it half-way up the side of a hill.

H3/2 Clach an't Sagairt; NF 880760; dec – 8°.8. (RCAHMS 1928: no. 170). This is a most unusual stone, quite unlike any other standing stone I have visited on the Outer Hebrides. It is quite possibly a free-standing natural boulder. Thom (MSB: 130) describes it as "orientated". It is not, however, orientated on Boreray, whose azimuth from the stone is 287°.6. Thom (1966: fig. 8d) shows it as orientated at 270°. The Royal Commissioners describe it as "set on edge facing the south-east", i.e. orientated south-west.

Clach an't Sagairt stands at about 100 ft. (30 m) above sea-level. Across the proposed sighting line to Boreray lies a long promontary of sand-dune-covered land, the Machair Leathann at Sollas, 3 miles (5 km) from the stone. Part of the sand-dune area directly on the sighting line includes the Udal, which recent excavation (Crawford & Switsur 1977) shows to have been occupied in Neolithic times. So there can be little doubt that the sand-dunes were there when the alignment was in use. These dunes are about 40-50 ft (12-15 m) above sea-level and would thus seriously interfere with observations of the sun setting on the distant horizon. One wonders why a stone was not erected on the promontary—thus securing a clear view out to sea, or on the side of Beinn Mhor (623ft; 190 m) a hill behind the stone—thus gaining sufficient height to see over the promontary. Moving to the southern side of Beinn Mhor would hardly affect the indicated declination.

H4/4 Rueval Stone; NF 814533; dec = 16°.9. This stone is within ½ mile (1 km) of two chambered cairns (Henshall 1972: UST1 & UST 29) and is situated about 100 ft (30 m) above sea-level. It protrudes about 4 ft (1.2 m) from the ground. On the day I visited the stone, Boreray was clear on the horizon and the flat side of the stone certainly indicated this general direction. Some nearby ground, about ½ mile (1 km) away on the sighting line, is also at the level of the horizon. This ground is at present being extensively quarried—and the possibility that, before quarrying began, the ground was slightly higher cannot be ignored. Any small increase in the height of the foreground would seriously interfere with, or completely obscure, the sighting line. As with the 2 previous stones considered, nearby higher ground—the southern slopes of Rueval (408 ft; 124 m)—would allow uninterrupted views of the horizon. The change in indicated declination would be less than 1/4°. Alternatively, the land at present being quarried would provide an equally clear view.

H5/1 An Carra; NF 770321; dec = 21°.9. (RCAHMS 1928: no. 407.) This is a fine standing stone, about 16 ft (5 m) tall. It is located about 200 ft (60 m) above sea-level and commands good views westwards over the great expanses of the South Uist machairs. It is thus well-positioned as an observing station for the setting sun. However, An Carra is orientated roughly NNE-SSW, not on Boreray. Thom does not claim that it is. His measured azimuth of Boreray from An Carra is 315°.4. I could see no evidence at the site that this direction was intended to be marked by the erectors. In order to see the sun setting over the top of the stone it is necessary to stand well back from it, and at present there is no indication of where to stand.

Thus of the 4 stones on the Outer Hebrides claimed as backsights for Boreray, 3 are not orientated on that island, and 3 are badly situated for scientific observations. There are 3 other calendrical sites on the Long Islands in MSB table 9.2.

<u>H3/11 Leacach an Tigh Chloiche/Unival; NF 800699; dec = $-21^{\circ}.7$.</u> (Scott 1947; Henshall 1972: UST 34.) Thom's failure to recognise the true nature of this site has been discussed fully (Moir 1980b) and an independent, accurate site plan given (Ruggles & Norris 1980). The site is, in fact, the chambered tomb Unival, and it, or the standing stone next to it, is the supposed backsight. The sighting line is indicated by 2 chambered cairns in line, Thom's sites H3/19 and H3/20, i.e. UST 23 and UST 25 (Henshall 1972). However these two tombs are well below the horizon, thus rendering them unsuitable for scientific use. Moreover I could not see these sites with the unaided eye from Unival, though admittedly they are now very ruinous. Interestingly, the stones of Unival stand out on the horizon when viewed from H3/20, but the indicated declination has no obvious astronomical interpretation.

<u>H3/18 Sornach Coir Fhinn/Loch a Phobuill; NF 829630; dec = $21^{\circ}.6$.</u> (RCAHMS 1928: no. 249; Burl 1976b: 147). The backsight is some part of of the stone circle H3/18 (Thom 1966: fig. 14). The foresight is H3/14, about a mile distant on the mound Cringraval, which forms the skyline. The nature o of the foresight is difficult to ascertain. In Henshall's view (1972: 534) the cairn marked at the site is probably a house. But, as she explains, this does not account for the 2 stone slabs next to the structure. Presumably these slabs are to be taken as the foresight. (I do not know if they are intervisible with the stone circle.)

Thom describes the foresight as "indicated". The indication (Thom 1966: fig. 14) is provided by one or other of 2 stones within the eastern sector of the stone circle. The whole eastern half of the monument is buried in peat. It is impossible to be sure what relation the 2 stones (particularly the low one) have to the ring. It could be that H3/18 is an example of a flattened circle (the Royal Commissioners describe it as slightly oval, dimensions 130 by 115 ft (40 by 35 m); in which case the 2 stones might be part of the flattened arc. The situation is further complicated by another of Thom's lines at the site. This is to Unival (Thom 1966: fig. 8b), showing midsummer sunset. It is also described as "indicated". The azimuths of the two lines differ by only $5\frac{1}{2}^{\circ}$. Which stone indicates which line cannot be decided with certainty. If the stones are standing stones the major portion of each lies buried in the peat. How then can we be certain of the direction in which it is pointing?

If the astronomical explanations for the 2 stones are accepted, what then are we to make of the stone circle? It is irrelevant to the astronomy. (If the centre of the circle is taken as the observing point, then the foresights are not indicated.) Were the stones erected long after the circle went out of use (or vice-versa)? If astronomical "explanations" are wanted, surely easier ones can be formulated. The standing stone at Unival is in the general direction of midsummer sunset from the ring. (It cannot be used for precise observations.) If the 2 stones are the remaining part of a flattened circle, then we can envisage a midsummer festival taking place within the circle: perhaps a dance with the culmination for the participants being the silhouetting of the standing stone and cairn at Unival against the glow in the sky after the sun has set. Such and explanation is just conjecture. The site has not been excavated, and there are very many intervisible sites on the western side of North Uist. But it is simpler than Thom's, and does demonstrate that at ritual sites (stone circles, cairns, etc.) ritual explanations will almost always be possible.

<u>H4/2 Gramisdale S./ Suidheachadh Sealg; NF 825552; dec = 16.°2.</u> (RCAHMS 1928: no. 352; Henshall 1972: UST 31.) The backsight is the almost-destroyed circle, H4/2, the centre of which would now be difficult to determine. The foresight is, I assume, the fallen stone at Hacklett (RCAHMS 1928: no. 355). There is nothing at the backsight to indicate the direction.

<u>L1/1 Castle Rigg; NY 292236; dec = -8°.1.</u> (Dymond 1881; Morrow 1910; Burl 1979: 220.) Here the "accurate" line is to the outlier, which lies close to a hedge. However as Burl has pointed out (1979: 70) the outlier has been re-erected and almost certainly is not in its original position, for it is plough-scratched, and so must have lain somewhere in the field in which the circle stands. The outlier is also well below the horizon.

Evidence that the outlier has been recently repositioned is given by Morrow (1910), a follower of Sir Norman Lockyer, who surveyed Castle Rigg at the beginning of this century with a view to elucidating its astronomical functions. Morrow assumed that the outlying stone was the backsight, the centre of the circle forming the foresight. He gives the azimuth of this line as 64° 54½' and the distance of the stone from the circle centre as 198 ft (60 m). Further, he describes the outlying stone as "being now almost completely buried". Those who have visited the site will know that the stone now stands to a height of 4 ft (1.2 m). From Morrow's data, the azimuth of the outlier from the circle centre is 244° 54½'. Thom gives the azimuth as 251°.5, and its distance from the circle as 296 ft (90 m). Clearly, the stone has been re-erected, on a different line and further away.

Interestingly, Morrow concludes a calendrical use for his outlier to circle line. He finds the indicated declination —13° 25'—satisfactorily close to the theoretical value for his agricultural calendar, with its important dates of May 6th and August 8th. Dymond (1881), in a comprehensive study of antiquarian references on Castle Rigg, does not mention any outlier.

<u>L1/7 Long Meg and her Daughters; NY 570372: dec = 16°·7.</u> (Dymond 1881; Burl 1979: 230). The backsight is the (unmarked) centre of the stone ring. The foresight is Fiends Fell (2080 ft; 635 m) about 5 miles (8 km) away. The line is "indicated" by Little Meg (½ mile (1 km) away), otherwise known as the Maughanby cairn. This cairn contained a cist with an urn and burnt bones (Taylor 1883). In view of its diameter and the large size of the kerb stones, it may well be a kerb-cairn. Whatever its nature, it lies far below the horizon and so cannot indicate a rising point with any great precision. The horizon undulates in a fairly gentle fashion, and includes no obvious notches or other striking features.

I cannot accept that either this site or the previous one were erected with astronomy as the primary purpose. The line to Fiends Fell does not explain the existence of a circle 360 ft (110 m) in diameter. The alignments, if they were ever intended, can only be ritualistic. Indeed Thom describes Castle Rigg as "symbolic, ritualistic" (MLO:12).

N2/1 Learable Hill; NC 893234; decs = 0°.3, 9°.1 and 16°.6. (RCAHMS 1911: no. 381; Moir 1980b.) The site consists of many long rows of small, flat-sided stones (most less than 1 ft (0.3 m) high), low mounds and a standing stone. Almost all the stones shown in the Royal Commission's plan (RCAHMS 1911: fig. 54) can be fairly easily located on the ground. Thom's plan (MSB: fig. 12.13) shows only a portion of the site—about one-third of the stones—and the menhir is wrongly positioned. In particular, two of the rows which are supposed to indicate calendrical declinations extend beyond the area shown by Thom, down a slope leading one's eye away from the foresights (notches in distant hills). The proposed explanations only account for some of the rows. What of the others? Why have 4 rows all indicating the same declination? Such small stones in long rows are unsuitable for precise observations. Would they not trip up an observer as he moves about when trying to get into the correct position?

W5/1 Moel ty Ucha/Moel Ty-uchaf; SJ 057371; dec = 16°.9. (RCAHMW 1921; no. 225.) (NOT VISITED). The backsight, Moel Ty-uchaf, is a "kerb-circle" (Lynch 1972), i.e. a form of ring monument. The horizon is at about the same height as Moel Ty-uchaf (1400 ft; 425 m). The line is indicated by W5/2 Tyfos, which lies at about 550 ft (170 m) in the valley bottom opposite. Tyfos (RCAHMW 1921: no. 226) is probably a "cairn-circle" (Lynch 1972); another type of ring monument. Neither site is a stone circle, both are forms of cairns.

W11/5 Ynys Hir; SN 921383; dec = 21°.7. (Dunning 1943.) (NOT VISITED.) The data for this site are confusing. An azimuth of 126°.5 cannot correspond to a declination of 21°.7. Presumably the azimuth is 306°.5. It is not clear what the foresight is, nor how it is indicated. Thom remarks about it "visibility unchecked".

This site is important as it is one of the few stone circles to have been completely excavated (Dunning 1943). Many of the stones had fallen. Dunning located the original stone holes and re-erected the stones where possible. This monument is thus today probably one of the few rings where we have a faithful representation of what the original builders erected. Thom records the diameter of this site as 58.6 ft (17.9 m)—almost exactly 0.5 megalithic yards different from a whole number of megalithic yards in diameter. Also Dunning discovered a post-hole a little way inside the southern sector of the circle. Any future astronomical hypotheses at this site should include consideration of this post-hole.

All 21 of the 81 calendrical lines in MSB table 8.1 selected by Thom for inclusion in MSB table 9.2 have now been reviewed. It can be seen that from a sample of only 21 lines Thom uses for backsights standing stones, stones positioned within stone circles, long rows of low stones, the centres of stone circles, chambered cairns, kerb-cairns and kerb-circles. For foresights in addition to notches and peaks on the horizon, the direction of view is indicated by standing stones, cists, chambered cairns and various other types of cairns. This leads to a quite multitudinous variety of combinations for sighting lines. Many of the lines are unsuitable for scientific astronomy. The sites are badly positioned or ill-adapted for scientific use; stones do not indicate the required direction; foresights are well below the horizon (and, sometimes, doubtfully visible); the horizons contain more than one distinctive feature. At some sites alternative interpretations in terms of ritual astronomy are feasible. For me, the inescapable conclusion is that various features at many different sites were selected by Thom purely because they coincided with his preconceived astronomical views.

A detailed discussion of MSB table 9.2, such as given above, refutes Thom's conclusion that in table 9.2

> "... we have conclusive proof that the erectors succeeded to a remarkable degree in getting a reliable calendar ...", (MSB: 114).

The remaining lines in MSB table 8.1 are either orientations or have no astronomical interpretation.

6.11 Conclusions

I have considered the evidence for scientific astronomy and find it is extremely weak. Prof. Thom or his supporters may fairly argue that no archaeologist using archaeological arguments would ever secure a conviction in a court of law. Archaeology cannot require proofs beyond all reasonable doubt and I trust I do not demand this of Thom's work. But even on a balance of probabilities, Thom's evidence in MSB is too weak to sustain his hypotheses. And surely for such radical ideas as Thom has proposed we can expect a high standard of evidence.

Sites such as cairns and stone rings are essentially ritualistic in nature. Any possible astronomy connected with them can always be interpreted as part of this ritual—even if in fact they were also being used as part of a system of precise scientific observations. In order to demonstrate a possible scientific astronomy in Megalithic Britain it is necessary to investigate classes of sites that are (or at least at present seem to be) unconnected with ritual. Standing stone alignments provide a suitable example. But ethnographic work on standing stones in Indonesia (Crystal 1974) should be noted here. Crystal studied groups of megaliths and showed that they were only visible remains of complex funerary ceremonies (with no astronomical components), the other material elements of which disappeared within a year or two.

Groups of standing stone alignments are often found in well-defined geographical areas, apparently unconnected with cairns or stone rings; for example: pairs of standing stones in West Perthshire (Stewart 1966) and stone alignments in SW Wales. A thorough literature search should be undertaken prior to field surveys: to identify all the sites within the region of interest; eliminate bogus sites; ascertain if possible where sites had been re-erected or altered; and to incorporate the results of any excavations or the evidence provided by chance finds, particularly where these offer a possibility of independently dating the site. In short, the systematic approach advocated by Ruggles (this volume) needs to be adopted. Only in this way can the interesting possibilities opened up by Prof. Thom's work be fairly assessed. And the approach suggested above applied to standing stone alignments would at the very least provide more information on sites that are, at present, virtually archaeologically incomprehensible.

NOTE

1. These points may be invalid if abnormal refraction conditions existed at my visit.

ACKNOWLEDGEMENTS

In 1970 I first read "Megalithic Sites in Britain". This book stimulated me into visiting some of the more remote and most beautiful parts of Britain. It led me, an applied mathematician, to the study of archaeology—a subject whose approach and content is very different from mathematics; and to that most enjoyable outdoor activity—excavation. For these reasons, though I may be critical of the contents of MSB, I am very grateful to the author.

I should like to thank Dr. Graham Ritchie, of the Royal Commission on Ancient and Historical Monuments (Scotland) for information on certain Scottish sites; and Mr. David Leighton, of the Welsh Royal Commission on Ancient and Historical Monuments, for information on certain Welsh sites. Also I must record my thanks to Adam Welfare and Colin Burgess of the department of Adult Education, Newcastle University, for many stimulating discussions.

REFERENCES

Aaboe, A., 1974. Scientific astronomy in antiquity. Phil. Trans. R. Soc. Lond., A276, 21-42.

Balfour-Browne, W. A. F., 1948. Field meetings. Trans. Dumfries & Galloway Nat. Hist. & Ant. Soc., 25, 185.

Bibby, J. S., 1974. Petrofabric analysis. Phil. Trans. R. Soc. Lond., A276, 191-194.

Burl, A., 1976a. Intimations of numeracy in the Neolithic and Bronze Age societies of the British Isles. Arch. J., 133, 9-35.

Burl, A., 1976b. The stone circles of the British Isles. London: Yale University Press.

Burl, A., 1979. Rings of stone. London: Frances Lincoln.

Burl, A., 1980. Science or symbolism: problems of archaeo-astronomy. Antiquity, 54, 191-200.

Burgess, C. B., 1980. The age of Stonehenge. London.

Cardona, D., 1979a. The stones of Ballochroy. Kronos, 4(3), 23-37.

Cardona, D., 1979b. The cairns of Kintraw. Kronos, 4(3), 38-55.

Crawford, I. & Switsur, R., 1977. Sandscaping and C-14: the Udal, N. Uist. Antiquity, 51, 124.

Crystal, E., 1974. Man and the menhir. Contemporary megalithic practice of the Sa'dan Toraja of Sulawesi, Indonesia. In Donnan & Clewlow (eds.), Ethnoarchaeology. Los Angeles: Monograph IV, Arch. Survey, Inst. of Arch.

Dunning, G. C., 1943. A stone circle and cairn on Mynydd Epynt, Brecknockshire. Arch. Camb., 97, 169-194.

Dymond, C. W., 1881. A group of Cumberland megaliths. Trans. Cumberland & Westmorland Ant. & Arch. Soc., 5 (old series), 39-57.

Glob. P. V., 1974. The mound people. London.

Hadingham, E., 1975. Circles and standing stones. London.

Heggie, D. C., 1977. Megalithic astronomy—fact or fiction? Quart. J. R. Ast. Soc., 18, 450-458.

Henshall, A. S., 1963. The chambered tombs of Scotland, Vol. 1. Edinburgh.

Henshall, A. S., 1972. The chambered tombs of Scotland, Vol. 2. Edinburgh.

Lynch, F., 1972. Ring-cairns and related monuments in Wales. Scot. Arch. Forum, 4, 61-80.

McCluskey, S. C., 1977. The astronomy of the Hopi Indians. J. Hist. Ast., 8, 174-195.

McCreery, T., 1979. Megalithic lunar observatories—a critique, Part I, Kronos, 5(1), 47-63.

McCreery, T., 1980a. Megalithic lunar observatories—a critique, Part II. Kronos, 5(2), 6-26.

McCreery, T., 1980b. The Kintraw stone platform. Kronos, 5(3), 71-79.

MacKie, E. W., 1974. Archaeological tests on supposed astronomical sites in Scotland. Phil. Trans. R. Soc. Lond., A276, 169-194.

MacKie, E. W., 1977. Science and society in prehistoric Britain. London: Paul Elek.

Maistrov, L. E., 1974. Contribution to discussion on ancient astronomy: the unwritten evidence. Phil. Trans. R. Soc. Lond., A276 267-268.

Moir, G., 1978. Standing stones and astronomy in the Outer Hebrides. Newsletter of the Northumberland Arch. Group, 2(1), 13-23.

Moir, G., 1979. Hoyle on Stonehenge. Antiquity, 53, 124-128.

Moir, G., 1980a. A review of megalithic lunar lines. Northern Arch., 1, 14-22.

Moir, G., 1980b. Megalithic science and sone Scottish site plans: Part I. Antiquity, 54, 37-40.

Morrow, J., 1910. Sun and star observations at the stone circles of Keswick and Long Meg. Proc. Durham Phil. Soc., 3, 71-83.

O'Connell, D. J. K., 1958. The green flash and other low sun phenomena. Rome: Vatican Observatory.

O'Kelly, C., 1971. An illustrated guide to Newgrange. (2nd edn.) Wexford. John English.

Patrick, J., 1974. Midwinter sunrise at Newgrange. Nature, 249, 517-519.

Patrick, J., 1979. A reassessment of the lunar observatory hypothesis for the Kilmartin stones. Archaeoastronomy (supplement to J. Hist. Ast.), 1, 78-85.

R.C.A.H.M.S., 1911. Second report and inventory of monuments and constructions in the County of Sutherland. Edinburgh: H. M. S. O.

R.C.A.H.M.S., 1914. Fifth report and inventory of monuments and constructions in Galloway, Vol. 2: County of the Stewartry of Kirkcudbright. Edinburgh: H.M.S.O.

R.C.A.H.M.S, 1928. Inventories of Ancient Monuments, Vol. 9: Outer Hebrides, Skye and the Small Isles. Edinburgh: H.M.S.O.

R.C.A.H.M.S., 1956. An inventory of the Ancient and Historical Monuments of Roxburghshire (2 vols.) Edinburgh: H.M.S.O.

R.C.A.H.M.W., 1921. An inventory of the Ancient Monuments in Wales and Monmouthshire, Vol. 6: County of Merioneth. London: H.M.S.O.

Rideout, E. H., 1932. Standing stones and other antiquities in Jura. Proc. Soc. Ant. Scotl., 66, 146-151.

Ritchie, J. N. G. & Maclaren, A., 1972. Ring cairns and related monuments in Scotland. Scot. Arch. Forum, 4, 1-17.

Ritchie, J. N. G. & Thornber, I., 1974. Small cairns in Argyll: some recent work. Proc. Soc. Ant. Scotl., 106, 15-30.

Ruggles, C. L. N. & Norris, R. P., 1980. Megalithic science and some Scottish site plans: Part II. Antiquity, 54, 40-43.

Scott, W. L., 1947. The chamber tomb of Unival, North Uist. Proc. Soc. Ant. Scotl., 82, 1-49.

Simpson, D. D. A., 1966. Excavations at Kintraw, Argyll. Proc. Soc. Ant. Scotl., 99, 54-99.

Spence, M., 1893. Standing stones and Maes Howe of Stenness. Scottish Review, October. Reprinted by RILKO, London, 1974.

Stewart, M. E. C., 1966. Excavations of a setting of standing stones at Lundin Farm, near Aberfeldy, Perthshire. Proc. Soc. Ant. Scotl., 98, 126-149.

Taylor, M. W., 1883. On a cup-marked stone found at Redhills, near Penrith. Trans. Cumberland & Westmorland Ant. & Arch. Soc., 6 (old series), 110-118.

Thom, A., 1958. An empirical investigation of atmospheric refraction. Empire Survey Review, 14, 248-262.

Thom, A., 1966. Megalithic astronomy: indications in standing stones. Vistas in Ast., 7, 1-57.

Thom, A., 1967 ("MSB"). Megalithic sites in Britain. Oxford: Oxford University Press.

Thom, A., 1971 ("MLO"). Megalithic lunar observatories. Oxford: Oxford University Press.

Thom, A. & Thom, A. S., 1978. Megalithic remains in Britain and Brittany. Oxford: Oxford University Press.

Welfare, S. & Fairley, J., 1980. Arthur C. Clarke's Mysterious World. London.

7. "BY THE LIGHT OF THE CINERARY MOON": CHAMBERED TOMBS AND THE ASTRONOMY OF DEATH

Aubrey Burl

7.1 Introduction

There were astronomers in prehistoric Britain. People observed the heavens, the sun and the moon, perhaps even the stars as they moved along their ordained courses. The dispute is not whether men ever watched the skies but the reasons for their watching, not whether a chambered tomb faced the rising moon but whether the orientation of its passage was decided by an astronomer-priest or a clan magician, one a proto-scientist investigating the complexity of the lunar cycle, the other a shaman performing rituals of death and regeneration beneath the cold brilliance of the moon. The motives of such men would not have been the same and the standing stones that were the mechanisms through which they worked would have had a different significance to each of them.

> 'Two men look out through the same bars:
> One sees the mud, and one the stars'.
>
> Frederick Langbridge

Referring to the stone-using natives of the Trobriand Islands Malinowski (1927:203) wrote that 'it is also easy to see that an elementary astronomical calendar would not be beyond the mental range of the simplest savages', and it is true that the more obvious movements of the moon and sun would soon be known to country dwellers whose lives were regulated by dawn, daylight, dusk and the nights of the full moon. To align upright stones or a megalithic tomb upon midwinter sunset would not have been difficult, especially if the alignment did not have to be exact. We should ask not whether there were such alignments but whether they were created to facilitate the prediction of eclipses or to satisfy social needs that were innocent of any spirit of scientific enquiry.

Malinowski (ibid.:205) went on, 'The moon plays a far greater part in the lives of the natives than either the sun or the stars. But here also there are no traces of the pseudo-scientific curiosity frequently ascribed to the natives. They do not worry about the cause of the lunar changes or phases; they have no tendency, practically or mythologically, to interpret the lunar cycle in any symbolic or cryptic form'.

This is very different from Wood's (1978:113) picture of 'an individual or group of people, with inspiration and brilliant mental capability [who] found a geometrical method of working out, from observations made on either side of the maximum, where they would have had to stand to see the moon set at maximum declination'. A study of chambered tombs may reveal the origins and purpose of astronomical observations in the British Isles, showing whether these were inspired by intellectual interest or emotional symbolism.

Knowing that primitive people elsewhere had belief-systems that often included a relationship between death and the sun anthropologists and archaeologists have found no difficulty in agreeing that the orientations in many British prehistoric monuments may have been symbolic, constructed for the same reasons that the unseeing corpses of Beaker and Food Vessel people were laid in their graves to face the rising or setting sun (Tuckwell 1975:115) just as some Iron Age and Christian burials were disposed to associate them with the sun (Wells & Green 1973; Rahtz 1978). Acknowledgement of these ethnological parallels, however, is not the same as accepting that early communities in these islands instituted a programme of astronomical research, meticulously recording solar and lunar events and perpetuating them in megalithic alignments. Archaeological evidence for dispersed peasant societies is not easily reconciled with models of a priesthood who 'were motivated by the same urge to study phenomena that drives the scientists of today' (Thom & Thom 1978:182). Sceptics might ask whether the precision of the supposed alignments has emerged simply from the over-precision and over-discrimination of modern investigators.

Faced with the increasing body of astronomical material accumulated especially from the punctilious fieldwork and planning of the Thoms the reactions of archaeologists have differed. Some, persuaded by the conclusions inherent in acceptance of the archaeoastronomical case and by the logic of the statistical data upon which it is founded, have modified their views of ancient British societies (Atkinson 1975; MacKie 1977a). Others have disagreed, believing that the traditional framework of prehistory should not be dismantled merely because it did not easily accommodate an astronomical hypothesis based on alignments that might have been misinterpreted (Fleming 1975; Daniel 1980). More archaeologists have vacillated or just hoped that it would all go away. Many simply ignored the controversy. The argument, however, is so fundamental to our understanding of early prehistory that it is essential both to recognise the limitations of the material available for examination and to define the approaches best suited to a successful enquiry.

Considerable problems confront anyone attempting to identify what prehistoric communities intended astronomically. Because it is easy to 'find' alignments in any group of standing stones investigators have quite properly applied rigorous standards before accepting any theoretical line (Wood & Penny 1975:207). Yet this procedure itself inevitably creates a corpus of ultra-accurate orientations that may be misleading. If megalith builders, as William Stukeley (1743:20) remarked, laid out sightlines 'with a sufficient, tho' not a mathematical exactness' then the alignments rediscovered today with an accuracy of a few seconds of arc may be a small proportion only of the originals and a proportion, moreover, which is deceptively precise and 'scientific'.

It is a difficulty that has been exacerbated by the practice of examining monuments individually rather than in groups. The passages of two chambered tombs at Balnuaran of Clava, Inverness-shire, face midwinter sunset but the significance of this is diminished because none of the nine other related passage-graves in the region indicates a solar declination. Possible alignments on midsummer sunrise in the two Scottish recumbent stone circles of Loanhead of Daviot and Esslie the Greater become less persuasive when it is realised that both of these rings and 48 similar rings in the vicinity were oriented on the moon (Burl 1979a: 33-7; 1980). Only when it is possible to study a fair-sized group of associated monuments which contain comparable lines can one be reasonably sure that these were the lines set out by their designers.

As well as this need to analyse groups rather than individual sites the archaeoastronomer must accept two further difficulties. One is the need to be able to recognise the type of structure being studied. Failure to do this has on occasion led to unfortunate conclusions being reached about the astronomical properties of ruined Iron Age settlements and 19th century cattle-stalls. Another is the chance that an alignment may exist and yet not be astronomical, directed towards a natural landmark such as a hill once considered sacred or towards a man-made structure such as a chambered tomb containing ancestral relics. Lewis (1892: 148-51) listed 100 sites 'in which outlying stones, prominent hills and other special objects are seen from certain circles in England and Wales', and the possibility of alignments having alternative explanations must always be remembered. For such 'landmark' sightlines to be acceptable they too, of course, would have to be tested as rigorously as the astronomical ones.

The challenge facing archaeoastronomers could be summarised in the following four questions:-

i. Did prehistoric people incorporate alignments into some of their monuments?

ii. If so, were these alignments astronomical or were some related to other targets?

iii. If there were astronomical alignments, how can they be distinguished from the possible non-astronomical lines?

iv. What was the purpose of an astronomical alignment?

In this paper it is suggested that a study of the bearings of the passages in chambered tombs may resolve some of these questions. The paper also demonstrates, as the writer well knows, the considerable intricacies involved in archaeoastronomical research.

7.2 The origins of orientations in the British Isles

It is not at all clear when orientation became an important factor in the beliefs of people in this country. Despite the occasional evidence from the continental mainland of Palaeolithic burials in which the bodies were laid out east-west, perhaps to face the setting sun as was the case with five of the six Neanderthal corpses at La Ferrassie in France, nothing similar is recorded from Britain until early Neolithic communities constructed their earthen long barrows.

That these were specifically aligned has long been known. 'We find more generally one end of these barrows broader than the other, and that broad end pointing towards the east' (Hoare 1812: 20), but this is an observation that needs qualification. The orientations of the long mounds cover a wide arc. Although amongst them there is a concentration of bearings towards the east many are aligned as far from this as NE or SE and there are some, like Belas Knap with an azimuth or compass-bearing of about 19°, that face NNE or SSE. This range of orientations has led to speculation that the barrows were crudely aligned on sunrise at various times of the year, a belief that was explicitly related to the megalithic court-cairns of northern Ireland.

'Alignment on the rising or setting sun would, in these latitudes, give a splay of some 80 degrees between summer and winter positions. A general facing towards the side on which the sun rose may have been all that was required' (de Valera 1960: 29).

Neolithic funerary structures obviously did have alignments built into them and as some such as Lambourn in Berkshire were put up as much as six thousand years ago (3415±180 bc or about 4270 BC in calibrated years) they were probably the earliest examples of oriented structures in these islands, predating even the lunar stakeholes at the entrance to Stonehenge I (Newham 1972: 15) by over a millennium. In consequence, study of the bearings of these monuments may show why prehistoric societies became concerned with such matters. The axis of a weathered and worn-down earthen long mound, however, is difficult to determine with any precision whereas the stone-lined passage of a chambered tomb does provide a well-defined and unequivocal sightline. It is chambered tombs, therefore, that have been chosen for this preliminary examination.

Perhaps because archaeologists have been distrustful of implications that peasant groups might have been attentive to matters of scientific accuracy hardly any serious work has been done on the orientation of chambered tombs and it is rare to meet anything more than a brief reference to the subject. As a result it is not widely known that these megalithic monuments, as well as rows and circles of stone, have been considered to contain alignments on the sun or moon, a suggestion that may weaken the scientific and strengthen the symbolic interpretation of astronomy because a line down a passage can never have been as exact as an alignment along the sharp edges of a row of pillars. Seventy years ago Lockyer wrote that the stone tombs in Cornwall were not sepulchres but observatories in which watchers of the sky would have been protected from rain, wind and wild beasts. The alignments were calculated not by the locals but by 'peripatetic astronomer-priests who went from place to place establishing and orienting the circle and the priest's house [the tomb] and then leaving subordinate priest-druids—curates—in charge' (Lockyer 1909: 335).

Although he was mistaken about these burial places the limited range of orientations in many groups does imply that specific targets determined the direction in which the tombs should face. Recently it has been suggested that the alignments of long mounds were connected with the need to establish an agricultural calendar for early Neolithic farmers and that in later centuries these mounds were incorporated into more elaborate structures such as cursuses—not cursūs any more than two circuses are circi—to make even more comprehensive observatories (Penny & Wood 1973; Wood 1978: 81).

Such a transformation is plausible. Proof, however, depends in part on the hypothetical astronomy of the Neolithic burial places in the British Isles. Writers have commented on the uniformity of alignment in them, in the easterly-facing earthen long barrows (Ashbee 1970: 28-30) amongst which three-quarters of those on Cranborne Chase faced south-easterly directions because of their siting on NNW-SSE ridges (ibid.: 23-4); in Cotswold-Severn tombs (Daniel 1950: 80); and in Clava Cairns of which Walker (1965: 88-91, 96-8) somewhat pessimistically observed that 'the orientation of prehistoric monuments is a subject deserving of more serious study than it has hitherto been given, but in many cases it must have reflected nothing more than a tradition—a tradition that may very well have been as little understood as that of the orientation of Christian churches and burials'. Fifteen years later we are a little closer to testing the truth of this.

The realisation that the roofbox over the entrance to Newgrange passage-grave was perfectly aligned on midwinter sunrise (O'Kelly 1971: 94; Patrick 1974) seems confirmation that Neolithic people intentionally and intelligently constructed some ritual monuments to be in line with selected celestial events. Maeshowe, the Orcadian passage-grave architecturally rather similar to Newgrange, was apparently aligned to admit the rays of the midwinter sunset (Brown 1975) through the 'letter-box' above its blocking stone (Burl 1981). Years ago O'Reilly (1898: 605) stated that in some Irish tombs 'due orientation was a condition observed by the builders'. Somerville (1909: 191; 1923) agreed with this and Lockyer (1909: 434-41) claimed solar alignments at Bodowyr, Bryn Celli Ddu, Lligwy and other megalithic tombs in North Wales. Thom (1966: 18; 1967: 97, A10/3 et al.) has given other instances in which tombs were seemingly aligned on solar, lunar or stellar risings and settings.

If these interpretations are correct the implication is that astronomical observation had long been practised in the British Isles, beginning not with the megalithic rings of the Late Neolithic but far earlier with the mortuary houses, the long mounds and the passage-graves whose entrances were so placed that the rays of the sun or moon could shine into their interiors on to the expectant bones of the dead. Such an association might explain other aspects of these places. Some carvings on the orthostats of chambered tombs have been considered to be solar or lunar motifs both in Brittany (Pequart & le Rouzic 1927: 57-60; Allibert et al. 1978) and in Ireland where 'evidence of the solar cult is abundant at Loughcrew; rayed suns and wheel-like figures are plentiful' (Coffey 1912: 102).

The conjunction of art and astronomy in these tombs is striking. 'A tradition had long existed that the rising sun, at some unspecified time, used to light up the 3-spiral stone...in the end chamber of the tomb' at Newgrange (O'Kelly 1978: 111), suggesting that the spiral itself may have symbolised the sun or the journey that the dead were believed to follow. If this were so then Newgrange was aligned upon the sunrise not for the benefit of astronomer-priests but for the ghostly spirits whose bones lay inside the blocked tomb.

It would seem a simple matter to analyse the azimuths (clockwise compass-bearings from True North) in each group of tombs to see if they related to anything astronomical but, as in every other facet of archaeo-astronomy, the task is not easy. Not only have so many of the monuments been so badly damaged that it is no longer possible to be sure of the exact direction of the passage but in other instances stones have shifted since prehistoric times or the monument has been reconstructed wrongly. Prehistoric people themselves sometimes altered the positions of stones when they were enlarging or adapting a tomb.

Where a tomb has been backfilled after excavation, blocking the passage, the situation is even more frustrating. Its plan, drawn up to a century ago, may be inaccurate or too small for any hope of recovering a precise compass-bearing from it. Even if drawn to an adequate scale one cannot rely upon the north point because usually there is no information as to whether the arrow indicates True, Magnetic or merely approximate North. Even where surveyors have worked to high standards there can be disagreement. At least five good plans have been published of the passage-grave at Balnuaran of Clava SW near Inverness and every one of them differs slightly so that five conflicting azimuths for the direction of the passage exist. Fraser (1884: 343) showed it facing SW, 219°; Somerville (1923: 221), 213.5°; Piggott (1956: 195), 209°; Henshall (1963: 365), 212°; and Thom (1966: 27), 216.5°. Jolly (1882: 310) casually planned it at 242°; Somerville recorded the altitude of the skyline as 3°30', Thom as 1.7°. Such contradictions emphasise the uncertainties attendant upon examining individual sites where an error of even 1° must seriously distort any astronomical analysis.

7.3 The variety of orientations, celestial and non-astronomical

Another, perhaps even more considerable difficulty for the archaeo-astronomer is the question of whether there were orientations which were directed towards non-celestial targets.

Ashbee (1974: 71) has pointed out that the entrance-graves on the Scilly Isles pointed to many different parts of the compass and although the dominant orientation was easterly the reason did not seem to be astronomical. 'The chamber tombs are generally set on slopes and entrances were seemingly contrived, frequently, but not always, to face down the slope and give an optimum view of the structure. From the facts of orientation all that could be said is that the dominant choice of axial and entrance direction might reflect some specific desire of the builders'. A comparable wish for the monument to be conspicuous may have determined the situation of tombs on Rousay Island in the Orkneys (Davidson, Jones & Renfrew 1976: 360).

Another factor that sometimes affected alignments was the lie of the land. The long barrows of Cranborne Chase have already been mentioned. Their builders were compelled to lay out the mounds NW-SE because of the ridges upon which they were erected (Ashbee 1970: 24) although even here 'the long barrow incorporated in one side of the Dorset Cursus is roughly at right angles to the normal trend. This, like the long barrow at the eastern end of the Stonehenge Cursus, provides evidence that when circumstances demanded, drastic departures could be made from a uniformity dictated, to some extent, by nature'. As these long barrows were almost certainly built before the cursuses their unusual orientations were probably determined by factors not now apparent to us.

In Yorkshire similar barrows were positioned to overlook river-valleys or to stand by streams, often 'arranged in lines along the elevated ground on the brow of these escarpments' (Mortimer 1905: xxiii).

With such complications it might not be possible to separate and distinguish the factors that affected the layout of an individual monument. Orientation may have been only one of several considerations that determined the location and design of a tomb and conflict between these needs may often have ended in compromise. It is quite possible that absolute accuracy was sometimes relaxed to admit equally important requirements such as a conspicuous position or a situation near water. Such modifications will not always be obvious in a single cairn. Only study of a large group will reveal departures from the norm.

Dealing not only with a variety of structures but with structures that were built centuries apart during which time beliefs may have been affected by the coming of new people, by social crises, by technological and climatic change and by the increasing complexity of society it would be naive to detach a monument from its cultural environment and 'explain' it as though it had existed in solitude. Group study is vital. And even where a group has a homogeneous cluster of azimuths care must be taken not to attribute an astronomical explanation for them without considering other possibilities.

The Irish passage-grave cemetery of Carrowkeel in the Bricklieve Mountains of Co. Sligo contains tombs whose passages face NNW (fig. 7.1). Cairn G, a cruciform-celled tomb, has a bearing of 328° so close

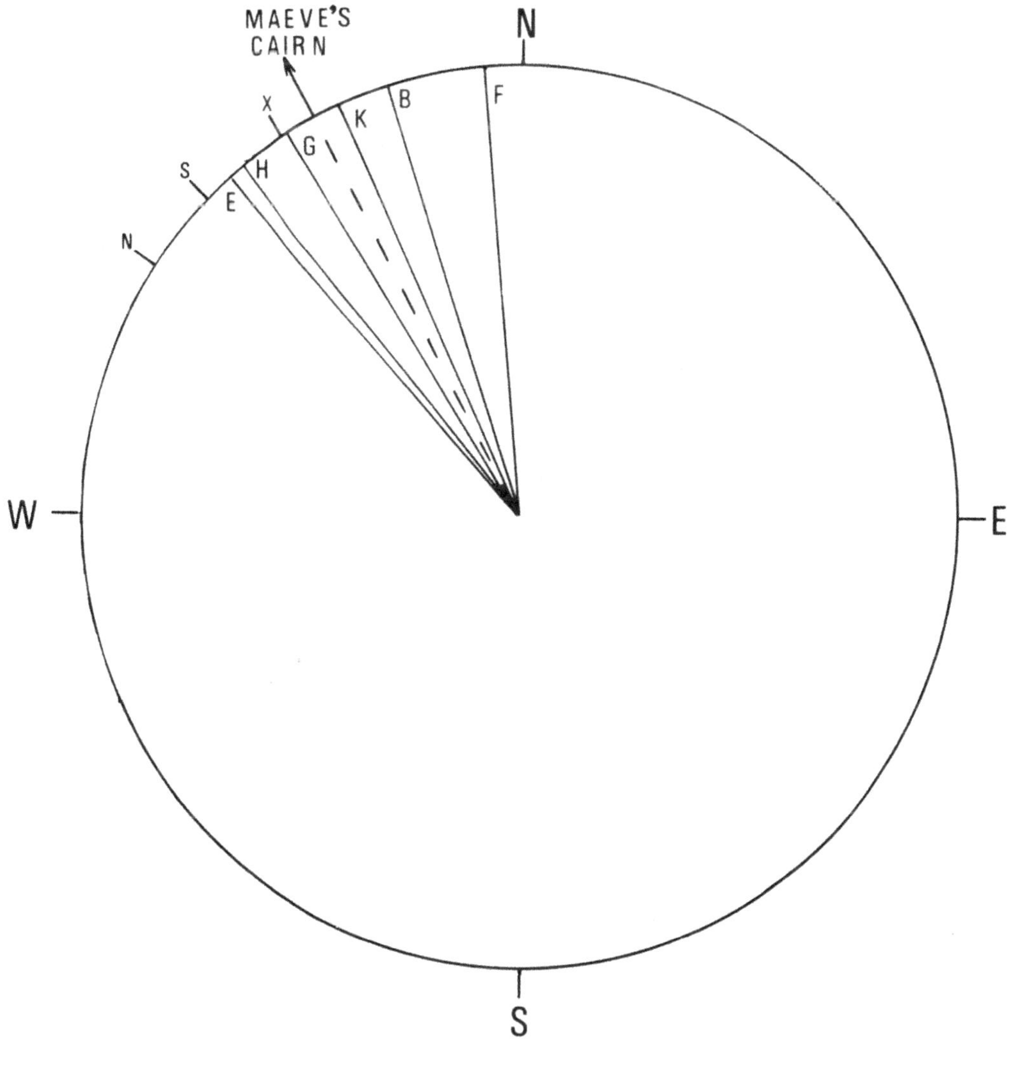

Fig. 7.1 Azimuths of Cairns B, E, F, G, H & K in the Carrowkeel cemetery. X = major lunar standstill; N = minor lunar standstill; S = midsummer sunset.

to an alignment on the maximum moonset at about 325° that one might assume that this had been an attempt on the part of the builders to align their tomb on a lunar event.

The azimuths of the other Carrowkeel tombs, however, including the dual-court long cairn (Herity 1974: 273), do not conform to a lunar pattern. At least three of them (Cairns B, F, K) are aligned too far to the north and Patrick (1975: 13) has shown that all of the sites were probably built to look towards the enormous kerbed passage-grave of Maeve's Heap on Knocknarea Mountain 17 miles (27 km) to the NNW. This gigantic flat-topped mound at G 626346 is indeed visible from the Carrowkeel necropolis at G 755115 but although no simple astronomical explanation is tenable for their alignments the tomb-passages are far from precisely aligned on Misgaun Mebn, the cairn of Maeve, an erotic goddess who was never 'without one man in the shadow of another'. Cairn F's passage bears 23° too far to the north and the angled passage of Cairn H was as much as 32° misaligned to the south.

Such approximations imply a far less demanding attitude towards orientation and, perhaps, far less visual ability amongst the builders of these tombs than has been claimed for the numinous astronomer-priests who designed the unerring observatories of the Bronze Age. Yet the Carrowkeel alignments should not be disregarded because of their imprecision. There is evidence elsewhere that tombs built later in the Neolithic period may also have been laid out to look towards the places where ancestors were buried.

In the Orkneys the passage-graves of Cuween and Wideford Hill face each other across three miles of low-lying, fertile country. Cuween, perhaps the earlier and not quite aligned upon Wideford, may have been oriented towards the equinoctial rising of the sun but Wideford has an atypical westerly azimuth of 274° and looks directly towards Cuween. Other 'secondary' tombs in a group may also have been built to face non-astronomical targets such as this, their axis determined by social constraints that are now obscured by time and lie almost undetectable amongst the scatter of azimuths that confront us.

There is a corollary to this which adds to the complicated possibilities that have to be considered in the interpretation of tomb orientations. If some tombs were built so that their entrances were aligned upon the burial places of ancestors then it is arguable that the earliest were planned to face the direction from which the first settlers had approached the territory. This has been suggested for some of the Clyde megalithic long mounds of south-western Scotland. 'If the cairn were the means whereby the dead returned to join their ancestors, then the tail of the cairn might be taken to point in the direction from which those ancestors were assumed to have come. Thus the carrying of the body through the entrance into the burial chamber could be the start of the spirit's journey in the direction of the land of the ancestors' (Scott 1969: 233).

Scott gives several instances to support this idea which also involves the depressing possibility that alignments might have been laid out not only from the chamber outwards but also from the entrance and along the passage into the chamber. Indeed, bodies were sometimes oriented 'back to front' because of this belief. 'It used to be claimed by ethnographers that the corpse was always oriented in the direction in which the spirit was to travel, a direction which correlated both with the position of the land of the dead and the home of the people's forefathers' (Ucko 1969: 272). But as if to terrify would-be researchers with the permutations of funerary custom recorded amongst primitive societies Ucko (ibid.: 273) went on, 'I fondly imagine that Ashanti practices would contribute to an archaeologist's nightmare, for in Ashanti there is a general rule that the body should not face the village; however, there are some Ashanti who say that immediately after burial the body turns itself round to face the village. Some, but not all, Ashanti therefore bury the dead facing the village knowing that the body will turn itself round and will therefore eventually conform to the rule of facing the forest'. Such human perversity cannot commend itself to the archaeoastronomer whose faith has been nurtured and shaped by the constant and dependable motions of the sun and moon.

In Britain there are tombs whose orientation may not have been the most obvious one looking out from the entrance but was reversed. Cotswold-Severn barrows such as Nympsfield and Hetty Pegler's Tump were built at the very edge of the escarpment that overlooked the lush Vale of Evesham, perhaps so that their mounds would be conspicuous but maybe so that the dead could more easily find their way down into that green and well-watered landscape.

On Arran the passage at Torlin looks beyond the end of the chamber towards the pyramidal island of Ailsa Craig. At Newgrange, although the entrance faces the midwinter sunrise, the opposite alignment down the passage leads NW towards the decorated kerbstones 50, 51 and 52 on the far side of the mound in line with the midsummer sunset. These stones are marvels of megalithic art, their position emphasised by the carvings of 'serpents', axe-polishings, zigzags and the lavish wonderment of the motifs on K52, an abundance of composition made all the more noticeable by the uncarved plainness of the kerbstones on either side of the group (O'Kelly 1978: 23).

At Maeshowe, the elegant passage-grave on Orkney, the passage facing SW may have been aligned on the midwinter solstice although the major axis of 221° is more truly in line with sunsets in late November. It is just as feasible that the builders laid out the passage to look towards the NE where as Spence (1894: 12) remarked, 'it points unequivocally to the point where the sun rises on midsummer morning'. Lockyer's calculations here (1909: 128) of a horizon height, allowing for refraction, of 1.26° and an azimuth of 41.2° produce a declination of 23.98°. This corresponds closely with the solar declination of 23.979° in 2500 BC (Hawkins 1966: 17), a date which itself agrees quite well with the two radiocarbon determinations obtained from peat in Maeshowe's ditch of 2185±65 bc and 2020±70 bc (2770-2570 BC).

Such examples show the need to examine alignments in both directions. Even the famous row at Ballochroy, so often mentioned as a solar observatory, may have had a lunar significance. Before the construction of the cairn at the SW the row did point roughly towards the midwinter sunset but to the NE where the sightline was never interrupted by any man-made structure the line just as neatly indicated the moon rising at its major standstill, appearing dramatically above the steep slope of the hill nearby (Thom 1967: 142).

So many factors conditioned the orientation of chambered tombs and stone rows that it is not always the one most apparent to us that supplied the original impulse. Maeshowe, already referred to as a possible example of a tomb whose passage may have been aligned on the midsummer sunrise, offers an insight into the subtleties with which we may be working. Spence (1894) was the first to suggest that its passage and entrance faced the midwinter sunset. Lockyer (1909: 35) considered Spence's computations had 'no pretension to scientific accuracy' but Spence may have been partly correct. Brown (1975) claimed to have seen the midwinter sunset inside Maeshowe although this was possibly as a reflected glow rather than a direct beam.

From the back of the chamber to the end of the 3 ft (1 m) wide passage is 52 ft (16 m), but although the passage is 4 ft (1.2 m) high it slopes gently downwards and is a foot (0.3 m) lower at the entrance thus reducing the observable altitude for anyone in the chamber to 0.5°. The skyline, therefore, would have been hidden. This stretch of passage, moreover, is aligned 221° and the resulting declination would have been about -22.4°, quite a long way from that of the solar solstice.

Yet at the entrance the finely-shaped portal slab is a little too low to block the opening completely. When dragged into place from its recess in the passage it would have left a rectangular aperture 3 ft wide and 1 ft 6 ins high (1 m x 0.5 m) just below the lintel rather like the roofbox at Newgrange. Beyond it an uncovered short length of passage was skewed slightly to the SW (216°) towards some low hills over which the rays of the midwinter setting sun would have shone through the 'letter-box' and down the passage to a point where two pillars like jamb-stones stood at the entrance to the chamber (Henshall 1963: 220). It is to be wondered whether such refinement was intended by the builders of this majestic tomb and shrine.

There is no space to discuss in detail what other elements affected the ground-plans of these early burial places and centres of ritual. Some may have been aligned towards prominent parts of the landscape. Hills such as the Paps of Anu in Co. Kerry and Mither Tap in Aberdeenshire were named after Iron Age goddesses as were Ballochroy's Paps of Jura and it would not be surprising if some megalithic alignments had been directed towards natural features such as these. Lynch (1975: 124) has commented on the visual effect of Carn Meini, the source of the Stonehenge bluestones in the Preseli Mountains, 'an amazing coronet of stone which dominates the landscape for miles around. Whenever you are in sight of it your eye is drawn towards it with a fascination that is uncannily compelling'.

Years ago Lilian Chitty (1926) observed that three gaps in the Black Marsh stone circle in Shropshire were in line with the positions of three tall hills and she speculated that the builders of the ring might have regarded these hills as sacred places. There is nothing improbable in this but even here there is a complication. From Thom's accurate plan (1967: 65, D2/2) it can be calculated that the SW gap towards the high saddle-back of Corndon Hill was also in line with the setting of the moon at its major standstill (latitude 52.59°; horizon, allowing for refraction (Thom 1967: 26), 2.8°; azimuth 210°; declination -29.1°). It suggests that the builders of Black Marsh may have combined the landscape and the sky in the design of this unobtrusive flattened ring with its central pillar. Here where the broad hump of Corndon glowers over the flatness of the plain there may be yet another instance of the niceties contained in what were once called 'rude stone monuments'. In his excavations at Cultoon stone circle on Islay MacKie noticed how the axis of the ellipse was in line both with the mountain of Slieve Snaght in Co. Donegal and with the midwinter sunset (MacKie 1977b: 104).

If the countryside did sometimes provide targets for alignments in chambered tombs and other megalithic monuments then this should appear

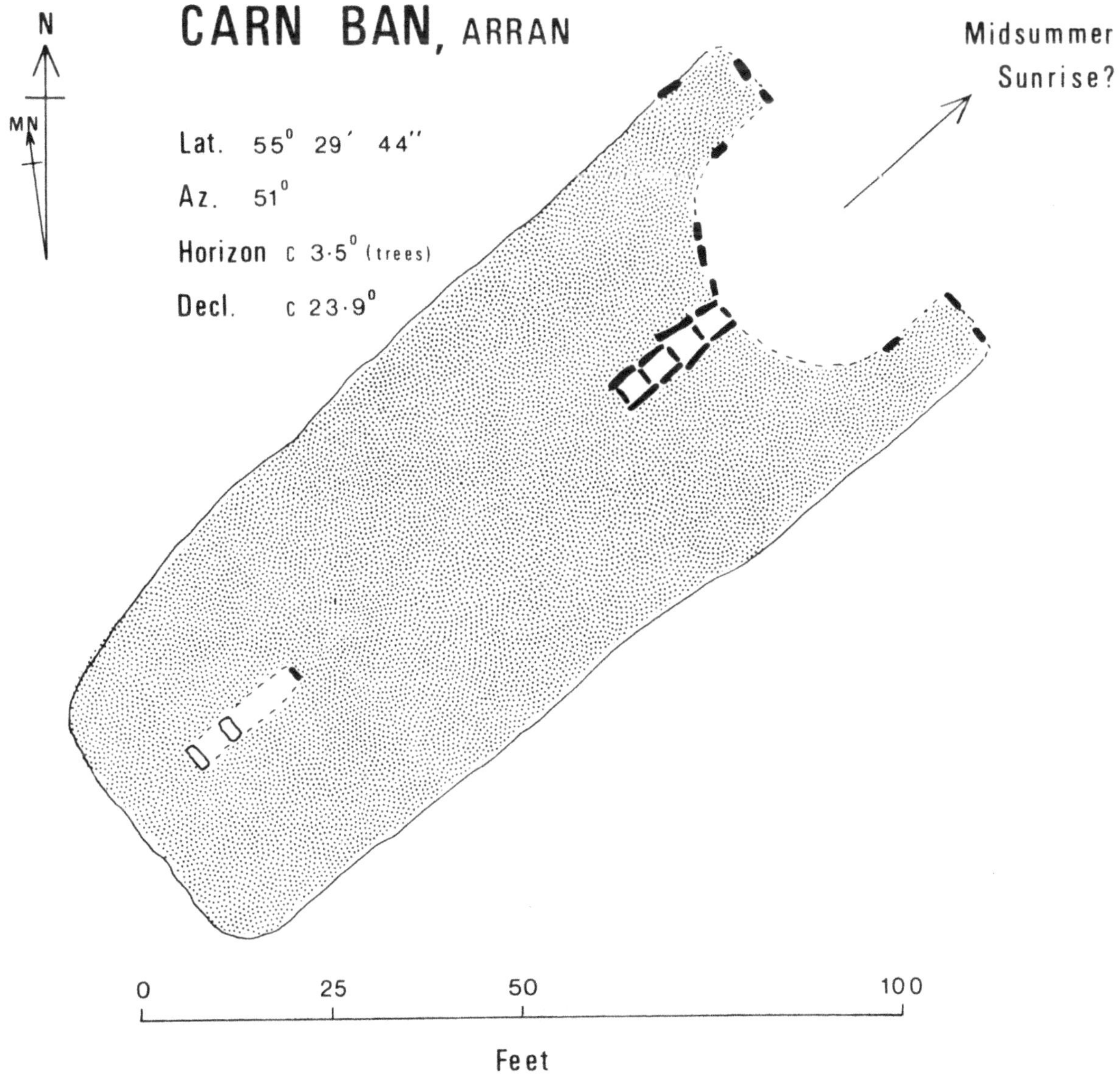
CARN BAN, ARRAN
N
MN
Lat. 55° 29′ 44″
Az. 51°
Horizon c 3·5° (trees)
Decl. c 23·9°
Midsummer
Sunrise?
0
25
50
100
Feet

Fig. 7.2

Table 7.1 Azimuths of Arran Tombs, based on fieldwork in 1979. (Arran (ARN) numbers taken from Henshall (1972).)

Site			Azimuth	Site			Azimuth
ARN	1	Sannox	297	ARN	11	Giants Graves I	5
ARN	2	Moinechoill	100	ARN	12	Dippen	2
ARN	4	Tormore I	36	ARN	13	Baile Meadhonach	109
ARN	5	Tormore II	54	ARN	14	East Bennan	297
ARN	6	Glenrickard	164	ARN	15	Torlin	344
ARN	7	Dunan Beag	18	ARN	16	Clachaig	336
ARN	8a	Dunan Mor A	169	ARN	17	Sliddery Water	327
ARN	8b	Dunan Mor B	272	ARN	18	Carmahome	256
ARN	8c	Dunan Mor C	26	ARN	19	Giants Graves S	294
ARN	9	Monamore	232	ARN	21	Moss Farm	99
ARN	10	Carn Ban	51	ARN	22	Machrie Water N	157

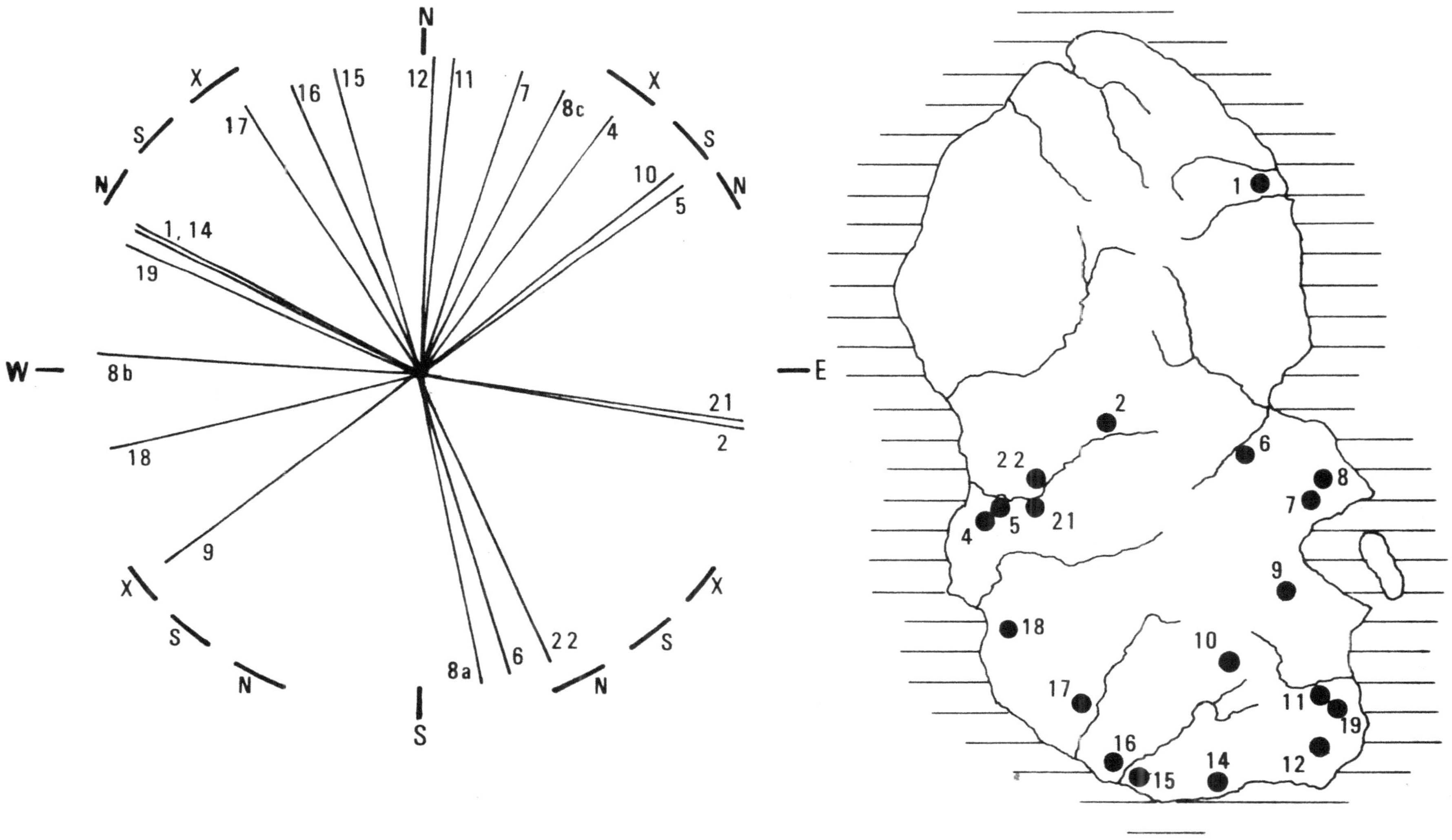

Fig. 7.3 The azimuths of chambered tombs on Arran.

in as consistent a pattern as the astronomical orientations. There should be groups of sites that point towards a particular hill or look down a valley or to the source of a river or spring. The writer does not know of such a group although many of the Clava Cairns in north-east Scotland do seem to look towards mountain peaks. In the jagged skylines of their territory this might be inevitable. Again, only the detailed study of a whole group would show whether such orientations were accidental or intended.

7.4 The need to study groups

With so many possibilities available to explain why a chambered tomb faced in a particular direction, including the dismaying thought that there never had been a special reason, it is clear that any single site might afford a misleading result. Carn Ban, a Clyde chambered long cairn on Arran, is an excellent example of this (fig. 7.2). It was excavated in 1902 and although virtually nothing was found in its segmented chambers the excavator made the intriguing suggestion that its deeply crescentic forecourt might have been extended to form a stone circle by the addition of a semi-circle of standing stones to its NE (Bryce 1903: 39), a tantalising hint of the transition from one type of ritual monument to another.

The rectangular passage of this cairn faces uphill to the NE, 51°, and this azimuth conjoined with a latitude of 55° 29' 44" and a horizon altitude, allowing for refraction, of about 3.5°, the skyline now being obscured by trees that cover the sharply rising ground, produce a declination of $+23.96^{\circ}$. This is so close to that of midsummer sunrise that it is tempting to visualise the builders aligning the sideslabs of the passage on the sun, aware that every year at midsummer the rays of the dawning sun would shine onthe entrance to the tomb and into the sacred stone circle in front of it.

It is tempting but it may be wrong. Nineteen other Clyde tombs exist on Arran and not one of them has an azimuth facing midsummer sunrise which, depending on the latitude and horizon, would vary between 44° and 51° (see table 7.1). In fig. 7.3 all the azimuths are shown, three being given for Dunan Mor because this round cairn contained three separate chambers, each aligned differently. The respective rising and setting positions for the sun (S) and the moon at its major (X) and minor (N) standstills are drawn as arcs to show how their azimuths would vary according to a horizon altitude from 0° up to 4° (for a nearby hill).

At least half these tombs did not face any significant solar or lunar position. Indeed, the azimuths of ARN 7, 8a, 8c, 11, 12 and 15 are so far to the north or south that the sun and moon could never have risen or set there. From this it would seem that Carn Ban's 'alignment' may be fortuitous just as there may have been an illusory orientation towards the minimum setting moon at Monamore.

The Arran azimuths are unusually dispersed for a group of chambered tombs with no obvious concentration towards any part of the horizon and it is possible that their entrances were laid out to point towards passes through the hills or nearby sources of water or even to face another tomb. Nor should Scott's suggestion that their mounds indicated the direction of the first colonists be forgotten.

This does not invalidate the necessity for the archaeoastronomer to study groups. Instead, it highlights the uncertainties that accompany the attempt to understand the prehistoric mind. Fortunately, examination of another, more tightly-related group is more rewarding.

7.5 The Clava Cairns, north-east Scotland

These megalithic monuments consist of passage-graves and ring-cairns both of which are unusually surrounded by stone circles in which the stones are graded in height. The cairns also have an uncommon orientation, the entrances of the passage-graves and the biggest kerbstones and tallest pillars in the circles of the ring-cairns virtually always occurring in the SW quadrant.

Thom (1966: 18) and Somerville before him (1923: 220) suggested that midwinter sunset could have been observed down the passages of the two chambered tombs at Balnuaran of Clava near Culloden battlefield. Somerville added that at each site and at the ring-cairn that lies almost between them there is a circle-stone standing exactly at 233° marking the place of sunset at Samain or Hallowe'en at the end of October, a night which later was to be the occasion of the most important of the four Celtic festivals and the beginning of the new year. In his own survey Thom noticed that the complex of three cairns seemed 'to contain pointers to all four solstitial rising and setting points'. He also deduced that midwinter sunset was indicated at some other Clava sites, namely the Daviot ring-cairn (Thom 1967: 76) and at the passage-grave of Druidtemple (ibid.: 69, 143) although at the latter the passage is a good 28° to the east of the circle-stone concerned. With such research already undertaken the Clava Cairns were attractive candidates for further investigation.

Even with their uncharacteristic architecture these cairns are convenient to study because a corpus of plans is available in Henshall (1963: 341, 358-91) and the accuracy of individual sites can be checked against Beaton (1882), Cash (1906; 1910), Coles (1907), Fraser (1884), Piggott (1956), Somerville (1923), Thom (1967) and Woodham (1956). From these plans the probable azimuth for each tomb has been calculated, taken from Thom where possible. Although for the reasons stated in section 7.2, and because of plan-distortion during printing each azimuth can be considered accurate only to within ±2° this should be adequate for an overall view of the astronomy of the group (table 7.2).

Assuming that the majority of these cairns were built between 4000 and 3000 BC (3225-2325 bc) the relevant declinations for the sun and moon are: sun, 24.112° to 24.027°; maximum moon, 29.262° to 29.177°; minimum moon, 18.962° to 18.877° (Hawkins 1966: 17). In this paper the intervening declinations for 3500 BC have been used and converted into azimuths for the mean Clava latitude of 57° 42'.

Figs 7.4a (passage-graves and 7.4b (ring-cairns) show the azimuths of these tombs. They range from 184° to 233° for the setting of a celestial object and from 172° to 178° for its rising. From these azimuths horizon heights can be estimated from the O.S. 1:50,000 maps and allowance made for refraction. The declinations consequently obtained present a consistent pattern (table 7.3).

Particularly amongst the chambered tombs whose entrances define the orientation much more precisely than is allowed by the passage-less ring-cairns there is a very clear correlation between the declinations and solar and lunar settings (fig. 7.5). Even in the ring-cairns this pattern is repeated albeit with some aberrant results at sites 16, 22, 23, 27 and 28. Examination in the field might remove these anomalies.

The effects of horizontal parallax have not been allowed for, nor has the question of whether it was the last gleam or the full orb of the sun and moon that was observed. The good fit of the solar declinations quoted

Table 7.2 The Azimuths of Clava Cairns in Inverness-Shire and Other Counties

		Henshall	Other Surveyors	Azimuth Chosen
(a)	Passage-Graves			
1	Avielochan	178	Cash (1910) 175	178
2	Balnuaran of Clava NE	207	Fraser 221, Piggott 208, Somerville 215, Thom 216.5 (B7/1,3)	216.5
3	Balnuaran of Clava SW	212	Fraser 219, Piggott 209, Somerville 213.5, Thom 216.5 (B7/1,2)	216.5
4	Carn Daley	226	—	226
5	Carn Urnan, Ross	200	Beaton 187, Woodham 193	200
6	Corrimony	228	Piggott 232	228
7	Croftcroy	191	Fraser 177, Thom 184 (B7/17)	184
8	Dalcross	227	Piggott 227, Thom 232 (B7/6)	232
9	Druidtemple	174	Fraser 175, Piggott 175, Thom 172 (B7/18)	172
10	Kinchyle of Dores	179	Beaton 193, Fraser 192, Piggott 184, Thom 192 (B7/19)	192
11	Upper Lagmore, Banff	174	Coles 175	174
(b)	Ring-Cairns			
12	Aviemore	209	Cash (1906) 197, Thom 203 (B7/12)	203
13	Bruaich	191	—	191
14	Balnuaran of Clava Centre	195	Fraser 204, Piggott 197, Somerville 201, Thom 200 (B7/1,1)	200
15	Culburnie	208	—	208
16	Culdoich	228	Fraser 222, Piggott 224, Thom 228 (B7/2)	228
17	Culduthel	169	—	169
18	Daviot	204	Fraser 184, Thom 197 (B7/5)	197
19	Delfour	231	Cash (1906) 218, Thom 219 (B7/10)	219
20	Doune of Dalmore, Banff	-	Davidson 204 (pers. comm.)	204
21	Gask	249	Fraser 198, Thom 215 (B7/15)	215
22	Grenish	230	Cash (1906) 214, Thom 221 (B7/13)	221
23	Little Urchany, Nairn	232	Thom 221 (B6/1)	221
24	Marionburgh, Banff	202	Coles 204	202
25	Moyness, Nairn	213	Thom 203 (B6/2)	203
26	Newton of Petty	233	Fraser 214	233
27	Raigmore, Stoneyfield	208	Fraser 183	208
28	Tordarroch	217	Fraser 204, Thom 205 (B7/16)	205

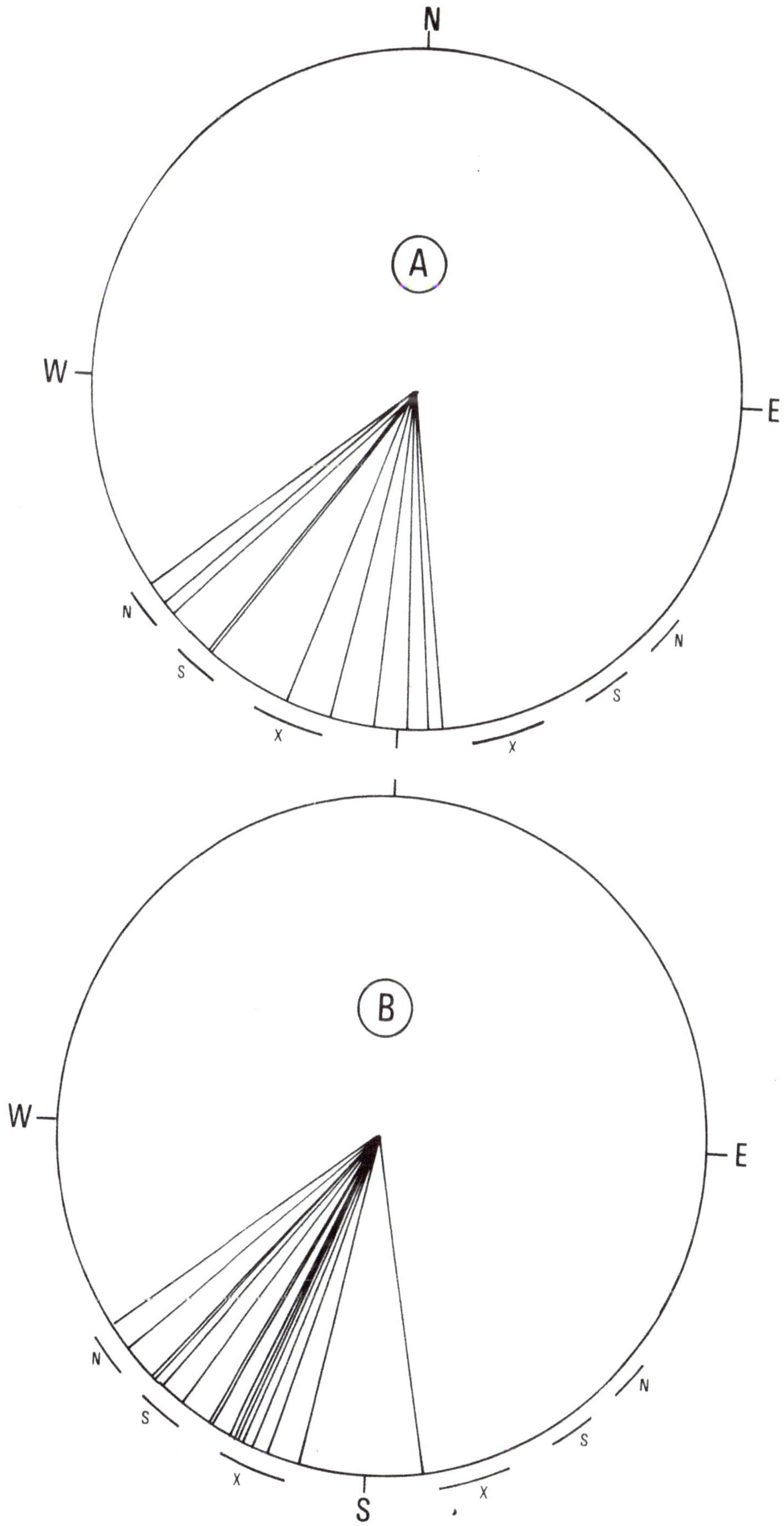

Fig. 7.4 The azimuths of Clava Cairns. (A), passage-graves; (B), ring-cairns. X = major lunar standstill; N = minor lunar standstill; S = midwinter sunset.

Table 7.3 Declination in Clava Cairns

		Latitude	Horizon	Azimuth	Declination
(a)	Passage-Graves				
1	Avielochan	57.23	4.0	178	-28.75
2	Balnuaran of Clava NE	57.47	1.7	216.5	-24.02
3	Balnuaran of Clava SW	57.47	1.7	216.5	-24.02
4	Carn Daley	57.35	3.3	226	-19.01
5	Carn Urnan	57.54	0.6	200	-29.7
6	Corrimony	57.34	2.5	228	-18.9
7	Croftcroy	57.37	3.4	184	-29.14
8	Dalcross	57.50	0.5	232	-18.87
9	Druidtemple	57.44	2.5	172	-29.71
10	Kinchyle of Dores	57.42	1.8	192	-29.98
11	Upper Lagmore	57.40	4.0	174	-28.4
(b)	Ring-Cairns				
12	Aviemore	57.20	0.7	203	-29.23
13	Bruaich	57.44	3.3	191	-28.6
14	Balnuaran of Clava Centre	57.47	3.9	200	-28.54
15	Culburnie	57.44	4.0	208	-24.5
16	Culdoich	57.47	1.0	228	-20.18
17	Culduthel	57.44	3.6	169	-28.3
18	Daviot	57.45	2.1	197	-28.9
19	Delfour	57.16	2.0	219	-23.07
20	Doune of Dalmore	57.35	0.5	204	-29.04
21	Gask	57.39	1.9	215	-24.4
22	Grenish	57.22	3.5	221	-20.89
23	Little Urchany	57.52	1.8	221	-22.2
24	Marionburgh	57.41	6.0	202	-24.1
25	Moyness	57.56	1.2	203	-28.4
26	Newton of Petty	57.52	1.0	233	-17.96
27	Raigmore, Stoneyfield	57.48	1.2	208	-27.19
28	Tordarroch	57.37	2.6	205	-26.74

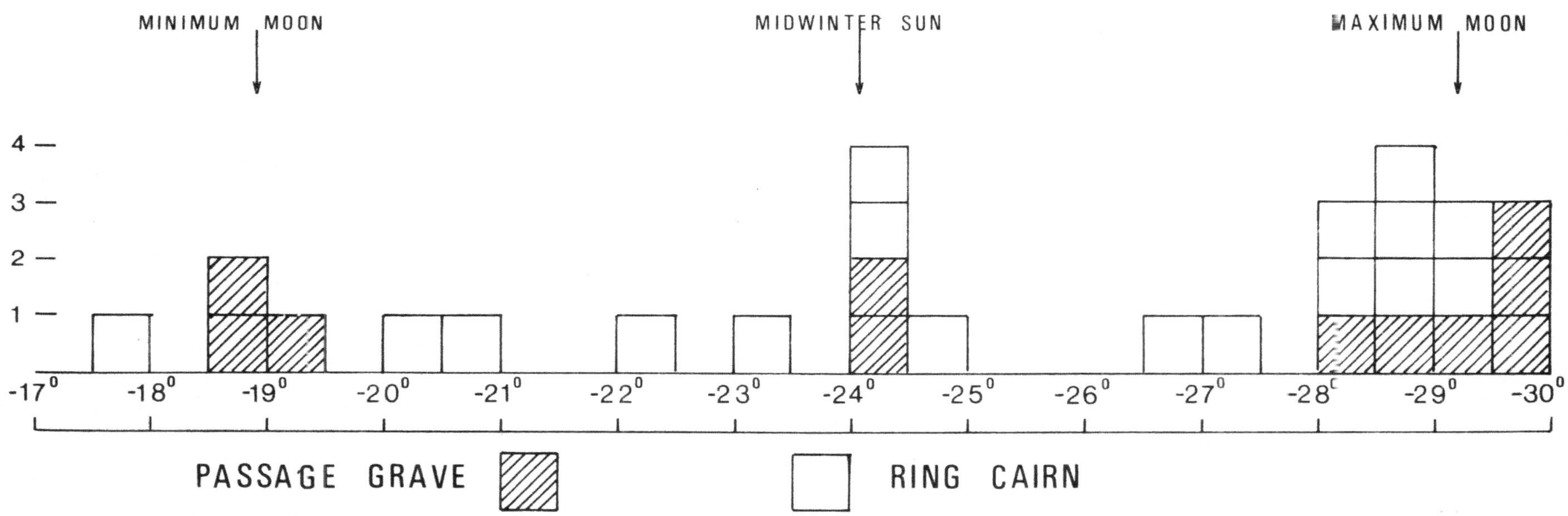

Fig. 7.5 Histogram of the declinations of Clava Cairns.

here suggests that it might have been the last gleam that was preferred but only fieldwork at each site can provide data accurate enough for these problems to be considered properly.

The three passage-graves which seem to be aligned not on the setting but the rising moon—nos. 1, 9, 11, Avielochan, Druidtemple and Upper Lagmore—are not typical of the group. Avielochan had no stone circle around it. Druidtemple is the only Clava tomb whose megalithic ring and cairn were of different shapes (Burl 1972: 35), and Upper Lagmore is far to the east, unusually situated on a hillside and separated from the main group by over twenty miles of mountain and forest. Little can be said about a fourth site, no. 17, Culduthel, with its azimuth of 169°. It was so badly wrecked in the 19th century that today only a few of its kerbstones remain. One might remark, however, that the astronomical significance of these anomalous sites only really becomes apparent in the context of all the other cairns.

Several points arise from this study. The first and perhaps the most important is that these Clava declinations suggest that the major-minor cycle of the moon was known in Neolithic times, something already recognised from the Stonehenge I stakeholes but not previously suspected for megalithic tombs some of which must have predated the Late Neolithic henge on Salisbury Plain by several centuries. The discovery of the moon's most southerly rising and setting positions, its major standstill, would not have been exceptionally difficult to establish 'but the second extreme direction of the moon, at its "minor standstill"...would be far more difficult to identify. It is neither the most northerly nor the most southerly direction in which the moon ever sets. In theory it could have been found as the most southerly of those directions at which the full moon sets at midwinter' (Thatcher 1976: 144) but if its alignment really were discovered in this way the builders 'would have had to make a very long series of observations indeed'. If this demanded the patience of several generations then the moon must have been a very compelling object indeed.

The second point is that the Clava passage-graves, and perhaps also the ring-cairns, were not observatories. Although Dalcross (no. 8) and Druidtemple (no. 9) were too badly damaged for reconstruction the remaining nine chambered tombs were in good enough condition for their excavators to record the heights and lengths of their passages. In six the roof was too low for the alignment to have been seen by anyone inside the chamber even if he had been sitting with his eye-level, as is likely with a Neolithic observer, about 2 ft (0.6 m) above the floor (table 7.4).

In six cairns (nos. 1, 4, 5, 6, 7, 11) the horizon was completely obscured so that no one inside the tomb would have had even a glimpse of the setting moon or sun. Only at the Balnuaran sites (nos. 2, 3) and at Kinchyle of Dores (no. 10) was the passage high enough to show the skyline. Even in these three cairns, like the others, the entrances were too low and narrow for anyone to follow the descent of a celestial body down the sky and it would have been quite impossible for these places to have been used as observatories for the calculation of eclipses or lunar perturbations. This suggests that, like Newgrange, the orientations in them were intended to bring the dead into conjunction with the moon or the sun.

It is also worth noting that the declinations in these Clava tombs are quite different from those in the Camster passage-graves in the same region. Although the majority of these cairns with stalled side-cells occupy regions as far north as the Orkneys there are some southerly outliers in

Table 7.4 Entrance Altitudes in Clava Passage-Graves

Site no.	A	B	C	D
1	19' 0"	2' 7"	4.0°	1.8°
2	26' 3"	4' 0"	1.7°	4.4°
3	25' 9"	4' 0"	1.7°	4.4°
4	18' 5"	2' 3"	3.3°	0.8°
5	20' 3"	2' 0"	0.6°	0.0°
6	29' 0"	3' 0"	2.5°	2.0°
7	13' 0"	1' 6"	3.4°	-2.2°
10	15' 5"	3' 6"	1.8°	5.6°
11	21' 6"	3' 0"	4.0°	2.7°

Key

A Length of passage to centre of chamber
B Height of passage
C Altitude of skyline
D Altitude observable from centre of the chamber

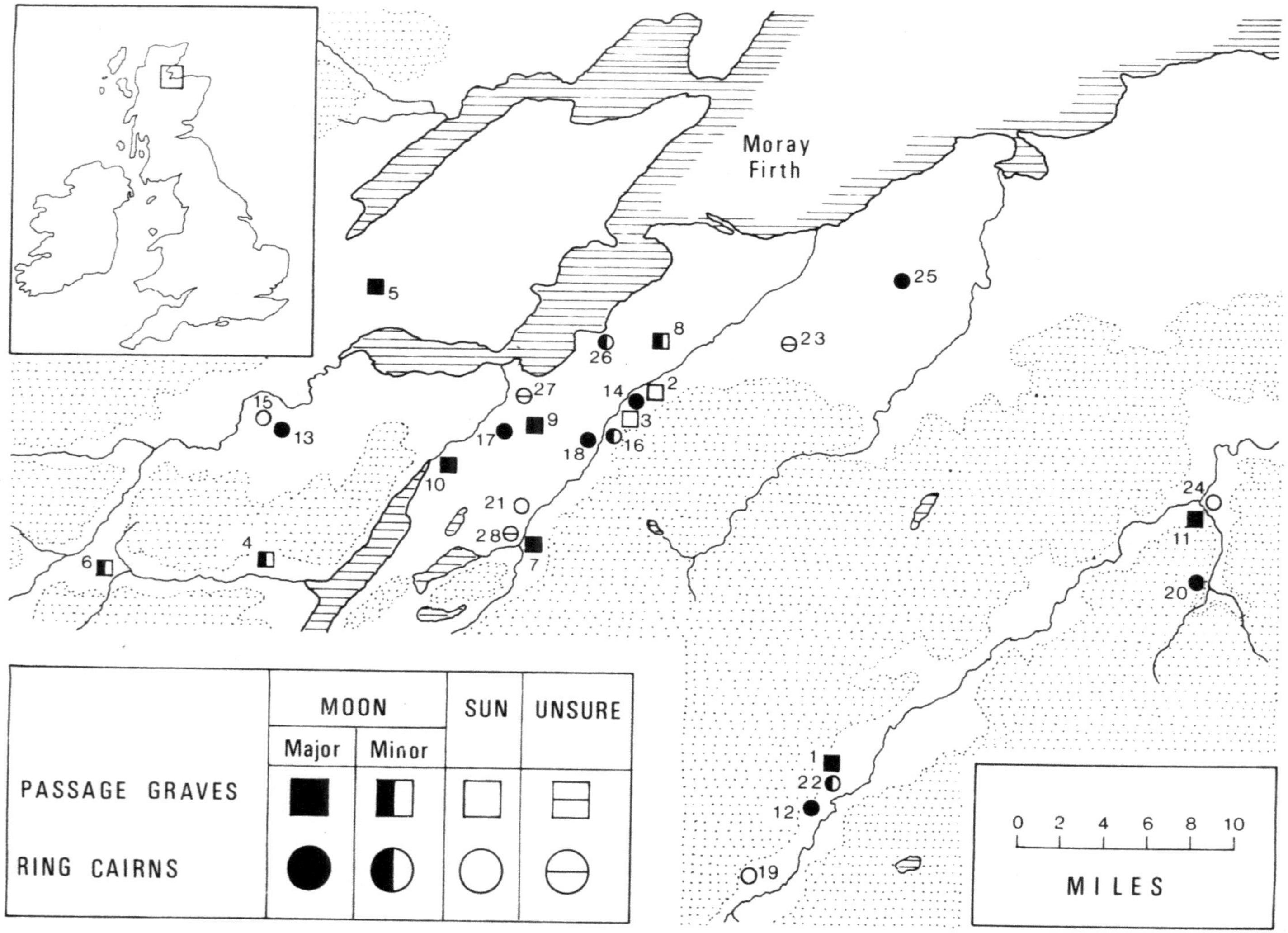

Fig. 7.6 Clava Cairns. The distribution of lunar and solar sites.

Ross-shire and in them the passages face eastwards (Henshall 1963: 333-57) with declinations ranging from +18.5° to -8.2°. Further south still, in Inverness-shire, the easterly orientation is even more pronounced. Leachkin (ibid.: 381) has an azimuth of about 92° and a declination of -0.6°; Reelig (ibid.: 384), 89° and +2.6°; and even in the long-multiphase mound of Essich Moor (ibid.: 376) the central passage faces 112° with a declination of about -10.3°. The significance of such declinations must be the subject of another paper. Here it is sufficient to remark on the very distinct differences in the orientations of two overlapping groups of chambered tombs that were, presumably, of roughly the same date and which to some extent shared the same territory.

At present there is no direct radiocarbon date for any Clava Cairn although there are two assays from materials which predate the ring-cairn of 3033±130 bc and 2782±90bc (c. 3690 BC) from Raigmore. In the absence of any chronological framework there is little value in making deductions about the distribution of lunar and solar sites in this group (fig. 7.6). No area contains cairns of only lunar or solar interest and there are several instances in which there are examples of both with major and minor lunar and also solar oriented cairns quite close together. Whether there is any significance in such clusters is a problem for further research. It is a great pity that because of the destruction of cairns in the last two hundred years it is not possible to analyse the astronomical relationships of a three-quarter mile long line of eight cairns from the northerly Mains of Clava SW down to Culdoich (Henshall 1963: 21). They form a linear cemetery akin to those in Wessex and in the Kilmartin Valley of Argyll. With the famous Balnuaran of Clava sites at their centre it would have been interesting to know the proportion of lunar to solar orientations and of passage-graves to ring-cairns in this complex.

The writer, incidentally, was relieved to find that a SW extension of this rather irregular alignment would bypass a castle with the unfortunate name of Leys by a good two miles. It would also miss the four adjacent Leys farms by the same comforting distance before reaching the mountain of Creag a Clachain that stood in line with the midwinter sunset, a fusion of landscape and astronomy noticed in other regions.

The Clava Cairns are most closely related in architectural features and in astronomical alignments to the recumbent stone circles of Aberdeenshire the earliest of which were probably put up around 2500 BC when the Clava passage-graves must already have been old. These stone circles seem to have been aligned on the moon at its major standstill not when it was rising or setting but when it was up in the sky (Burl 1980). As only seventy miles of easily travelled coastlands separate these delightful rings from the Clava Cairns it is possible that in them we may perceive one of the many transitions from Neolithic to Bronze Age, from chambered tomb and burial place to open-air ritual centre, from a charnel-house enriched by the life of the moon to a sacred enclosure in which priests or magicians invoked the power of the moon for their own purposes.

7.6 'Something haunting in the light of the moon' (Joseph Conrad)

The Clava Cairns are not the only chambered tombs with a restricted range of azimuths and a thorough investigation might discover the reasons for the alignments in other groups. The Cotswold-Severn tombs (Daniel 1950: 80; Corcoran 1969) were oriented mainly towards the east. In the Shetlands the heel-shaped cairns (Henshall 1963: 156-82) were built with entrances between east and SSE and this was also largely true of the stalled cairns of the Orkneys (ibid.: 183-246). The builders of the Maeshowe tombs on the same islands (ibid.: 194-245) seemed to have had some interest in the cardinal points of east, south and west. On the Scottish mainland the Orkney-Cromarty Camster Cairns had passages facing anywhere between NE and SE (ibid.: 257-357) but, as has already been remarked, those in Ross shire and Inverness shire had a predominantly easterly orientation while, to the north in Caithness, the tombs looked to the SE. In south-west Scotland the Clyde long tombs (Henshall 1972: 98-100; Scott 1969) had passages aligned between NNE and ESE. 'The largest group, eight cairns, face between 41° and 50° (Henshall 1972: 99), perhaps aligned on the midsummer rising sun and this may be true also of the cognate court-cairns of northern Ireland (de Valera & O'Nuallain 1964: 110; 1972: 152). Rather differently, a slight majority of the double-ended dual-court cairns in the same Irish region had east-west alignments. The related centre-court cairns (O'Nuallain 1976) had similar east-west bearings. Unlike all these, the wedge-graves of western Ireland (de Valera & O'Nuallain 1961: 106; 1964: 113; 1972: 158) were almost invariably built to face westwards, anywhere between west and SSW.

It is likely that many of the chambered cairns in these widely separated groups were erected by people who were concerned to integrate the presence of the moon or the sun with the places that were to contain the bones of the dead, and, to the writer, it is this association with death that is significant about the astronomy of megalithic tombs. It is an association that really only becomes noticeable when a whole group has been studied. The same criterion may apply to rows of standing stones and other structures that have been interpreted as observatories for astronomer-priests but which, instead, may have been intended as links between the sun or moon and death.

This explanation may also be appropriate to other 'observatories' such as Kintraw where the larger cairn covered a peripheral cist, oriented NE-SW, containing burnt bone and wood and was surrounded by a concentration of white quartz (Simpson 1967: 57). Ballochroy also, with its cairn and cist, has obvious funerary associations and before accepting this row as an 'interesting solstitial site' of 'ample sensitivity' (Thom 1967: 151, 153) it would be instructive to learn what astronomical sightlines its many counterparts around the North Channel possessed (fig. 7.7).

As an aside one might add when discussing problems of interpretation that the variety of structures claimed as observatories seems incompatible with the hypothesis of a tightly-knit national priesthood whose measuring-rod and geometrical constructions were known and accepted from the Shetlands down to Land's End. With such a countrywide use of these standards a corresponding uniformity might be expected in the design of the observatories. If Britain were controlled by a theocracy whose members were required to have been 'meticulous observers of the night sky, to have calculated with numbers, and to have communicated sophisticated astronomical knowledge among themselves from generation to generation' (Hoyle 1977: vii) it seems incongruous that the observing-stations of these powerful astronomer-priests should be so dissimilar. Some consisted of a short line of stones (Ballochroy); of a cairn, a monolith and a platform (Kintraw); of fan-shaped stone rows (Mid Clyth); of parallel rows (Merrivale); an ellipse of standing stones (Cultoon); a stone circle and cairns (Ring of Brodgar); an arched circle (Stonehenge); pairs of kerb-cairns (Fowlis Wester); complexes of stones (Crois Mhic-aoida); chambered tombs (Balnuaran of Clava); timber buildings (Woodhenge); and megalithic temples (Stanydale). Indeed, except in

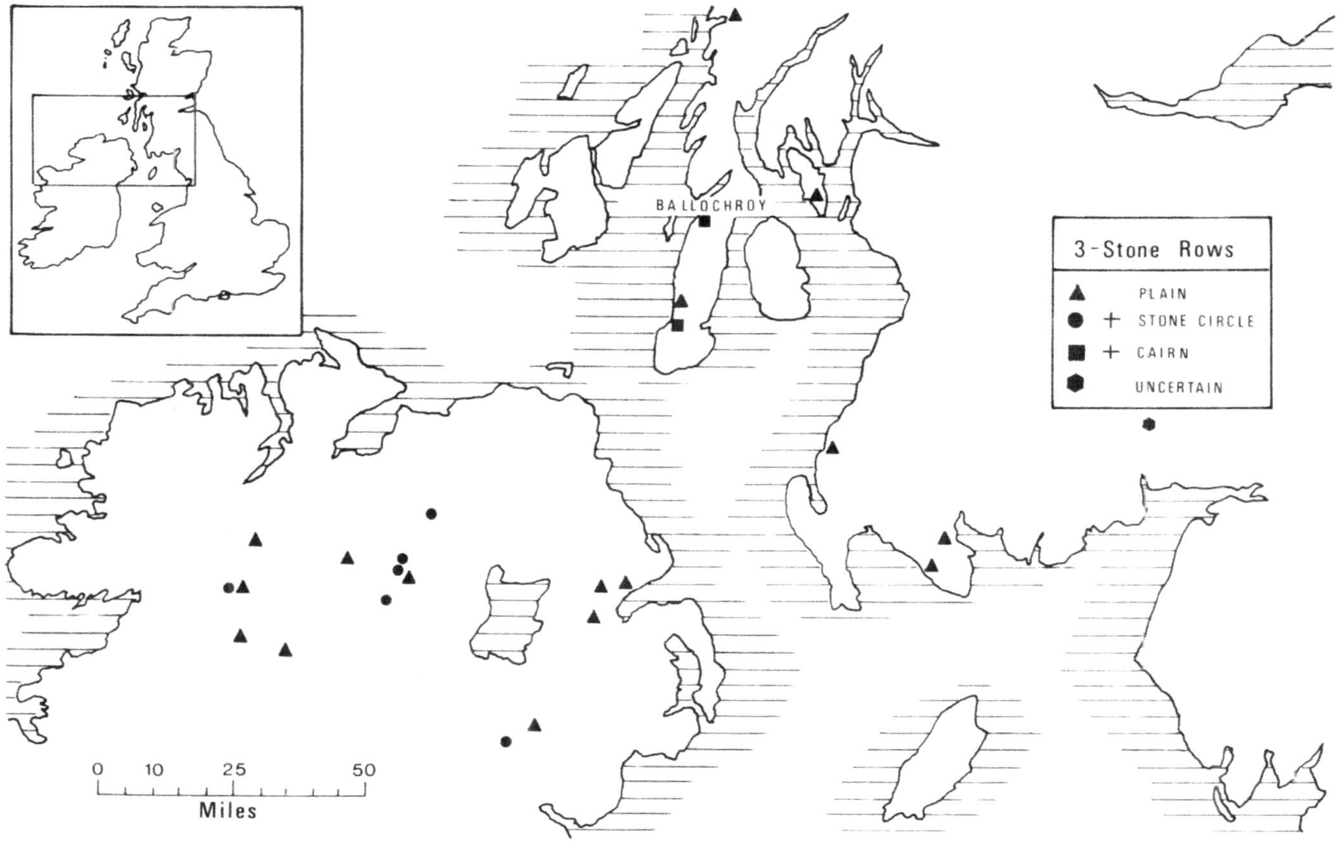

Fig. 7.7 The distribution of 3-stone rows around the North Channel.

the case of the multitudinous single standing stones it is rare to find even one 'observatory' that is like any other.

The same objection cannot be made against chambered tombs and when a group is found to possess consistent orientations towards the sun or moon it is probable that these alignments were deliberately fashioned by prehistoric people for reasons that we may now never recover completely but whose outlines can be deduced.

It could be argued that the astronomical analysis of Neolithic tombs is irrelevant to the question of Bronze Age observatories but such an objection implies a lack of continuity that is belied by the links between the Clava Cairns and the succeeding recumbent stone circles; between the chambered tombs, Bronze Age house-plans and the Stanydale and Yoxie 'temples' in the Shetlands; between the terminal entrances of chambered tombs and the Late Neolithic 'Coves' of Wessex; and by the preservation of a NE-SW axis at Stonehenge from the Late Neolithic to the erection of the sarsen circle nearly a thousand years later. A similar conservatism appeared in the orientation of the henges at Llandegai in North Wales put up a millennium apart.

Cemeteries of round barrows huddle around Neolithic earthen long barrows in Wessex, and at Winterbourne Stoke their line actually respects the alignment of the long mound. Continuity of tradition is likely. This is not to claim that there was no change of belief between 4000 and 2000 BC. What alignments there were in the Bronze Age may well have been more precise, constructed for different purposes but such developments would be evidence of gradual evolution rather than for discontinuity. The rituals by moonlight of the much defamed Iron Age druids add some support to this belief in an unbroken cultural tradition.

The study of chambered tombs illuminates the much greater attention paid by prehistoric people to the moon than to the sun. It has been assumed, particularly by archaeologists, that it was the sun that attracted early societies but the moon, in fact, would have been easier to observe. Although the minutiae of its cycle were complicated it did possess two enormous advantages, first that it could be looked at directly with the naked eye, and secondly that its movements were repeated more rapidly and were more easily noticed than the sun's slow shifting along the horizon. The full moon rose and set monthly, conspicuous against the blackness of night. Its very association with night may have symbolised darkness, cold and death in a way that appealed vividly to the allusive imagination of prehistoric man.

Many primitive societies have used the moon for their annual calendar, dividing the year up into ten or twelve lunar months not because they could count no further but 'because the number 10 or 12 corresponds more exactly to their practical and pragmatic interests. The tail end of the year, the uneventful level time after the harvest has started and before the festivities begin, remains vague to such an extent that moon 13 is entirely dropped out, even in esoteric accounts' (Malinowski 1927: 215). Harrison (1913: 32), referring to societies nearer to Europe than the Trobriand Islanders, wrote that 'round the moon, whose cycle is long, but not too long, ritual very early centred, but probably only when its supposed influence on vegetation was first surmised. The moon, as it were, practises magic herself; she waxes and wanes, and with her, man thinks, all the vegetable kingdom waxes and wanes too...'

So influential was the moon that the Iron Age Celts started their year in October/November by the full moon then. So did the natives of

Malekula in the New Hebrides and it is strangely coincidental that these societies, half a world apart, both also believed that this was the time when the dead returned to threaten the living. The Malekulans reckoned their year by lunar months that were tallied by clan magicians whose role passed from father to son. During the year the bones of ancestors were kept in a communal charnel-house but at the time of the harvest festival whose date was determined by the first appearance of the moon the ghosts awakened and had to be appeased by offerings of food and other gifts.

In several ways these beliefs mirrored those of the Celts and perhaps even those of the people who built the Neolithic chambered tombs. Their late October moon began the new year, provided the occasion for an annual feast and maintained the links between the living and the dead (Deacon 1934: 644). Of the same people Layard (1942: 288) remarked that the New Hebridean harvest feast marked 'the beginning of the calendar in the October-November moon. This is also the period when the ghosts of the dead are believed to revisit their former homes'. Amongst the Berawans of North Borneo rice was offered to the dead on the first full moon after the harvest when the moonlight ensured that people would be able to recognise their ancestor's ghost (Huntington & Metcalf 1979: 79).

These people also believed that the soul was released from the corpse only when the bones were dry and that between death and the ultimate decomposition of the flesh the spirit was suspended in a kind of purgatory, lingering resentful and malevolent near its earthly home (ibid,: 80). It might survive like this for as long as a year while the living anxiously performed the rites necessary to release it, counting the full moons to the time when the ghost could travel to the land of the dead. Some South Sea Islanders believed that there had once been a time when human beings were able to die and revive themselves at will 'following the cycle of the moon' (Diolé 1977: 73) and that it was not until a corpse had rotted in the open air, decayed into a skeleton and been buried that it was harmless.

Excarnation and the secondary interment of bones has been suspected amongst the builders of chambered tombs and earthen long barrows in the British Isles and it is possible that the rituals in these monuments were also regulated by the times of the moon. Whatever its significance to Neolithic and Bronze Age people there is no doubt that the moon was important to Iron Age communities. Their calendar that they used to compute the times of their four great festivals was based on lunar observations (Powell 1963: 120). In his Natural History (XVI: 249) Pliny the Elder wrote of the Celts that 'mistletoe...is gathered with great ceremony, and particularly on the sixth day of the moon (which for these tribes constitutes the beginning of the months and the years), and after every thirty years of a new generation, because it is then rising in strength and not one half of its full size' (Rackham 1968: 549-51).

The moon determined the time of Samain and the start of the cold half of the year at the end of October. 'It is well known that numerous peoples throughout the world have attached a very special significance to the juncture or interstice between two distinct temporal periods, regarding it in some sense as time outside time or as a temporary resumption of mythic or primordial time...All divine beings and the spirits of the dead move freely among men and interfere, sometimes violently, in their affairs' (MacCana 1970: 127). In such supernatural contexts the moon and death were closely joined and it is noteworthy that the lunar associations of Neolithic chambered tombs, of Bronze Age standing stones and cairns, and of the Iron Age festival of Samain were all based on the theme of death and the ghosts of the dead. The moon was a symbol rather than an object of scientific curiosity. Even in the latest prehistoric period when the druids were strong in the land there is 'not much evidence that they possessed any remarkable learning, astronomical or otherwise' (Kendrick 1927: 182). Classical writers such as Pliny who initially had believed these priests and law-givers to have considerable arcane knowledge later came to regard them as wizards and charlatans.

The druids may well have understood something of the moon, sun and stars, quite possibly inheriting this learning from the observations of earlier societies, but it is likely that they used this information, as Strabo said, in 'sacrifices and divinations that are opposed to our usage'. Pliny recorded that animals were sometimes killed as offerings on the sixth day of the moon and Tacitus (Annals, XIV: 26) remarked in disgust that the druids thought it necessary 'to drench their altars in the blood of prisoners and consult their gods by means of human entrails'.

Such sanguinary superstitions are not easy to reconcile with the objectivity expected of scientific researchers. These Iron Age sacrifices, carried out by the light of the full moon, are of the same astronomical background as the sacrifice of the girl at Woodhenge two thousand years earlier, her body 'turned towards the north-east, i.e. to the rising sun at midsummer' (Cunnington 1929: 13; Thom 1967: 74). They are in keeping with the death of another young girl, half-buried beneath a stone at the Sanctuary circle near Avebury, lying with her face towards the equinoctial sunrise (Cunnington 1931: 313; Burl 1979b: 197). They fit with the fusion of sun and human bones at Newgrange and Maeshowe, with the moon and human cremations in the Clava Cairns, all of them expressions of symbolism that have little to do with intellectual enquiry and the exploration of the heavens.

The assiduous fieldwork of Alexander Thom and his family over the past four decades has provided us with an invaluable set of plans for megalithic rings and stone rows without which much of the current research into archaeoastronomy would be greatly retarded. Thom's work has stimulated archaeologists to consider problems which have been all too casually ignored in the past and whose solution is essential to our better understanding of antiquity. Prehistorians owe the Thoms much gratitude for the results of their determined and sadly unencouraged labours and this debt should be acknowledged even by those who are not able to agree with the Thoms' interpretations about the design and purpose of prehistoric stones and tombs. Much more research will have to be done before there can be general agreement about the function of these megalithic structures.

A catholic approach is needed by anyone contemplating the problems of archaeoastronomy. A single-minded belief in shamans, magicians and witch-doctors may be unhelpful. Equally, an obsession with astronomer-priests and the miniscule perturbations of the moon may lead an investigator a long way from the truth.

As Gloucester complained in King Lear, 'These late eclipses in the sun and moon portend no good to us'.

ACKNOWLEDGEMENTS

I am exceptionally grateful to Professor R. J. C. Atkinson of University College, Cardiff and to Dr. D. C. Heggie of Edinburgh University, both of whom offered pertinent advice about the astronomical calculations in this paper. What errors and omissions remain are testimony only to the writer's semi-numeracy.

REFERENCES

Allibert, M. et al., 1978. L'interpretation astronomique de la Pierre Ogivale et le symbolisme solaire. In: Le dolmen de la Table des Marchands. Fontenay-le-Fleury. 32-37.

Ashbee, P., 1970. The earthen long barrow in Britain. London.

Ashbee, P., 1974. Ancient Scilly. Newton Abbot.

Atkinson, R. J. C., 1975. Megalithic astronomy—a prehistorian's comments. J. Hist. Ast., 6, 42-52.

Beaton, A. J., 1882. Notes on the antiquities of the Black Isle. Proc. Soc. Ant. Scotl., 16, 477-492.

Brown, G. M., 1975. Letters from Hamnavoe. Edinburgh.

Bryce, T. H., 1903. On the cairns of Arran. Proc. Soc. Ant. Scotl., 37, 36-37.

Burl, A., 1972. Stone circles and ring-cairns. Scot. Arch. Forum, 4, 31-47.

Burl, A., 1979a. Rings of stone, London.

Burl, A., 1979b. Prehistoric Avebury. London & New Haven.

Burl, A., 1980. Science or symbolism: problems of archaeoastronomy. Antiquity, 54, 191-200.

Burl, A., 1981. Rites of the gods. London.

Cash, C. G., 1906. Stone circles at Grenish etc. Proc. Soc. Ant. Scotl. 40, 245-254.

Cash, C. G., 1910. Archaeological notes from Aviemore. Proc. Soc. Ant. Scotl., 44, 189-203.

Chitty, L. F., 1926. The Hoar Stone or Marsh Pool circle. Trans. Shropshire Arch. Soc., 10, 247-253.

Coffey, G., 1912. New Grange and other incised tumuli. Dublin.

Coles, F. R., 1907. Stone circles in Banff and Moray. Proc. Soc. Ant. Scotl., 41, 130-172.

Corcoran, J. X. W. P., 1969. The Cotswold-Severn group. In: T. G. E. Powell, J. X. W. P. Corcoran, F. Lynch & J. G. Scott, Megalithic enquiries in the west of Britain. Liverpool. 13-106.

Cunnington, M. E., 1929. Woodhenge. Devizes.

Cunnington, M. E., 1931. The "Sanctuary" on Overton Hill, near Avebury. Wilts. Arch. Mag., 45, 300-335.

Daniel, G., 1950. Prehistoric chamber tombs of England and Wales. Cambridge.

Daniel, G., 1980. Megalithic monuments. Scientific American, July, 78-90.

Davidson, D. A., Jones, R. L. & Renfrew, C., 1976. Palaeoenvironmental reconstruction and evaluation: a case study from Orkney. Trans. Inst. Brit. Geog., 1(3), 346-361.

Deacon, B., 1934. Malekula. A vanishing people in the New Hebrides. London.

de Valera, R., 1960. The court-cairns of Ireland. Proc. R. Irish Acad., 60C, 2-140.

de Valera, R. & O'Nuallain, S. P., 1961. Survey of the megalithic tombs of Ireland, I: County Clare. Dublin.

de Valera, R. & O'Nuallain, S. P., 1964. Ibid., II: County Mayo. Dublin.

de Valera, R. & O'Nuallain, S. P., 1972. Ibid., III: Counties Galway etc. Dublin.

Diolé, P., 1977. The forgotten people of the Pacific. London.

Fleming, A., 1975. Megalithic astronomy: a prehistorian's view. Nature, 255, 575.

Fraser, J., 1884. Stone circles of Strathnairn etc. Proc. Soc. Ant. Scotl., 18, 328-362.

Harrison, J., 1913. Ancient art and ritual. London.

Hawkins, G. S., 1966. Astro-archaeology. Cambridge (Mass.).

Henshall, A. S., 1963. The chambered tombs of Scotland, Vol. 1. Edinburgh.

Henshall, A. S., 1972. Ibid., Vol. 2. Edinburgh.

Herity, M., 1974. Irish passage graves. Dublin.

Hoare, R. C., 1812. The ancient history of south Wiltshire. London.

Hoyle, F., 1977. On Stonehenge. London.

Huntington, R. & Metcalf, P., 1979. Celebrations of death. Cambridge.

Jolly, W., 1882. On cupmarked stones. Proc. Soc. Ant. Scotl., 16, 300-401.

Kendrick, T. D., 1927. The Druids. A study in Keltic prehistory. London.

Layard, J., 1942. Stone men of Malekula. Vao. London.

Lewis, A. L., 1892. Stone circles in Britain. Arch. J., 49, 136-154.

Lockyer, N., 1909. Stonehenge and other British stone monuments astronomically considered (2nd edn.). London.

Lynch, F., 1975. The impact of the landscape on prehistoric man. In: The effect of man on the landscape: the Highland zone. 124-126.

MacCana, P., 1970. Celtic mythology. London.

MacKie, E. W., 1977a. Science and society in prehistoric Britain. London: Paul Elek.

MacKie, E. W., 1977b. The megalith builders. Oxford: Phaidon.

Malinowski, B., 1927. Lunar and seasonal calendar in the Trobriands. J. R. Anthrop. Inst., 57, 203-216.

Mortimer, J., 1905. Forty years' researches. London.

Newham, C. A., 1972. The astronomical significance of Stonehenge. Leeds.

O'Kelly, C., 1971. An illustrated guide to Newgrange. (2nd edn.) Wexford: John English.

O'Kelly, C., 1978. Ibid. (3rd edn.) Wexford.

O'Nuallain, S. P., 1976. The central court-cairns of the north-west of Ireland. J. R. Soc. Ant. Ireland, 106, 92-117.

O'Reilly, J. P., 1898. On the orientation of some cromlechs in the neighbourhood of Dublin. I & II. Proc. R. Irish Acad., 20, 589-605.

Patrick, J., 1974. Midwinter sunrise at Newgrange. Nature, 249, 517-519.

Patrick, J., 1975. Megalithic exegesis: a comment. Irish Arch. Res. Forum, 2(2), 9-14.

Penny, A. & Wood, J. E., 1973. The Dorset Cursus complex. A Neolithic astronomical observatory? Arch. J., 130, 44-76.

Péquart, M., Péquart, St. J. & le Rouzic, Z., 1927. Corpus des signes gravés des monuments megalithiques du Morbihan. Paris.

Piggott, S., 1956. Excavations in passage graves and ring-cairns of the Clava group. Proc. Soc. Ant. Scotl., 88, 173-207.

Powell, T., 1963. The Celts. London.

Rackham, H. (translator), 1968. Pliny's "Natural History", Vol. 4. London & Cambridge (Mass.).

Rahtz, P., 1978. Grave orientation. Arch. J., 135, 1-14.

Scott, J. G., 1969. The Clyde cairns of Scotland. In T. G. E. Powell, J. X. W. P. Corcoran, F. Lynch & J. G. Scott, Megalithic enquiries in the west of Britain. Liverpool. 175-222.

Simpson, D. D. A., 1966. Excavations at Kintraw, Argyll. Proc.Soc. Ant. Scotl., 99, 54-59.

Somerville, B., 1909. Ancient stone monuments near Lough Swilly. J. R. Soc. Ant. Ireland, 39, 192-202, 215-233, 343-349.

Somerville, B., 1912. Prehistoric monuments in the Outer Hebrides and their astronomical significance. J. R. Anthrop. Inst., 42, 23-52.

Somerville, B., 1923. Instances of orientation. Archaeologia, 23, 193-224.

Spence, M., 1894. Standing stones and Maes Howe of Stenness. Paisley.

Stukeley, W., 1743. Abury, a temple of the British Druids. London.

Thatcher, A. R., 1976. The Station Stones at Stonehenge. Antiquity, 50, 144-146.

Thom, A., 1966. Megalithic astronomy: indications in standing stones. Vistas in Ast., 7, 1-57.

Thom, A., 1967. Megalithic sites in Britain. Oxford: Oxford University Press.

Thom, A. & Thom, A. S., 1978. Megalithic remains in Britain and Brittany. Oxford: Oxford University Press.

Tuckwell, A., 1975. Patterns of burial orientation in the round barrows of East Yorkshire. Bull. Inst. Arch., 12, 95-123.

Ucko, P. J., 1969. Ethnography and archaeological interpretation of funerary remains. World Arch., 1, 262-280.

Walker, I. C., 1965. The Clava cairns. Proc. Soc, Ant. Scotl., 96, 87-106.

Wells, C. & Green, C., 1973. Sunrise dating of death and burial. Norfolk & Norwich Arch. Soc., 34, 435-442.

Wood, J. E., 1978. Sun, moon and standing stones. Oxford: Oxford University Press.

Wood, J. E. & Penny, A., 1975. A megalithic observatory on Dartmoor. Nature, 257, 205-207.

Woodham, A. A., 1956. Survey of prehistoric monuments in the Black Isle. Proc. Soc. Ant. Scotl., 88, 65-93.

8. ETHNOASTRONOMY: ITS PATTERNS AND ARCHAEOLOGICAL IMPLICATIONS

I. J. Thorpe

The most nagging doubt which archaeologists have had about the suggestions of great sophistication in megalithic astronomy is that first expressed by Jacquetta Hawkes (1967); that things simply were not like that. The suggestions of rule by astronomer-priests in Neolithic and Bronze Age Britain simply do not accord with our view of the nature of Neolithic and Bronze Age Society. This is not in any way an ethnocentric belief that prehistoric societies are not capable of considerable achievements, but rather that the views of some archaeoastronomers (for instance MacKie 1977) on the nature of megalithic astronomy, if correct, would necessitate an extreme degree of imbalance in one particular direction that would have resulted in a completely different society from the one reflected in the archaeological record. It is the main aim of this paper to see whether the results of a brief preliminary ethnographic survey suggest that we need assume that a high level of expertise in astronomical techniques is needed to create an accurate calendar, and whether, where such sophistication does exist, it is accompanied by a commensurate degree of complexity in other aspects of society.

It is not my intention to discuss the well-known astronomies of Egypt (Parker 1974), Babylon (Sachs 1974), the Maya (Thompson 1974), or China (Needham 1974), or even that of the Druids (Burl 1976: 79-80; Piggott 1968: 122-5). This is not only because these have been covered far better than I could have done, but also because by concentrating on them we ignore the mass of data which may be less exciting, but is potentially far more informative. It is easy to be so dazzled by the complexities of these better-recorded later civilisations that we fall into the trap, warned against by Turton and Ruggles (1978), of over-estimating the sophistication of prehistoric time-reckoning. Not only this, but it is also extremely easy to use, or rather misuse, the ethnographic data in presenting an ethnographically recorded society as an analogue of a prehistoric one. Any attempt to find exact parallels for Neolithic and Bronze Age societies (as in MacKie 1977, or Renfrew 1973) is doomed to failure from the start. As Binford has said so rightly (1968: 13), merely attempting to fit archaeological remains into ethnographically known patterns of life adds nothing to our understanding of the past. However, to ignore the ethnographic evidence, in fear of becoming bogged down in an illusory search for exact parallels with prehistoric Britain, is unnecessarily to restrict the range of knowledge at our disposal. Used with proper caution, ethnographic evidence can help to formulate hypotheses which may be used to attempt to answer some of the seemingly intractable problems posed by archaeoastronomy.

The literature of ethnoastronomy is a vast one, and thus I can do nothing more than skim the surface of the evidence. Extremely valuable surveys were made by Fraser (1912; 1936), Cope (1919), and especially Nilsson (1920); while more recently, Baity (1973) has usefully pulled together some of the more recent work, although done in support of a highly dubious diffusionist paradigm, and Turton and Ruggles' (1978) case study of the Mursi of Ethiopia has pointed the way to a much more rigorous and explicit methodology. Evidence is to be obtained from all quarters of the globe, although, for obvious reasons, little documentation exists for Europe. There is unquestionably a geographical bias to this evidence, with a considerable shortage of material from northern Asia, especially China, and an abundance of material from Polynesia and Micronesia, stimulated by the debate over Oceanic navigation and settlement. Obviously the quality of the evidence varies considerably, for while first-contact accounts are clearly based on first-hand knowledge of societies unaltered by the impact of contact with western societies, they are all too frequently scanty and ambiguous. Similarly, more recent work will clearly probe deeper into the society under study, but the nature of the society may have changed so much that its pre-contact patterns may have been completely lost. However, this is not a cause for despair, merely a reminder that the ethnographic record is no more untouched than is the archaeological record.

Perhaps the first important task in looking at any relatively unexamined field is to attempt to produce a meaningful system of classification. This is especially true of a field of such bewildering complexity as that of ethnoastronomy. Leach (1954) has proposed that there are five main modes of thinking about time. These are: time as a recurring cycle without chronology; time as history, not cyclic, but sequential; time as magic, capable of being controlled; time as duration, a simple dimension analogous to length; time as a political tool, as in the calendars of various revolutionary movements. Of these five modes of thinking, the last two, times as duration, and time as a political tool, have only been developed in the modern industrial era. Leach suggests (1954: 120) that time as history, and time as magic, develop out of thinking of time as a recurring cycle. This increasing complexity of ways of thinking about time, in Leach's view, is correlated with the emergence of a priestly class of time-keepers and more systematic observation of the heavens. So it is my intention to test Leach's hypothesis by a preliminary survey of the ethnographic evidence, to see whether we can indeed see astronomy developing on its own, or whether such developments only take place in the context of other social changes also occurring.

Before examining the ethnographic record we must ask the question, why astronomy? There are various ways in which astronomical knowledge can be put to use. Perhaps the most obvious is the connection between the use of a calendar and the agricultural year, so that it may be known when to plant crops and when to harvest them. There is also the question of the use of astronomy for navigation, whether through the use of stars as navigation markers, or by producing a calendar which will predict the likely periods of bad weather in the year. Finally, there is the question of the timing of ritual activities by astronomical observation, at its most sinister the timing of human sacrifices.

Those societies which pay no great attention to the heavens are not necessarily those which one would expect to be unconcerned because of their relatively simple social organisation; for instance the Bushmen (Bleek 1928), the Hottentots (Frazer 1912) and the Australian aborigines (Parker 1905; Frazer 1912; Nilsson 1920) are all recorded as evidencing interest in following the movement of the heavens, in order to schedule their food-gathering cycle. Even the Tasmanian aborigines, generally regarded as one of the most simple of societies, with a considerable lack of technological equipment, observed the stars because they told them when to burn over the forest, when to walk to the coast for shellfish, and when to swim to offshore islands to catch the nesting sooty petrel (Coon 1972: 287). There are, however, hunter-gatherers that show no interest in the stars, as for instance, the Andaman Islanders (Radcliffe-Brown 1922: 141). There are also those agricultural societies which show no evidence of being interested in the heavens. Thus the Ibo-speaking tribes are described as being singularly incurious about heavenly bodies and occurrences, naming only the Pleiades, the belt of Orion, the Great Bear, and Venus (Thomas 1913: 127-8). Again, the Abaluyia of Kenya take little interest in the stars, naming no stellar constellations, and not remarking any difference between the planets and the fixed stars, only distinguishing Venus (Wagner 1954: 32). A more surprising member of this group are the Fijians, who are reported (Deane 1921: 29) to name only the belt of Orion, the Southern Cross and the Pleiades. This seemingly astonishing ignorance, when compared to the considerable knowledge possessed by other Polynesian societies, is ascribed to the fact that they undertake only very limited voyages, and are thus bad navigators in the technical sense although good sailors.

The most obvious phenomenon which the heavens afford is that of the lunar month, and consequently there are very few peoples that have no knowledge of a regular lunar month. It is also a fairly easy matter to make of these lunar months a calendar of twelve lunar months to the year, and Cope (1919) and Nilsson (1920) have collected large numbers of examples. These can be divided into numerical calendars, each month simply being given a number from one to twelve; and descriptive calendars, where each month is given a name appropriate to its time of year. The Kaffirs of South Africa possess such a descriptive calendar (Macdonald 1890: 291), naming months after flowers, birds, and the presence or absence of rain; a similar calendar can be observed among the Nyangatom of Ethiopia (Tornay 1978: 598), with months called 'mud,' 'maturation', 'whiteness' and so on. Greater interest in the movements of the moon is shown by the Sandawe of Tanzania (Raa 1969: 43), who watch the point on the horizon where the moon rises, the skyline of the Sandawe hills being serrated by large boulders and rocks which towards the end of the dry season are clearly visible between the bare trees. When the moon reaches a certain point in its annual southward movement, the Sandawe know that the beginning of the short rains is at hand and therefore it is time for the hoeing season to begin.

There is also a considerable amount of evidence for the observation of the stars and the planets. A great number of societies are recorded as observing the planet Venus, but fewer seem to have taken notice of the movements of the planet Mars. One society that did take a considerable interest in the stars in general, and Mars in particular, are the Pawnee Indians. They recorded the positions of the stars on charts of tanned hide (Buckstaff 1927), which were kept with the sacred equipment of the priesthood. Of greater interest to the more sensation-seeking among us was the Pawnee Morning-Star sacrifice (Linton 1926), performed in those years when Mars was the morning star. It usually originated in a dream in which the Morning Star appeared to some man and told him to capture a suitable victim, the victim being dispatched with an arrow, shot by the man who had caught her, at the moment when the Morning Star rose. The Morning Star sacrifice was last performed in 1838 (Weltfish 1965: 84). On a more mundane level, it is very easy to time an event by observing the stars. Simply by rising at dawn and seeing the stars that are shining in the brightening east, and then going to bed at dusk and seeing the stars that appear in the west, it will be noticed that the stars' positions vary throughout the year and that this variation keeps pace with the phases of Nature. So there lies ready to hand a means of determining the time of the year, a method much more accurate than that of observing the phases of Nature (Nilsson 1920: 128-9). Thus among the Thonga of South Africa (Junod 1927: 309) the rising of the Pleiades in July/August indicates the resumption of tilling. The Mandaya tribe of Mindinao in the Philippines observe the Pleiades at the start of November to know when to start cleaning land by their appearance, and plant their crops when they are straight above at the beginning of December (Frazer 1936: 394). The Indians of north-western Brazil plant their manioc crop when the Pleiades set, because it is then that the heavy rains set in (Frazer 1912: 309; Levi-Strauss 1964: 217). Closer to home Pliny the Elder records ('Historia naturalis,' xviii, 223) that in Italy, Greece and Asia the movements of the Pleiades dictated the time of sowing grain (Rackham 1950). Nilsson (1920: 130) suggests that the popularity of the Pleiades is not merely a result of the fact that their appearance coincides with important phases of the vegetation, but also because they form themselves into an easily recognised group without any aid from the imagination. Although the Pleiades are observed most frequently, other constellations have their users, thus the Sema Nagas determine the time of sowing by the ascendancy of the belt of Orion (Hutton 1921: 260); while the Bangala of the Upper Congo observed a group of five stars called kole, in our constellation Lepus, planting when kole reaches its highest point (Weeks 1909: 129).

It is clear that these two calendrical systems, that produced by moon-observation, and that produced by star-observation, are in conflict, because the year produced by twelve lunar months is shorter than the solar year by a few days. There are obviously two possible methods of correcting this imbalance, either by observing the rising or setting of a particular star for the first time in the year; or by observing the sun, either watching for the solstices, or noting the annual appearance of the sun at a particular point on the horizon. However, as Turton and Ruggles warn (1978: 590) we cannot assume that this conflict between lunar and solar reckoning will necessarily be seen as a problem requiring an urgent solution. Thus the Mursi of Ethiopia have a lunar calendar, but fix the date for planting by observing the successive setting of four stars, and also observe the annual course of the sun; however, they do not use these observations to adjust their calendar, the correct month only emerging very gradually as a consensus view. There are, however, a number of unequivocal instances of the stars being used to regulate the calendar. Thus among the Blackfeet Indians (L'Heureux 1886: 301-02) the Pleiades mark the beginning of the agricultural year, overruling the lunar calendar. The Bafioti of Loango (Frazer 1936: 360) corrected their lunar calendar of twelve months by observing the rising of Sirius, inserting an extra month when Sirius failed to rise to coincide with the new moon. Among the Kaffirs of South Africa (Macdonald 1890: 291) the confusion which frequently arises as to the identity of the month is always settled by the first appearance of the Pleiades just before sunrise.

As mentioned above, the error caused by the difference between the lunar and solar year may be rectified by observing the sun. So, among the Kwakiutl Indians the winter solstice serves as a period of adjustment (Cope 1919: 138). The Thompson Indians (Boas 1938: 274) carefully observe the summer and winter solstices and readjust their calendar accordingly. Among the Haida (Murdock 1934: 222-3), if the new moon on which the first or seventh month is due to begin appears before the sun has reached the solstice the old moon is made to extend over two

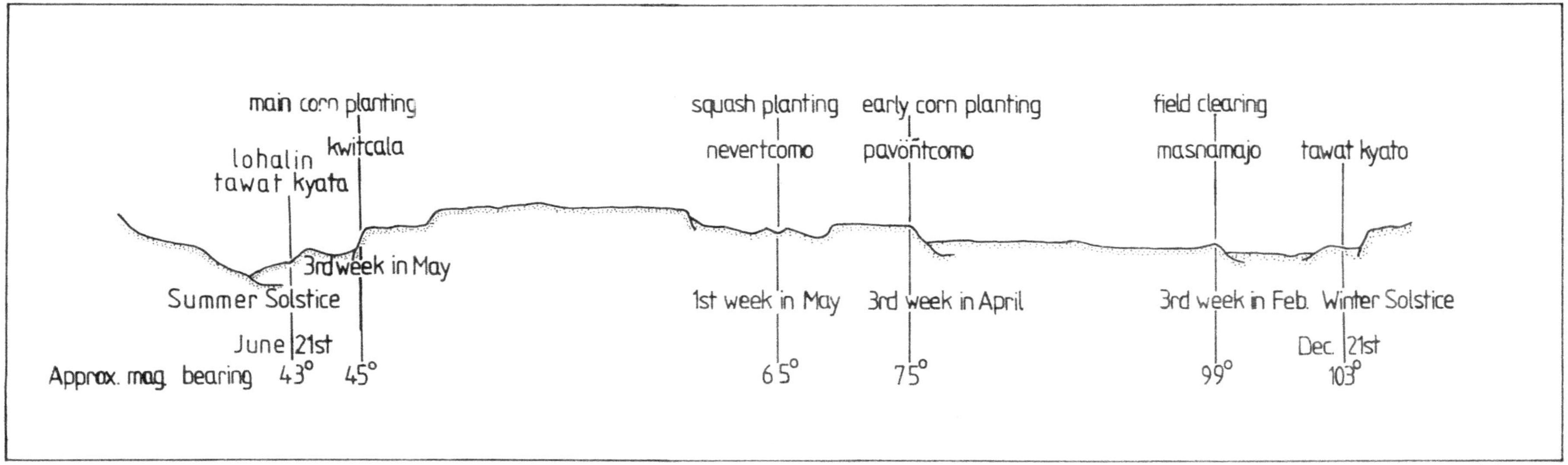

Fig. 8.1 Horizon calendar, December to June, Walpi

moons. However, observation of the sun can produce an accurate calendar by itself, and there are a number of societies that determine the time of year by observing the movement of the sun against the horizon as seen from a fixed point. Lindblom (1920: 338) reports that the Akamba of East Africa have a method of calculating the time of year by observing from a level piece of ground with an open view a number of points along the horizon, checking in particular the beginning of the rainy season. For Russia Maistrov (1974) records that some 40-50 village holidays in the year were fixed by observing the movement of the sun against a horizon of mountain ridges. In the Philippines (Scott 1958) important dates for planting were fixed by observing notches cut in the mountainside, but Scott feels (1958: 569) that this practice is probably a result of the penetration of western ideas in the mid-nineteenth century, and therefore outside the scope of this paper.

The most well-known calendars, as a result of Renfrew's popularisation (1973), are those of the Pueblo-dwelling Indians of the south-west United States. Information exists on the sun calendars of the Tewa (Parsons 1926; Forde 1931), the Zuni (Stevenson 1901-02; Parsons 1917; 1933), and the Tiwa (Parsons 1939) as well as the more extensively discussed Hopi. The Hopi are an agricultural people, their staple food being maize, and therefore climate is an all-important factor to them. They have accordingly devised a method of accurately determining the correct times for planting, and the period of the rainy season, as well as the solstices, by almost continuous observation of the position of the sun on the horizon. Parsons (1933) has preserved a record of the series of plantings and ceremonials associated with particular points which the sun touches on the horizon; while Forde (1931) has recorded two horizon maps maps of the Hopi, one from Walpi (fig. 8.1), and one from Shimopovi (fig. 8.2). The observation of up to twenty sunrise positions is skilled work and it is the responsibility of one religious official: the Sun-watcher is chosen from the members of a clan that has a hereditary right to fill that position, but he is chosen from those available by the heads of the other clans. However, at the most important festival of the year, the winter solstice ceremony (soyal), the Sun-watcher plays a relatively minor role after he announces the arrival of the ceremony. The important ceremonies are performed in secret, in the kiva (ritual clan headquarters) of the Sun clan, which does not provide the Sun-watcher; these ceremonies being performed in order to 'call back the sun' from its southerly movement (Fewkes 1920: 497). This is in direct contrast to the astronomical knowledge and abilities of the Sun-watcher which are freely discussed and commented on (Parsons 1925). The facet of Hopi astronomical expertise which has most impressed McCluskey (1977) in his researches is that Hopi have managed to maintain an agricultural and ceremonial calendar with a striking degree of precision, with individual festivals seldom falling more than two days from the norm (1977: 191).

These observations of the sun do not, however, mark the level of greatest interest in the movements of the heavenly bodies. For this we must turn to Oceania. Here, considerable controversy over the question of deliberate voyaging has produced a vast literature on the subject of navigation, with which astronomy is intimately connected. I do not propose to spend long on the question of navigation, for it is of but tangential importance to the subject at hand; but for those who do, the most important writings on the controversy are Sharp (1957), Golson (1962) and Lewis (1972). Sharp's strongly expressed doubts notwithstanding, Parsonson (1962: 42) has recorded sailing directions for an eighteenth-century voyage from Niuatoputapu to Tonga which involves aiming for

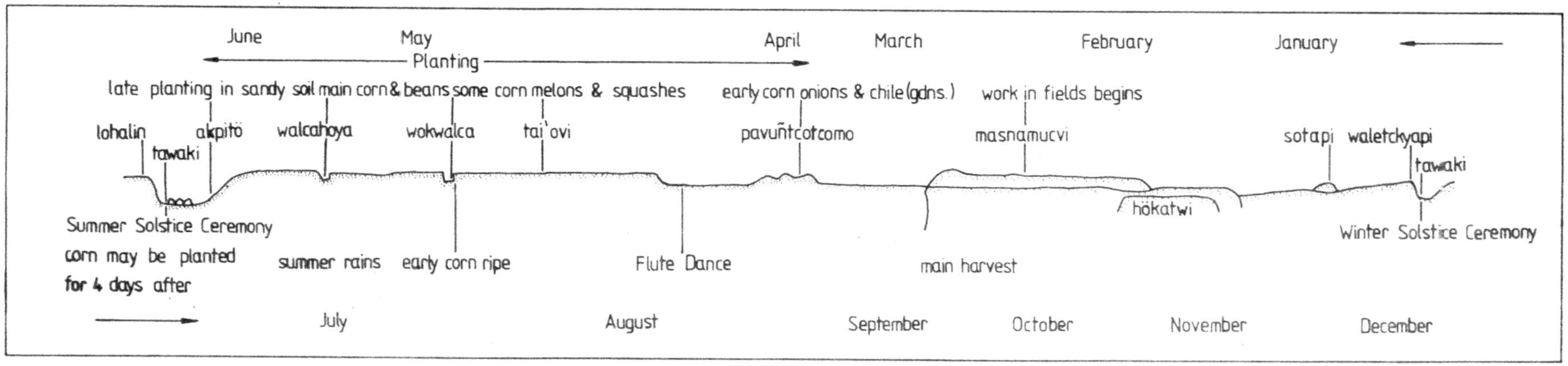

Fig. 8.2 Horizon calendar, Shimopovi (from a native drawing).

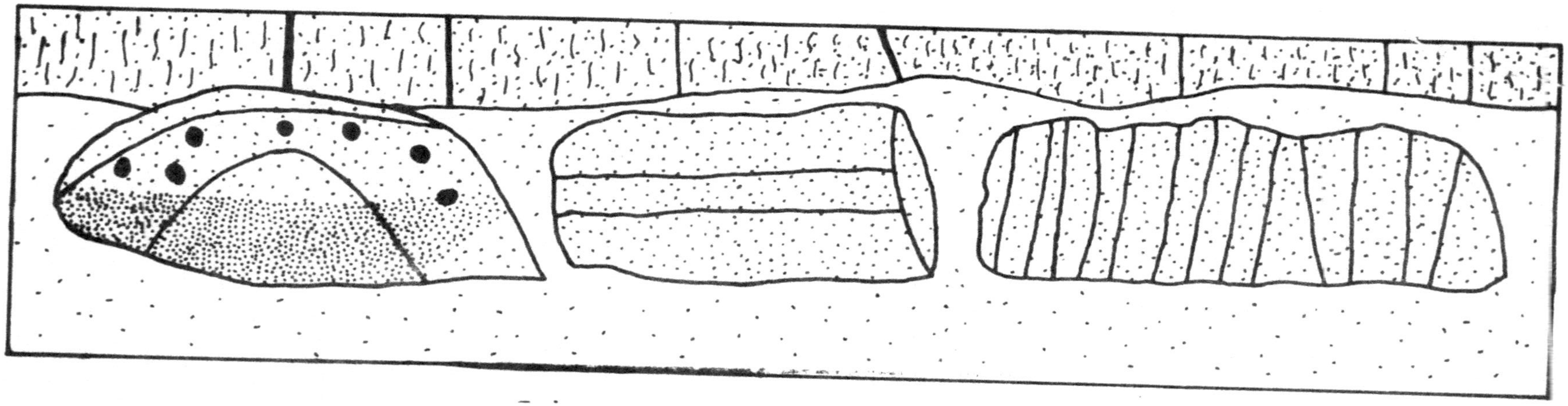

Fig. 8.3 Hupa calendar stones.

a succession of at least eight stars acting as markers. Certainly such sailing directions have been followed in recent times (Gladwin 1970; Lewis 1972), and have been found to work, so that most writers now accept their validity. It is only by constant observation of the stars that such directions can be formulated because every day the sun falls four minutes further behind a given star, so that the star used must be regularly changed so that it does not become misleading by apparently turning from East to West.

Less well known, however, are the land-based astronomical skills of the Polynesians, probably because in many places astronomy was regarded as merely an adjunct of navigation, as in Tonga (Collocott 1922: 157). Thus in the Marshall Islands, while a number of stars are closely observed (Nilsson 1920: 125-6), this is primarily in order to predict weather-patterns for the navigators. However, on many islands the heavens were observed for other purposes, such as agriculture; this observation was at a very intensive level, as in the Gilbert Islands, where the navigator, or tiaborau, noted the point of the sun's rising every ten days, and further checked its annual progress by observing the eastern constellations just before dawn (Makenson 1938: 377). The small island of Mangareva (Hiroa 1938: 414-15) had four observation posts for the study of the summer and winter solstices. Makenson (1938: 382) gives some idea of the techniques that may have been used, in the account of a native who watched an astronomer observing wero-i-te-ninihi (probably a comet) by putting sticks in the ground. These observations appear to have taken place from temple platforms, including the one recollected in the tradition of Puna district, Hawaii, that a temple at Haahae, a hill on the east coast, had a large, flat, altar rock with a groove running diagonally across it, and that when the sun rose from the sea at a point directly in line with this groove, a victim was sacrificed (Makenson 1938: 378). Among the Maori all months were named after the star that rose during the month, and the year as a whole was regulated by observation of the Pleiades and Orion (Nilsson 1920: 248; Firth 1929: 46-48). Similarly, each moon is named by the Kiwai Papuans after a specific constellation, the one that sets during the month (Nilsson 1920: 207-08). This degree of interest in the heavens requires a sustained pattern of observation, not just for timing specific festivals or for regulating the lunar calendar once a year.

So what can we expect these instances of astronomy to leave behind in the archaeological record? We may expect there to be a great range in the field monuments which may have astronomical importance. Probably the least sophisticated type of monument is that represented by the Hupa 'Calendar Stones' (Goldschmidt 1940) which seem to be simple mnemonic devices for timing ceremonial activities by the passing of the lunar months (fig. 8.3). While the Hopi of Walpi Pueblo used no monument to mark their observation place, at the other Hopi village of Mishongnovi the date of corn planting is determined by observation of sunrise made at Tavavonpi, a shrine at the west edge of town (Parsons 1933: 59). Among the Zuni Indians the Pekwin, the Sun-watcher, watches the sun rise throughout the year from Alela (upright), a round stone in a field on the east side of town (Parsons 1933: 93), which Hagar (1910: 65) describes as an erect sandstone slab adorned with a solar effigy. Clearly the temple platforms of Polynesian astronomers will be left behind; Grimble (1931: 205-06) describes the platforms, or buatarawa, as being three or four feet square at the base, rising to a height of between two and twelve feet, and tapering gradually. Flat and smooth on top, they faced out to sea, allowing the observer an excellent view. It is possible that posts or 'pylons' were set up on them to mark lines of sight. From Africa, Frobenius reports (1913: 304) that the Yoruba of West Africa set up at Ife a triangular monolith as one of a group commemorating the dead with the additional use of being a sun-dial. When the shadow of the monolith reached a certain point drawn on the ground then it was time for a particular festival to take place. Among the Maori, Rout (1926: 293) records that their temples were oriented so that the central post cast its shadow along the ridge pole at each March solstice, when ceremonial meetings were held. Clearly, then, there are a large number of different astronomical methods which could leave traces in the archaeological record, but there are at least an equal number of similar methods which would leave no trace.

How, then, does Leach's hypothesis (1954: 120) of the emergence of a specialist priestly class together with close observation of the heavens stand up to the test provided by the evidence of this brief ethnographic survey? There is a clear difference between the evidence of the Pueblo Indians and that of the Polynesians on this point. Although the Hopi clearly see time as magic, with rituals held in the kiva being capable of affecting the movements of the sun, the Sun-watcher, despite being generally regarded as the most important figure in the village, is by no means all-powerful. Forde (1931: 386) says that many of the points in the horizon calendar are well known, and indeed the Sun-watcher's performance of his duties is freely commented on by members of the tribe. For instance Crow Wing, author of the 'Pueblo Indian Journal' (Parsons 1925), says in 1921, 'We think the Sunwatcher is not a very good man. He missed some places, he was wrong last year. And I think he is going to miss or take on some place (i.e. observe inaccurately). All the people think that is why we had so much cold this winter and no snow.' This passage illustrates clearly the responsibility borne by the incompetent or lazy astronomer when his operations are a matter of public knowledge. In Polynesia also, time is seen as magic, with the actions of priests being noticed by the stars who control the seasons, as among the Maori (Firth 1929: 47). However, among the Polynesians there is a much closer control of astronomical knowledge. Thus in Tahiti (Forster 1778: 528) the sciences of navigation and astronomy were known only to a few; in Tonga (Lewis 1974: 136) the secrets of astronomy were held by a hierarchy of navigator families of varying, but always chiefly, rank; and in the Gilbert Islands (Makemson 1938: 376) only some twenty out of 30,000 natives had any knowledge of astronomy. So here indeed we have the sort of specialists envisaged by Leach, and associated with the existence of a more complex astronomical system. The vital question to answer is why this difference arises. What are the factors which cause this distinct ranking to occur in Polynesian society, but not in Hopi society? It is in this field that ethnoastronomy can make its most important contribution, for an understanding of the emergence of ranked societies is a problem which has very wide implications, not just for astronomy. It may be that the importance of astronomy for navigation is a factor in this emergence of ranking, with control of navigation being the key to the control of the exchange of prestige goods.

So, a brief ethnographic survey shows that: (1) there is an almost universal concern with the heavens, even among hunter-gatherers; (2) it is extremely easy to produce an agricultural calendar, especially by use of the Pleiades; (3) where the stars are used, it is almost invariably in the form of constellations and not single stars, and therefore if alignments on stellar objects are sought in the archaeoastronomical record, despite the considerable difficulties caused by their movement over time, then constellations should be seen as the norm rather than the exception; (4) there are many astronomical activities which do not leave behind any archaeological evidence, and those that do, frequently leave behind only upright posts and stones; (5) certainly, stone circles are not a feature of the ethnoastronomical record, which lends support to Burl's suggestion (1976: 53) that a circle is by no means the best shape for an astronomical monument; (6) the need for a level observing platform is shown from Polynesia, which correlates with the practice of building up artificially raised earth platforms, something which Burl (1980) suggests from Scottish recumbent stone circles, and which may possibly explain the existence of some of the prominently sited barrows which do not contain burials; (7) the final, and most important, point to note is that there is no need to postulate the existence of a Mayan-type regimented society (MacKie 1977). Instead of of applying the principles of Occam's Razor (Clarke 1978: 79-81), unnecesarily complex hypotheses have been developed to explain what, to the non-western observer, is not a particularly surprising phenomenon.

ACKNOWLEDGEMENTS

I should like to thank Richard Bradley for reading over this paper in draft form. I have also benefited from discussion with Aubrey Burl. Finally I should like to thank the staff of the Pitt Rivers Museum Library, Oxford, and in particular Liz Edwards, for their assistance.

REFERENCES

Baity, E. C., 1973. Archaeoastronomy and ethnoastronomy so far. Curr. Anthrop., 14, 389-449.

Binford, L. R., 1968. Archaeological perspectives. In S. R. Binford & L. R. Binford (eds.), New perspectives in archaeology. Chicago.

Bleek, D. F., 1928. The Naron. Cambridge.

Boas, F., 1938. Invention. In F. Boas (ed.), General Anthropology. New York.

Buckstaff, R. N., 1927. Stars and constellations of a Pawnee sky map. Amer. Anthrop., 29, 279-285.

Burl, H. A. W., 1976. The stone circles of the British Isles. London.

Burl, H. A. W., 1980. Science or Symbolism: problems of archaeoastronomy. Antiquity, 54, 191-200.

Clarke, D. L., 1978. Analytical archaeology (2nd edn.). London. Revised by R. W. Chapman.

Collocott, E. E. V., 1922. Tongan astronomy and calendar. Occas. Papers of Bernice P. Bishop Mus., 8, 157-173.

Coon, C. S., 1972. The Hunting Peoples. London.

Cope, L., 1919. Calendars of the Indians north of Mexico. Univ. of Calif. Pubs. in Amer. Arch. and Ethnol., 16, 119-76.

Deane, W., 1921. Fijian Society. London.

Fewkes, J. W., 1920. Sun worship among the Hopi Indians. Annual Report of Smithsonian Inst. 1918, 493-526.

Firth, R., 1929. Primitive economies of the New Zealand Maori. London.

Forde, C. D., 1931. Hopi agriculture and land ownership. J. R. Anthrop. Inst., 61, 357-405.

Forster, J. R., 1778. Observations made during a voyage round the world. London.

Frazer, J. G., 1912. Spirits of the Corn of the Wild (3rd edn.). London.

Frazer, J. G., 1936. Aftermath. London.

Frobenius, L., 1913. The voice of Africa. London.

Gladwin, T., 1970. East is a big bird. London.

Goldschmidt, W. R., 1940. A Hupa "calendar". Amer. Anthrop., 42, 176-177.

Golson, J. (ed.), 1962. Polynesian navigation. A symposium on Andrew Sharp's theory of accidental voyages. Wellington.

Grimble, M., 1931. Gilbertese astronomy and astronomical observances. J. Polynesian Soc., 40, 197-224.

Hagar, S., 1910. Sun, moon and stars. In J. Hastings (ed.), Encyclopaedia of religion and ethics, vol. XII. Edinburgh.

Hawkes, J., 1967. God in the machine. Antiquity, 41, 174-180.

Hiroa, T. R., 1938. Ethnology of Mangareva. Honolulu.

Hutton, J. H., 1921. The Sema Nagas. London.

Junod, H. A., 1927. The life of a South African tribe. London.

Leach, E. R., 1954. Primitive time-reckoning. In C. Singer, E. J. Holmyard and A. R. Hall (eds.), A history of technology, vol. I. Oxford.

Levi-Strauss, C., 1964. The raw and the cooked. London.

Lewis, D., 1972. We, the navigators. Canberra.

Lewis, D., 1974. Voyaging stars: aspects of Polynesian and Micronesian astronomy. Phil. Trans. R. Soc. Lond., A276, 133-148.

L'Heureux, J., 1886. Ethnological notes on the astronomical customs and religious beliefs of the Chokitapia or Blackfeet Indians, Canada. J. R. Anthrop. Inst., 15, 301-304.

Lindblom, G., 1920. The Akamba. Uppsala.

Linton, R., 1926. The origin of the Skidi Pawnee sacrifice to the Morning Star. Amer. Anthrop., 28, 457-466.

McCluskey, S. C., 1977. The astronomy of the Hopi Indians. J. Hist. Ast., 8, 174-195.

Macdonald, J., 1890. Manners, customs, superstitions, and religions of South African tribes. J. R. Anthrop. Inst., 19, 264-296.

MacKie, E. W., 1977. Science and society in prehistoric Britain. London.

Maistrov, L. E., 1974. Contribution to discussion on ancient astronomy: the unwritten evidence. Phil. Trans. R. Soc. Lond., A276, 267-268.

Makemson, M. W., 1938. Hawaiian astronomical concepts. Amer. Anthrop., 40, 370-383.

Murdock, G. P., 1934. Our primitive contemporaries. New York.

Needham, J., 1974. Astronomy in ancient and medieval China. Phil. Trans. R. Soc. Lond., A276, 67-82.

Nilsson, M. P., 1920. Primitive time-reckoning. Oxford.

Parker, K. L., 1905. The Euahlayi Tribe. London.

Parker, R. A., 1974. Ancient Egyptian astronomy. Phil. Trans. R. Soc. Lond., A276, 51-65.

Parsons, E. C., 1917. Notes on Zuni. Memoirs of the American Anthropological Association, Vol. IV no. 3.

Parsons, E. C., 1925. A Pueblo Indian Journal 1920-1921. Memoirs of the American Anthropological Association, no. 32.

Parsons, E. C., 1926. The ceremonial calendar of the Tewa of Arizona. Amer. Anthrop., 28, 209-229.

Parsons, E. C., 1933. Hopi and Zuni ceremonialism. Memoirs of the American Anthropological Association, no. 39.

Parsons, E. C., 1939. Picuris, New Mexico. Amer. Anthrop., 41, 206-222.

Parsonson, G. S., 1962. The settlement of Oceania. In J. Golson (ed.), op. cit., 11-63.

Piggott, S., 1968. The Druids. London.

Raa, E. T., 1969. The moon as a symbol of life and fertility in Sandawe thought. Africa, 39, 24-53.

Rackham, H. (translator), 1950. Pliny's "Natural History". London.

Radcliffe-Brown, A. R., 1922. The Andaman Islanders. Cambridge.

Renfrew, C., 1973. Before civilisation. London.

Rout, E. A., 1926. Maori symbolism. London.

Sachs, A., 1974. Babylonian observational astronomy. Phil. Trans. R. Soc. Lond., A276, 43-50.

Scott, W. H., 1958. Some calendars of Northern Luzon. Amer. Anthrop., 60, 563-570.

Sharp, A., 1957. Ancient voyagers in the Pacific. London.

Stevenson, M. C., 1901-2. The Zuni Indians: their mythology, esoteric societies and ceremonies. Annual Report of Bureau of Amer. Ethn., 23, 1-608.

Thomas, N. W., 1913. Anthropological report on the Ibo-speaking peoples of Nigeria. London.

Thompson, J. E. S., 1974. Maya astronomy. Phil. Trans. R. Soc. Lond., A276, 83-98.

Tornay, S., 1978. Comment on: Turton & Ruggles (op. cit.), Curr. Anthrop., 19, 597-598.

Turton, D. A. & Ruggles, C. L. N., 1978. Agreeing to disagree: the measurement of duration in a southwest Ethiopian community. Curr. Anthrop., 19, 585-593 & 599-600.

Wagner, G., 1954. The Abalunia of Kavirondo (Kenya). *In* D. Forde (ed.), *African worlds*. London.

Weeks, J. H., 1909. Anthropological notes on the Bangala of the Upper Congo River. *J. R. Anthrop. Inst.*, *39*, 97-136.

Weltfish, G., 1965. *The Lost Universe*. London.

9. SOME NOTES ON WORK ORGANISATION AND SOCIETY IN PREHISTORIC WESSEX

Bill Startin and Richard Bradley

"How long
Do works endure? As long
As they are not completed.
Since as long as they demand effort
They do not decay..."

Bertolt Brecht, 'About the way to construct enduring works'.

9.1 Introduction

In 1973 Renfrew published an important analysis of social evolution in Bronze Age Wessex. His main purpose was to show that the rich Wessex Culture of the second millennium bc could be the culmination of a process of internal growth, rather than a response to contact with the Mediterranean world. There has been some discussion of Renfrew's concept of a 'chiefdom', but less attention has been paid to his attempts to illustrate social change in Wessex by the increasing investment of manpower expended in the construction of communal monuments. It is this second aspect of his case that we wish to consider here.

Renfrew attempted to define a series of social territories in Wessex, and related these to a sequence of public monuments. There were difficulties in assuming the contemporaneity of a number of these sites, but he did show that the ceremonial foci of his successive units involved a steadily increasing investment of communal labour. His main estimates are summarised below.

Type of site	Work investment per site	Number of sites	Date range
Long barrow	10,000 man-hours or less	132	c. 3400-2400 bc
Causewayed enclosure	c. 100,000 man-hours	7	c. 3300-2500 bc
Major henge monument	c. 1,000,000 man-hours	5	c 2200-1800 bc

The work needed to build the individual sites increased by a power of ten with each level of his hierarchy. These estimates are expressed in terms of man hours and are based on an empirical formula devised by Atkinson (1961) to calculate the effort expended in earthwork construction. In this paper we present alternative estimates for building these monuments and go on to consider the evidence which the sites themselves provide for the organisation of the work force and for its likely size.

Given these estimates, individual work effort can also be considered. Our figures are then discussed in the light of a different evolutionary scheme from that of Renfrew, and are compared briefly with the evidence for social organisation suggested by contemporary burials. The detailed calculations will appear elsewhere (Startin, forthcoming).

9.2 Work organisation

In this section the main emphasis is on the size of the labour force and that of the individual teams which made it up. The main tasks involved in building the monuments of prehistoric Wessex were stone-moving, timber-working, and earthwork construction.

On any stone-built monument, for instance a megalithic tomb or a stone circle, the minimum size of the labour force must be the smallest number of people required to move the heaviest stone (Startin, forthcoming). On the basis of Atkinson's (1960) experiments it is possible to estimate that a ton weight can be moved over level ground (on a sledge but without the use of rollers) by applying a force of 460 lbs. Nine or ten men could provide such a force throughout a day, only seven being required for a short-term effort. Careful adjustments can be made for sloping ground. Since we know the approximate source areas for the different stones used at Stonehenge, the minimum labour force needed in its second phase can now be set at 100 men. 600 men would have been needed to build Stonehenge IIIa but only about 60 men would have been required to rearrange the bluestones in phases IIIb and IIIc.

Estimates for timberwork have already been published in relation to European Neolithic long houses (Startin 1978), and the same principles can be applied to British monuments. Building a small timber-framed house, like that at Ballynagilly, Co. Tyrone (ApSimon 1976), would have required an estimated 550 man hours. Such a building would have housed one small family. The timber component of a long barrow like Fussell's Lodge (Ashbee 1966) might have required 2,900 man-hours, and the southern circle at Durrington Walls (Wainwright & Longworth 1971) would have taken 11,000 man-hours. The size of the labour force might have been determined by the minimum number of people required to manoeuvre the largest timbers.

Earthworks are more informative. Renfrew's estimates employed an empirical formula devised by Atkinson (1961), but this makes certain *a priori* assumptions and works better on small sites like burial mounds than on the large ceremonial centres where it gives rather high figures (Startin, forthcoming). Instead, the figures employed in this paper are based on engineering texts published before the advent of mechanical aids (Hurst 1886; Rea 1913). It is known from practical experiments that primitive excavating tools are less efficient than their recent counterparts (Jewell 1963) and the published figures are adjusted accordingly. Further adjustments are needed since the excavated spoil would have been removed by means of baskets rather than wheelbarrows. Two estimates are relevant here: on level ground a man carrying a 30 lb load and returning unladen would be able to average about 4 ft/sec; on slopes it can be shown that a basket chain would be more efficient. With monuments of any size a mixture of both methods might work best. This approach provides an estimate of 62,760 man-hours for the construction of all three rings of the Windmill Hill enclosure, compared with 64,800 man-hours by Atkinson's formula, and an estimate of 500,000 man-hours for the earthworks at Avebury (Smith 1965), compared with Renfrew's figure of 1,500,000 man-hours.

For smaller earthworks labour was apparently organised into one of two basic modes: 'pit-digging' or 'trench-digging'. Pit-digging may be used where a large number of people need to complete a task rapidly. This method requires a quite loosely organised work-force consisting of small teams, each digging separate pits. 'Pit alignments' are examples of this mode of work, although they are often interchangeable with ditches which may have been trench-dug (Simpson 1966). Rankine points out that an excavation team will need a working face of 5-6 ft in order to work efficiently (1865: 336); this agrees with the size of the pits at Windmill Hill (Smith 1965). Allowing for the space between pits, each team might occupy a length of ditch of about 10-12 ft. Thus at Stonehenge Hawley (1924-8) noted that the excavated length of ditch, some 550 ft, was composed of about 50 pits. The outer ring of Windmill Hill demonstrates that pits were sometimes dug side by side in order to create a wider ditch. As at Stonehenge, the baulks were rarely removed entirely.

With trench-digging the team works along a length of ditch from one end to the other. This method will generally leave a ditch with steep sides and a flat bottom, as for example at South Lodge Camp (Pitt-Rivers 1898). It is an efficient way of digging since each team works to a face, but as they occupy an entire length rather than a pit, it may limit the size of the labour force which can be used. For larger monuments labour organisation would necessarily have been more complex, but for practical reasons a form of trench-digging would generally have been more appropriate (Startin, forthcoming).

The two basic modes of working may have occurred sequentially. Whittle, referring to European Neolithic enclosures, remarks that "U-shaped ditches were not ... used in the very earliest enclosures....It may be that the appearance of the U-shaped ditch marks the appearance of more permanent enclosures with strong banks ..." (1977: 334). A similar sequence may be seen in the construction of British burial mounds and the gradual adoption of trench-digging can explain the use of flat-bottomed ditches on both henges and round barrows.

Examination of field monuments will often reveal that they were composed in several sections. For example, both the side ditches at West Kennet long barrow (Piggott 1962: fig. 2) can be divided into three lengths. At South Street long barrow, a pit-dug monument, the ditch lengths were apparently dug in sequence, one length on either side being completed before the next were begun (Ashbee, Smith & Evans 1979). Elsewhere the unfinished form of sites like Ladle Hill allows us to see how much of the circuit was being worked at one time (Piggott 1931). Since only so many people can fit into a given space, such information can be used to estimate the size of the labour force.

On this basis it is possible to reconsider the total work load represented by the monuments discussed by Renfrew. We can also estimate the maximum size of the labour force and the approximate contribution of each individual. The detailed figures are included in a separate paper (Startin, forthcoming).

Site	Estimated labour	Estimated labour	Individual work load
Long barrow (Fussell's Lodge)	6,900 man-hours	32-40	173-215 hours
Megalithic long barrow (West Kennet)	15,700 man-hours	100	157 hours
Causewayed enclosure (outer ring Windmill Hill)	48,100 man-hours	mode 1:80 mode 2:1,600	600 hours 30 hours
Large henge earthwork (Durrington Walls)	500,000 man-hours	250-500	1,000-2,000 hours
Stonehenge II	360,000 man-hours	at least 100	3,600 hours or less
Stonehenge IIIa	1,750,000 man-hours	600	2,917 hours

N.B. The estimates of labour force sizes for the henges can probably be considered as minimum figures. Two modes are offered for Windmill Hill. Mode 1 assumes that the total labour force was concentrated in one length of ditch and mode 2 assumes that the whole circuit was being excavated simultaneously. It should be noted that, apart from Hambledon Hill, this site is one of the largest causewayed enclosures and for this reason the estimate refers only to the outer ring.

9.3 Implications

The contrasts described by Renfrew are still apparent, but the overall range of the estimates has been greatly reduced. These contrasts now extend to the sizes of the labour forces and to the individual work loads. But they do not imply a gradual sequence, as Renfrew's explanation suggests.

With the exception of Hambledon Hill, the contrast between causewayed enclosures and long barrows is less important than it originally seemed, and the individual work effort was not necessarily greater for the enclosed sites. There is little evidence for a sharply graded hierarchy and both monuments were built using the same style of labour organisation.

The important contrast is between the labour requirements of causewayed enclosures and those of large henge monuments. Renfrew considered that these two types succeeded one another directly, a view which was widely held when his paper was written. For example, Atkinson had suggested continuous population growth throughout the Neolithic (1968). Again this interpretation must be challenged. More recent work would suggest that causewayed enclosures and henge monuments may belong to quite different traditions, separated by as much as 500 years. It now seems that small Neolithic ring ditches may be one intermediate form (Kinnes 1979). At the same time pollen analysis suggests that forest regeneration occurred widely in the mid-third millennium and that the Late Neolithic subsistence economy may have been less intensive than that of the preceding phase (Bradley 1978: 102-7). This is incompatible with Atkinson's model and may also undermine the assumptions on which Renfrew's interpretation depends. In place of steady evolution we must envisage a cycle of discontinuous growth.

In any case the large henges were not the first to be built. The earliest sites shed a new light on Renfrew's scheme. Among these monuments are the so-called 'pit-circle' henges, which may have acted as cremation cemeteries. As their name suggests, these sites were built by a pit-digging technique, rather than the trench-digging likely for the larger henges. The two most elaborate sites in Wessex were Stonehenge I and Maumbury Rings. The labour for their construction can be estimated as 11,000 and 35,000 man-hours respectively. These figures are actually rather limited and have more in common with estimates for earlier sites than those for the larger henges used in Renfrew's paper. Such low investments of manpower do not suggest significant growth from the preceding period.

Although these arguments seem rather negative, they do provide certain clues to a different understanding of the evidence. There is a massive contrast between the labour requirements of the major henges and those of earlier monuments and there is a corresponding difference in labour organisation, with pit-digging continuing in use into the later third millennium. The scale of the labour forces involved in the larger henges would suggest a far greater organisation of the work force, very probably extending over a longer period. Trench-digging was normally used and it may be relevant that Silbury Hill (Atkinson 1970; Burl 1979: 129-33), and possibly Stonehenge II and IIIa (Atkinson 1960: 206) were built continuously, despite several changes of plan. Both sites made considerable demands on manpower.

There is also some analogy between the evidence of work organisation and the structure of contemporary burials. The communal 'tombs' of the fourth and third millennia bc are themselves divided into 'segments' with different burial areas occasionally reserved for specific age or sex groups. Fleming (1972: 62ff.) has argued that the layout of these monuments is a reflection of the segmentary nature of contemporary society. In the same way, pit-circle henges could also act as communal cemeteries, although individual burials are certainly known from the same period (Kinnes 1979). It has been suggested that the internal divisions within some long barrows might relate to work-sharing and may not have a structural purpose (Benson 1971; Startin, forthcoming) and, except where a large labour force was attempting to finish a task quickly (see above), it may be that pit-digging was a form of work organisation appropriate to a segmentary society. The contribution of each separate group may have been emphasised in the final form of the monument. This would explain the care with which causeways were left intact when some of the enclosure ditches were recut. Patterns of this kind disappeared by the early second millennium when the principle of individual burial gained a far wider ascendancy, even where several deposits were placed in the same mound. The work was more tightly organised; trench-digging was widely adopted; the form of the monuments no longer emphasised the diverse character of the labour force; and the grave goods in the mounds seem to suggest a society with internal differences of authority and wealth. The layout of field monuments can be described as 'unitary' rather than 'segmentary'.

But such estimates of work effort will only be illuminating if they can be related to the other parts of the energy budget. Erasmus (1965) has produced an interesting study on these lines in which he compares the energy invested in the Maya ceremonial centres at Uxmal with the amount of time the local population might have spent in food production. He concludes that the task of construction was well within the capacity of contemporary society. It is hard to make similar estimates for Wessex. Our only indication of the amount of work invested in food production comes from ethnographic sources. Leach (1976) has brought together figures on energy expenditure in a sample of seventeen societies cultivating what he terms 'pre-industrial crops'. He estimates that a typical family of four adults would devote about 3,500 man-hours a year to subsistence farming. The full range of estimates is between 2,000 and 5,000 man-hours a year and on this basis the average contribution of each individual would be about 800 man-hours. These figures are more than the estimates for third millennium monuments but much less than those for the larger henges. There is too little evidence to show whether these early sites were built continuously, but clearly there was ample time for their construction. The second question concerns population density and remains important however the work was timed. Neolithic population density is unknown but two alternative estimates have been made for the Early Bronze Age—on the basis of the number of barrow burials or on analogy with ethnographic evidence. It is very unlikely that all of the population received barrow burial, but even Atkinson's very low estimate of the burial population would give a figure for Wessex of about 530 (Atkinson 1972) compared with a labour requirement of about 600 for Stonehenge IIIa. By contrast, the ethnographic evidence assembled by Green (1974) would suggest a population density of very roughly 10 per km^2. He argues that between 0.5% and 5% of the Bronze Age population received barrow burial and this argument would give a population of about 10,000 in the Marlborough Downs region and of about 50,000 in Wessex as a whole. On this basis between 1% and 5% of the population could have undertaken the building of Stonehenge or Avebury respectively. Given the scale of the individual contributions, work may now have run for longer periods and may have involved specialist labour. If this estimate is of the right order, the work would hardly have interfered with the business of food production, even allowing for the need to produce a surplus while building was in progress. Indeed, agriculture may have been less intensive between the Middle Neolithic and the Middle Bronze Age (Bradley 1978).

9.4 Summary

This paper, then, makes three suggestions. First, monument building in Wessex need not have made an enormous demand on the energy budget, unless the work force was drawn from a restricted area. Secondly, the labour requirements of the major monuments of the second millennium bc are different in kind from those of the preceding Neolithic. This applies both to the energy expended in construction and to the size of the workforce. Against this background, there is very little evidence of the continuous evolution suggested by Renfrew. Lastly, a contrast can be seen in terms of work organisation, and differences of this kind may be reflected by the differences in political structure suggested by contemporary burials. This last point may take the evidence too far, but the building of so many monuments has always invited speculation. Let us end with John Aubrey:

> "Round about Stonehenge one may count 43 or 45 barrows, some much bigger than others. I am not of the opinion as most believe that these were all made for burying the dead after some battles here: 'twill require a great deal of time and leisure to collect so many thousands of loads of earth...Soldiers have something else to do". (Monumenta Britannica)

REFERENCES

ApSimon, A. M., 1976. Ballynagilly and the beginning and end of the Irish Neolithic. Acts of the 4th Atlantic Colloquium, Ghent 1975.

Ashbee, P., 1966. The Fussell's Lodge long barrow. Archaeologia, 100, 1-80.

Ashbee, P., Smith, I. F. & Evans, J. G., 1979. Excavations of three long barrows near Avebury, Wiltshire. Proc. Prehist. Soc., 45, 207-300.

Atkinson, R. J. C., 1960. Stonehenge. London.

Atkinson, R. J. C., 1961. Neolithic engineering. Antiquity, 35, 292-299.

Atkinson, R. J. C., 1968. Old mortality: some aspects of burial and population in Neolithic England. In J. M. Coles & D. D. A. Simpson (eds.), Studies in Ancient Europe. Leicester. 83-93.

Atkinson, R. J. C., 1970. Silbury Hill, 1969-70. Antiquity, 44, 313-314.

Atkinson, R. J. C., 1972. Burial and population in the British Bronze Age. In C. Burgess & F. Lynch (eds.), Early Man in Wales and the West. Bath. 107-116.

Benson, D., 1971. Ascott-under-Wychwood. Curr. Arch., 24, 7-10.

Bradley, R. J., 1978. The prehistoric settlement of Britain. London.

Burl, A., 1979. Prehistoric Avebury. London & New Haven.

Erasmus, C., 1965. Monument building: some field experiments. S.W. J. Anthrop., 21, 277-301.

Fleming, A., 1972. Vision and design: approaches to ceremonial monument typology. Man, 7, 57-73.

Green, H. S., 1974. Early Bronze Age territory and population in Milton Keynes, Buckinghamshire and the Great Ouse Valley. Arch. J., 131, 75-139.

Hawley, W., 1924-8. Excavations at Stonehenge. Antiq. J., 4, 30-39; 5, 21-50; 6, 1-16; 7, 149-176.

Hurst, J. T., 1886. Architectural surveyor's handbook. London.

Jewell, P. A., 1963. The experimental earthwork at Overton Down, Wilts. London.

Kinnes, I., 1979. Round barrows and ring-ditches in the British Neolithic. London.

Leach, G., 1976. Energy and food production. Guildford.

Piggott, S., 1931. Ladle Hill: an unfinished hillfort. Antiquity, 5, 474-485.

Piggott, S., 1962. The West Kennet long barrow. London.

Pitt-Rivers, A. H., 1898. Excavations in Cranborne Chase, Vol. IV. London.

Rankine, W. J. M., 1865. A manual of civil engineering. London.

Rea, J. T., 1913. How to Estimate. London.

Renfrew, C., 1973. Monuments, mobilisation, and social organisation in Neolithic Wessex. In C. Renfrew (ed.), The explanation of cultural change. London. 539-558.

Simpson, W. G., 1966. Romano-British settlement on the Welland gravels. In C. Thomas (ed.), Rural settlement in Roman Britain. London. 15-25.

Smith, I. F., 1965. Windmill Hill and Avebury. Oxford.

Startin, D. W. A., 1978. Linear pottery culture houses: reconstruction and manpower. Proc. Prehist. Soc., 44, 143-160.

Startin, D. W. A., forthcoming. Manpower and labour organisation in prehistoric Wessex.

Wainwright, G. J. & Longworth, I. H. (eds.), 1971. Durrington Walls, Excavations 1966-1968. London.

Whittle, A., 1977. Earlier Neolithic enclosures in north-west Europe. Proc. Prehist. Soc., 43, 329-348.

10. LATER NEOLITHIC SOCIETY IN BRITAIN: A REALIGNMENT

Alasdair Whittle

Vain was the chief's, the sage's pride !
They had no poet, and they died.
Alexander Pope, Imitations of Horace, Odes

10.1 Introduction

Apart from its intrinsic scientific interest the question of astronomy in the Later Neolithic of Britain, like so many other aspects of the period, raises essentially social issues, whose resolution requires a broad attempt at reconstruction of social organisation in the country. This is hardly an original viewpoint since several authors have already addressed themselves to this general question (e.g. Renfrew 1973; 1979; MacKie 1977) and there is growing attention to the need for social reconstruction. How is this to be achieved? It is clear that what we may call the particularist approach of MacKie, while offering several interesting insights, is too limited, since it bases a specific social model (of dominant astronomer-priests) on a controversial view of astronomy and forces this is on to a partial range of the evidence. A more generalising approach is likely to be more successful, in which as many aspects as possible of social organisation are considered together with the relations between them which determine the functioning of a given social system (Rowlands & Gledhill 1977; Friedman & Rowlands 1977). A similarly broad but more static approach underlies Renfrew's chiefdom model for Later Neolithic Wessex (1973; 1974) and its succeeding counterpart model of segmentary societies in Earlier Neolithic Orkney and Western Europe as a whole (1976; 1979), with their clear definition of the kinds of factors that may be involved in differing sets of social relations. These models are open to criticism as leading to 'check-list' archaeology rather than to an analysis of the actual prehistoric cases in hand and their underlying regularities, since they offer generalised (almost universal) social forms derived from the ethnographic record. Indeed in the case of chiefdoms, Service's initial model (1962) was followed by later apostasy (1971), which has been reinforced by others who have pointed out the variety of kinds of leadership, for example, which might be forced into a catch-all chiefdom model (e.g. Dalton 1977; Kehoe 1974). Nonetheless the definition of factors such as ranking, redistribution and centralisation has been of great value, and can be easily used in a more flexible approach, which it was predicted anyway would be necessary (Renfrew 1973: 557). A more practical criticism is that all too rarely has the required breadth of approach actually been applied. Communal monuments for example were the focus of analysis in the case of Later Neolithic Wessex, but it is unlikely that one aspect of a society's activities need uniquely reflect its whole structure or organisation, and one such analysis deserves to be set against the other evidence derived from burial, exchange, redistribution and so on.

These aspects themselves may not individually tell the whole story. There is considerable danger, for example, in the view (e.g. Binford 1972; Tainter 1977) that social relations are most clearly expressed in mortuary practices, if this were taken to imply that other social activities may be dismissed from consideration. The aim of this paper therefore is to try to apply a broad approach in which more than one social activity is considered. The range offered is hardly exhaustive since I will offer some remarks on settlement and economy, with some analysis of communal organisation and centralisation, exchange and redistribution, status and ranking. I am thus open to the very criticism just outlined, but even from this summary treatment I hope that a more complicated model emerges both of social organisation in the Later Neolithic period and of social development before this. From the contrasts that result some clues to the relations between different social activities may emerge, but this will remain an ill-understood aspect.

The main period under discussion is the Later Neolithic, here taken to subsume the traditional Early Bronze Age, as already briefly outlined elsewhere (Whittle 1980), and thus to run from c. 2500 to after 1500 bc in uncorrected radiocarbon chronology. In this scheme, designed to circumvent the distractions from continuity which the Three Age System insidiously entails, I have envisaged pre-Beaker (2500-2200/2100) early Beaker (2200/2100-1800/1700) and late Beaker/Urn phases (1800/1700)–after 1500). More detailed attention will in fact be paid to the former two of these phases but both the latter and the Earlier Neolithic will also come into consideration.

The discussion requires first some lengthy scene-setting in order to argue against an ethnic interpretation of the Beaker phenomenon, still supported by some authors in one form or another (e.g. Clarke 1976; Case 1977; Kinnes 1979), though increasingly doubted by others (e.g. Shennan 1976; 1977; 1978; Burgess & Shennan 1976). It is important to deal with remaining doubts over the British case. The detail offered may be justified by this need and by the moral underlying this volume, that complex issues should be resolved by complex and detailed treatment.

10.2 Later Neolithic settlement: native economy and regional developments

It is important briefly to outline the nature of Later Neolithic settlement in the country as a whole, since its characteristics including regional developments provide a necessary setting for the analysis of social organisation which follows. Though the evidence points to diverse and regional trends, a picture emerges overall of active, well-organised native communities. At the risk of gross over-simplification this pattern may be seen to continue into the traditional Early Bronze Age so that this background can cover virtually the whole of the period under discussion from other points of view.

The altered nature of the landscape by the mid third millennium bc is likely to have affected subsequent developments. On the chalk of the south open grassland was widespread and perhaps permanent (Evans 1972). Such a stage may have been preceded by exploitation including cultivation on a considerable scale, seen in evidence for lynchets at South Street long barrow, Wilts (Ashbee et al. 1979) and at Bishopstone, Sussex (Bell 1977), for larger deposits of soil wash at Brook, Kent (Kerney et al. 1964), and for the possible degradation of brown forest soils to thin rendzinas (Limbrey 1975). In the Thames Valley there may be similar evidence for soil change (Limbrey 1975: 190) and elsewhere pressure of some kind is surely reflected in the closing of many Earlier Neolithic clearances (Bradley 1978; Whittle 1978). This need not be a universal development: the substantial quantities of turf in the mounds of Knowth, Co. Meath (Eogan 1974) at approximately this date may serve to underline this point. Nor is the scale of other environmental changes clear, since both blanket peat growth and podsolisation could be documented from earlier dates (e.g. Smith 1970). Many discussions of the Later Neolithic period have assumed unbroken development of settlement from the Earlier Neolithic with a concomitant steady rise in population but this is increasingly unlikely in the light of the above, the more so as suggestions of clearance and environmental change are pushed back into the Mesolithic period, altering our view of resources even at the beginning of agricultural settlement. There are therefore strong grounds for seeing Later Neolithic development in the context of an altered setting.

Within this setting settlement and economy was diverse and regionalised. I have noted several regional trends (Whittle 1980). A continuation of the previous pattern is apparent in some areas, as in the northern and western isles with mixed farming well represented, or on the southern chalk with grassland. In other areas a strong recovery can be seen in pollen diagrams with a renewal and perhaps extension of clearances in northern Ireland, Cumbria and East Anglia. Dated clearances have been seen to peak c. 2000 bc (e.g. Bradley 1978: fig. 6:2) but far fewer dates are available for the Later period as a whole than for the Earlier. Many come from northern Ireland, where blanket peat spread considerably from around 2000 bc, few from Cumbria or East Anglia and none from areas such as Dartmoor and the Cornish lowlands. Since these were definitely occupied in the earlier second millennium bc, the chronological trend may be biased. Clearance was also extended into higher parts of the uplands such as the Pennines or North Yorkshire Moors which had largely been avoided earlier (e.g. Turner *et al*. 1973; Tinsley 1975).

Two features are of special interest. First, mixed farming continued to be widespread while some areas were used perhaps predominantly for pastoral purposes such as on the fen-edge at Fengate (Pryor 1978), or on the southern chalk, though cultivation is documented in the form of ard marks under barrows and of cereal impressions on pottery. Secondly, the scale of land division evident at some sites is impressive. The best example is the pre-peat fields of Co. Mayo and other parts of western Ireland (Caulfield 1978), with areas at least 1500 by 900 m enclosed; similar examples may come from the north (e.g. Davies 1939). Native communities therefore had a well-organised landscape, and the scale of control and of specialisation suggest that the economy in many areas was on a more developed level than self-sufficient subsistence. This is hard to test further in the absence of settlement data, though faunal remains from two major henges, Durrington Walls and Mount Pleasant (Harcourt *in* Wainwright & Longworth 1971; Wainwright 1979) do provide some support in having pig as the numerically dominant species, though they were not the largest suppliers of meat. The absence of cereals at such sites is unproven (*pace* MacKie 1977) either as grain or processed food, in the absence of the application of appropriate recovery techniques. Cereals have been found in Grooved Ware contexts from the Thames Valley (David Miles, pers. comm.), Dorset (Martin Jones, pers. comm.) and Skara Brae (Clarke 1976).

10.3 Beaker assemblages

It has been the conventional view for a long time that Beaker assemblages constitute a Beaker culture which represents a distinctive ethnic grouping, and historically, greatest effort has been expended on tracing the origins of this culture (summarised in Harrison 1974). In this sphere there were occasional doubts expressed. The following modern-sounding quotation—

> "The problem of whether water, milk, blood, beer or mead was drunk out of the beakers is quite as interesting as whether their shape originated in Spain or Mongolia"

comes from a note by T. C. Lethbridge in the report on Fifty Farm, Suffolk (Leaf 1935). But the Beaker culture survived to be the last allowable immigration of the invasion hypothesis in British prehistory (Clark 1966). David Clarke's classic treatment of the British evidence (1970) marks the culmination of this process, with its numerous groupings even within the British Beaker culture, each with distinctive and separate continental origins.

There then followed a phase of modification. Lanting and van der Waals' review of Clarke's work (1972) reaffirmed the belief in an ethnic group with continental origins, but posited only one immigration at the outset and substituted four main areas of subsequent regional development for Clarke's complex and geographically diffuse groups. Clarke himself offered a radical new perspective on the problem on a European scale by distinguishing between types of 'Beaker presence', which suggested varying distribution of both Beaker people and Beaker users (1976). At the same time in more detailed fashion Shennan (1976; 1977; 1978) showed that in central Europe Beaker assemblages were only a specialised and varying addition (perhaps 'status kits') to local traditions. However Clarke was still inclined to see Beaker people rather than merely users in the British Isles, or at least the eastern part of it, which covers the bulk of the British distribution (1976: 474). As far as Britain was concerned it was left to Burgess and Shennan (1976) to suggest that Beaker assemblages represent an international 'package' associated perhaps with fashionable cult practices, rather than an intrusive ethnic grouping with a series of associated, wide-ranging innovations.

Published opinion is still divided about this. Case has envisaged a first phase of Beaker diffusion attributable to fashion followed by an immigration of a distinctive human group with distinctive economic practices (1977). Kinnes has also argued (1979: 75) for 'some component of population movement from across the North Sea'.

Since Beaker assemblages provide such an important part of the evidence which can be used to discuss social organisation in the British Later Neolithic but have been largely ignored in recent overall discussions of this, it is necessary to attempt to resolve the issues outlined above, for the interpretation chosen will radically affect any overall view of the period. I propose therefore to discuss a range of pre-Beaker innovations in the Later Neolithic, to present the lack of distinctive settlement evidence associated with Beaker assemblages before the late phase (starting c. 1800-1700 bc) in Case's terms (1977), and then to explore some of the implications contained in such a confirmation of the rejection of essentially ethnic interpretations of the British Beaker assemblages.

10.4 Pre-Beaker innovations

Several important innovations or changes of the Later Neolithic which have often been considered to be linked exclusively with the intrusive Beaker culture can now be seen to be either definitely or possibly pre-Beaker in appearance (Burgess & Shennan 1976). These include robust, bracycephalic skeletal types; burial in round mounds with individual distinctive grave goods; metallurgy; flat-based pottery; substantial communal monuments; and finally the renewal of clearance activity in the later third millennium bc.

The appearance in the burial record contemporary with Beaker assemblages of a more robust, broad-headed human physical type, noted since the nineteenth century and reaffirmed recently (Brothwell & Krzanowski 1974), has often been seen as the clinching argument for a Beaker ethnic intrusion. At best however the significance of this development may be an open question, and at worst an irrelevance. Burgess (Burgess & Shennan 1978: 321) argued that crucial gaps in the evidence, from the mid to the later third millennium bc, are missing caused by the end of collective burial in long barrows and chambered tombs from the earlier date. To this we may add the practice of cremation in native cemeteries of the later third millennium bc (Kinnes 1979: 67-9). In fact some of the evidence from the intervening period in which transitions could have occurred did survive, though it has been little commented on. Important sites like Duggleby Howe and Liff's Low, of uncertain absolute date, but probably of the later third millennium bc and classed by Kinnes as belonging to stages C-E (1979: 16-18), had a series of inhumations. The characteristics of the former are described at length by Garson (in Mortimer 1905: 30-42). Garson was struck by the dolicephalic character of the skulls in metrical terms but two of his eight skulls were 'mesaticephalic', and in qualitative terms the flat-vaulted, coffin shaped skulls of burials G, L, J and M had a superficial, though in his view illusory, resemblance to skulls of the 'Round Barrow' period. Interestingly, the range of types seems evenly spread through the phases of burial. The primary inhumation at Liff's Low had a 'fine and intellectual' skull (Bateman 1848: 42) which is ambiguous but may be reminiscent of the later practice of equating large broad skulls with persons of authority (cf. Keith in Cunnington 1929: 57). The primary inhumation at Linch Hill, Stanton Harcourt, Oxon (Grimes 1944; see also below) was a slightly dolicocephalic female, 'nearer the longheads than the roundheads'. The young man of the secondary burial, with an accompanying Beaker, had a cephalic index slightly closer still to the longheads (75 as opposed to 77). In the far north, metrical data were not obtained on the skull fragments in the recently excavated chambers of Quanterness on the Orkney mainland, where deposition spanned this transitional period; stature seems to have been comparable to that of the populations of the Early Neolithic (Chesterman in Renfrew 1979).

Within the Beaker-associated series the range of variability and the size of sample available for the re-study urgently needed are both unknown. It is likely that 'broad-headed' became a catch-all term in describing skeletons of this period. It is at least clear that the association between Beakers or Beaker assemblages and bracycephalic types was not exclusive. Examples of association with 'dolicocephalic' skeletons, in crude terms, include the burial in the Thickthorn Down long barrow (Drew & Piggott 1936) the secondary burial at Stanton Harcourt (Grimes 1944), and that at Wherstead, Thorington Hall, Suffolk (Clarke 1970: no. 960). Strikingly, the late cemeteries at Cassington and Eynsham had a range of types, with dolicocephalic skulls more numerous, and others in the Oxford region were of intermediate character, including that at Radley 4a with the famous earrings (Case 1956: 14, n. 36). There was some variation too in the Barnack multiple round barrow (Wells in Donaldson 1977: 223). The small number of these could to some extent be offset by the large number of

Beaker burials in which age, sex or any skeletal details are poorly recorded, if at all. This doubt can be set in a wider context by comparison with Gerhardt's study of Beaker skeletons centred on south-west Germany (1976). For the purpose here three results stand out—the regional variation within the study area, the variant physical types alongside the stereotyped (planoccipital Steilkopf) Beaker male, and the difference between males and females. One may also note in parenthesis that even this detailed study was based on a rather small sample. In a wider study Ricquet (1970) has indicated two trends in the skeletal record of west and central Europe in the third millennium bc, the first considerable variability both within and between regions, as measured perhaps rather crudely by cranial data, and the second what he calls a general 'bracycephalisation'. His 'floraison des éléments bracycrânes' and 'degracilisation générale' affected areas from the Paris Basin to Scandinavia and western Russia (183-6). He argued that Corded Ware populations were not uniform and though he saw Beaker populations as radically different, noted also that the sample was small (106). A further continental study has emphasised that there is still much to understand about the physical behaviour of populations at this time. Louwe Kooijmans (1974: 320) has shown that in the Netherlands the most abrupt physical change is associated not with Corded Ware but with Beaker assemblages; this contrasts with the essential continuity of pot styles, associated grave goods, and burial traditions established from the PFB horizon (Lanting & van der Waals 1976). There can therefore be no certainty that either here or in neighbouring countries physical and cultural developments marched hand in hand. The evidence is badly recorded, and the whole question of variability (cf. Henneberg et al. 1978) needs reinvestigating by means other than simply cranial measurements. Presumably evidence for deliberate alteration of head shape (cf. Morris 1977; Kiszely 1978) would have by now been detected but as this is generally intimately connected with individual status it is a relevant possibility.

The specific Beaker contribution to burial traditions is now unclear. Burgess and Shennan have pointed to the existence of individual inhumations in both long and round mounds from the Earlier Neolithic, as well as cremation cemeteries and multiple inhumations in round mounds from the pre-Beaker Later Neolithic (1976: 315-7). Kinnes has refined this wealth of evidence into six pre-Beaker stages (1979). From the earlier Neolithic one may note the considerable diversity of burial form, incorporating in Yorkshire for example cremation pyres, cremation pits, linear zones, wooden and stone boxes, trench chambers and enclosed cemeteries. Single burials with personal grave goods are placed as early as Stage C, represented by Aldro 94, and tentatively dated to the late fourth/earlier third millennium bc. However the proposed sequence envisages the replacement of such elements in the Later Neolithic by cremation cemeteries, as seen for example in the undated sequence at Duggleby Howe, and in dated or dateable contexts at Stonehenge and Meldon Bridge, probably before any Beaker presence in this country. Such a break however is ill at ease with the great diversity of burial forms outlined and the radiocarbon dates are too few for certainty, virtually avoiding the second half of the third millennium bc (ibid.: fig. 8.1). The seriation-based sequence (ibid.: fig. 6.2) may not do justice to individual pieces of evidence. Thus the Towthorpe bowl of the first phase at Duggleby Howe is put in stages B-C of the Earlier or Middle Neolithic, (cf. Burgess & Shennan 1976: 317) but elsewhere such pottery has Grooved Ware complex associations (Manby 1974: 100). It is hard to reconcile the sequence at Linch Hill Corner, Stanton Harcourt, Oxon.

(Grimes 1944; 1960: no. XXI, 1), in which a Beaker burial with a crude Northern-Middle Rhine vessel (Clarke 1970: no. 772) cut the only partially silted ditch of a ring-ditch with a jet belt-slider of Peterborough associations, with a substantial hiatus between two in essence similar traditions of single burial. Grimes concluded (1944: 40) that the secondary ditch 'was dug after the process of quick silting of the larger ditches had been completed and the process of slower silting commenced.' The secondary burial cannot therefore have followed immediately but neither can an inordinate period of time have elapsed before it was made. If a more mixed tradition is thus supported in the later third millennium bc, the renewed claim for the distinctiveness of Beaker small round barrows, flat cemeteries, mortuary ritual and disposition of the body (Kinnes 1979: 75) must be open to doubt.

In addition to the structural component further comment may be made on mortuary ritual. The first point is negative. The bulk of Beaker burial evidence must date from after 2000 bc. There is clear evidence from the neighbouring continent that body disposition and orientation varied considerably both in space and time in Beaker burials in the Netherlands and in the Saale area (Louwe Kooijmans 1974: fig. 123), though these were tightly controlled within specific Beaker groups with the exception of AOO and Maritime groups. By contrast Clarke's data (1970: appendix 3.10) suggest more catholic preferences in British Beaker graves. Therefore a distinctive Beaker innovation in this sphere in this country is difficult to substantiate. The second point is more positive. To the earlier native tradition of single burials with personal grave goods can be added the practice of excarnation represented in burials by disarticulated and incomplete skeletons. This practice clearly has a long history back into the Earlier Neolithic, but recurs also in Later Neolithic contexts, as again at Duggleby Howe (Burial J: Mortimer 1905: 29), Cowlam 57 (ibid.; Kinnes 1979: 17), Handley 27 (Pitt Rivers 1898: 140-2) and Aldwincle 4 (Jackson 1976). This is also an element of Beaker burial practice (and indeed subsequently in the Early Bronze Age). Examples include Stanton Harcourt Aerodrome (Hamlin 1963) and Dorchester, Oxon (Inventaria Archaeologica 1955) in the south, and Aldro 54 (Mortimer 1905: 63), Aldro 116 (ibid.: 54-6), Hanging Grimston 55 (ibid.: 100-102), Painsthorpe 4 (ibid.: 113-17), Garrowby Wold 104 (ibid.: 134-6), Garton Slack 163 (ibid.: 214-15) and Rudston 62 (Greenwell 1877: 235-45) in the north. A Wessex example from Amesbury G71 (Christie 1967: phase II, two examples in central grave) predates Food Vessel and Urn burials. Some of the Yorkshire examples such as Aldro 116, Painsthorpe 4 and Rudston 62 may represent earlier disturbed burials but cumulatively the evidence for the practice is striking. The presence of some sort of bier at Dorchester site XII also serves to reinforce the reality of movement of disarticulated and incomplete burials. As far as I am aware, this was not part of continental practice, but its representation from early Beaker burial onwards in this country is another detail underpinning the overall picture of continuity from the later third millennium bc and indeed earlier.

The deposition of personal grave goods with individual burials clearly precedes the Beaker horizon (Kinnes 1979). The appearance of flat based pottery, in Grooved Ware style, also predates this horizon, on the basis of available radiocarbon dates (see below), though this was not part of mortuary assemblages. It is difficult to claim native origins for all the particular items of the distinctive Beaker grave assemblages, besides missing the point, but the technology of some of them may have a pre-Beaker background as well as the concept itself of such specialised assemblages. One of the classic roles assigned to a Beaker 'people' has been that of mobile metallurgists moving in search of new ores and markets. This at times has been given general European expression (e.g. Childe 1929; Sangmeister 1972) and related also specifically to the British case (Case 1966). Case however recognised the difficulties for the argument that Beaker people were the first metallurgists in Ireland and the rest of the British Isles inherent in the low Irish Beaker presence and the lack of direct association between Irish Beakers and early copper products. The doubt was raised again by Harbison (1973: 103) and should remain in our minds, reinforced by two further points. The first concerns the preceding continental diffusion of copper metallurgy. Copper products, if not copper metallurgy itself, were widely though sporadically represented in third millennium bc contexts in adjacent parts of western Europe, from Seine-Oise-Marne tombs in the Paris Basin (Bailloud 1967), Pfyn and Cortaillod settlements in the Alpine foreland of Switzerland (Ottaway & Strahm 1975), Dutch hunnebedden (Buinen, Drouwen, etc.) to Danish TRB hoards, barrows and graves (Bygholm; Barkaer; Salten) (Ottaway 1973 and references). Products include <u>Blechstil</u> ornaments and heavy trapezoidal axes; crucibles were found at several Pfyn settlements (Wyss 1969). The British Isles are often seen as recipients of local metallurgy at the same horizon as the Low Countries and south Scandinavia. In fact being the last major untapped source of ores they may possibly have seen earlier developments, particularly in view of the continental background, and the similarity of Irish thick- and broad-butted axes to continental examples (cf. Case 1966). The second point leads on from this. Finds on settlements of early metalwork are rare, even comparatively so in the later Bronze Age, and settlements themselves of the mid to later third millennium bc are rare too. As will be argued below there was considerable constraint or selection on what artefacts appeared in mortuary assemblages, for example only a handful of axes in Early Bronze Age graves and none in Beaker graves. This selectivity has a pre-Beaker background too, particularly in Ireland in those assemblages represented in the passage graves. It is unrealistic therefore to suppose that evidence of a putative pre-Beaker metalworking phase is likely to have survived. The question must remain far more open than commonly considered.

Beaker pottery is quite widely represented at communal monuments of the Later Neolithic. There has not been a sustained claim that Beaker people were responsible for the creation of the henge and cursus traditions, both of which now seem to have origins in the mid-third millennium bc (summarised in Smith 1974). The major henges excavated in recent years by Dr. Wainwright, Durrington Walls, Marden and Mount Pleasant, seem also to have primary associations with Grooved Ware pottery and artefacts (Wainwright & Longworth 1971; Wainwright 1971; 1979), seen also at Avebury (Smith 1965). There are nonetheless sites with apparently exclusive Beaker primary associations (Wainwright 1969) and sites like Stonehenge which are seen to have been subject to a Beaker 'takeover' (Case 1977: 80, 82). These cases are largely illusory and cannot be taken to support separate activity by distinctive Beaker people. Candidates for Beaker primacy have been seen to include Gorsey Bigbury (ApSimon 1976), Ballymeanoch (Wainwright 1969), Cairnpapple (Piggott 1948), Fargo (Stone 1938), Rudston (McInnes 1964), Llandegai B (Houlder 1976) and Dorchester Big Rings (Site XIII: Atkinson <u>et al</u>. 1951). But the associations are largely secondary or indirect (Gorsey Bigbury as noted by the excavator, Rudston, Llandegai, Fargo), or based on the presence of very few sherds (Ballymeanoch, Cairnpapple, Dorchester). Indeed in the Oxfordshire case a neighbouring henge, the Devil's Quoits at Stanton Harcourt, had virtually no material in its substantial ditches (Margaret Gray, pers. comm.). 'Takeovers' too tend to be based on such insubstantial quantities as the thirty odd weathered sherds of Atkinson's phase 2 at Stonehenge. These cases can be put into perspective by the relative quantities of numerous Grooved Ware and scarce Beaker pottery at Durrington Walls (Wainwright & Longworth 1971), and the sequence at site IV at Mount Pleasant, in which Beaker pottery joins Grooved Ware in the secondary silts and becomes more numerous than it but by no means supplants it over the site as a whole (Wainwright 1979). Where henges were kept clean or largely so, and in a situation where both Grooved Ware and Beaker pottery were in contemporary use, the presence of a few sherds of the latter can hardly be diagnostic.

Finally, distinctive Beaker economic practices may be considered. It has been held that Beaker groups particularly were the 'energetic mixed farmers' of the Later Neolithic (Case 1977: 76), in the absence of cereal finds and cereal processing equipment in Grooved Ware contexts (Wainwright & Longworth 1971). Cereal impressions on Early and Middle Style Beakers, mainly from the chalk of southern England, are well known (Helbaek 1952), and conversely both long-lived grassland and woodlands offering pannage are suggested by molluscan and faunal studies at major henges like Durrington Walls and Mount Pleasant. The evidence however has been treated in a simplistic manner. The lack of cereal finds in Grooved Ware contexts is a product of traditional practice of not sieving for floral remains, even in major excavations of recent years. Where the technique has been applied barley has been recovered in Grooved Ware contexts in all levels of recent excavations at Skara Brae on the Orkney mainland (Clarke 1976), and at Barton Court Farm Oxfordshire (David Miles, pers. comm.) and at Down Farm, Dorset (Martin Jones, pers. comm.); Hunstanton, Norfolk yielded no floral remains at all (Ian Kinnes, pers. comm.). It remains now to be seen whether all stages of crop processing and cleaning are represented or whether cleaned grain or even bread was imported to high-status settlements (cf. MacKie 1977: 222), but the weight of evidence has shifted. Dennell has also pointed out that as far as cereal impressions are concerned the context of pottery manufacture and the possible association of Beakers with beer or other beverages derived from barley must be taken into consideration (1976). Season of manufacture, and the generally poor quality of Grooved Ware fabric may also be relevant factors. It would not be surprising for the manufacture of one pottery style not to reflect the full range of contemporary agricultural practice. Ploughmarks of this horizon have been assigned to Beaker ploughmen (Case 1977: 76) but this is again ill at ease with the evidence. Marks under the Avebury bank (Evans 1972) are unlikely to have Beaker associations, and those under Barrow G71, Earl's Farm Down, Amesbury Wiltshire (Christie 1967) were followed first by a probably unaccompanied inhumation in a ring ditch, then by a further unaccompanied inhumation under a small barrow with a C14 date of 2010 ± 110 bc (NPL-77). A phase of ploughing across the silted ditch of the South Street long barrow was accompanied by scanty sherds of three Beaker pots and one vessel of uncertain affinities, and was immediately underlain by fragments of at least three Ebbsfleet/Mortlake bowls (Smith *in* Ashbee *et al*. 1979: 273, layers 6 and 7).

The ambiguities of this evidence are further heightened by consideration of the results of dated pollen analysis (partially summarised in Bradley 1978; Whittle 1978). After some sort of standstill in the mid third millennium bc, clearances were renewed or extended from the later third millennium bc, from a pre-Beaker horizon. The quantity of dated sites varies regionally but the trend may be seen in East Anglia and Cumbria as well as in the best documented area, northern Ireland (Smith 1975). It is appropriate therefore that at the upland site of Ballynagilly, Co. Tyrone (ApSimon 1976) and in its neighbourhood, dated clearances of this horizon should begin before the appearance of dated Beaker contexts on the site. The latter are dated by UB-200, -309, -316, -356, -555, -556 and -558 from 2100 ± 50 to 1900 ± 55 bc (Radiocarbon 13, 1971 and 15, 1973) whereas at Ballynagilly renewed clearance activity seems to be bracketed by the dates UB-248, -249 and -250 (2005 ± 55, 2075 ± 65, 2390 ± 65 bc) (Radiocarbon 12, 1970), and elsewhere is dated by UB-111 (2250 ± 80 bc), at Ballyscullion and UB-440 (2230 ± 90 bc) at Sluggan, in Co. Antrim, and by UB-274 (2215 ± 80 bc) at Slieve Gallion in Co. Tyrone. Other dated sites (Whittle 1978: 35) are apparently contemporary with the Beaker horizon at Ballynagilly. Dated sites in other parts of the country (Whittle 1978: 35-6) fit the same pattern. The interpretation presented here perhaps relies on finer resolution than the radiocarbon method can yet supply, but on face value the pattern does seem clear particularly as in the northern Irish case the Beaker presence is strictly limited anyway and there are dated Sandhills contexts preceding it, as at Ballynagilly itself (ApSimon 1976: 21). Admittedly the Irish case need not reflect the situation of the rest of the country, but cumulatively the evidence supports a picture of renewed clearances for both grazing and arable purposes beginning before the appearance of Beaker assemblages in this country as a whole.

Together these arguments confirm the views cited earlier that Beaker assemblages, while distinctive and in some senses novel in themselves, have a limited range of associated innovations. This hypothesis is further strengthened by the following reconsideration of Beaker settlement, which has often been seen as a final argument in favour of separate Beaker groups in this country (Clarke 1976; Case 1977; Kinnes 1979), in line with the settlement evidence available in some neighbouring countries, especially on the Lower Rhine (Louwe Koojimans 1974; Lanting & van der Waals 1976: 71). The evidence is complicated, and involves as yet unresolved chronological problems, but this point alone should make the question more open than it is generally considered to be.

10.5 Beaker settlement: general considerations and chronology

The conventional view of Beaker settlement is that it is abundantly represented at all phases (Clarke 1970 passim; Kinnes 1979: 72) and that any gaps in the record, as with several of Clarke's groups, can be explained away by the accidents of discovery and excavation (Clarke 1970: 88, 122, 156, 167, etc.). Lamenting 'this extraordinary state of affairs' with his Primary Northern/Dutch (NI/D) group, Clarke continued 'although other unknown factors may be operating, this lamentable fact is probably in part accident, andin greater part bound up with the lack of application of modern settlement detection and excavation techniques, as well as the diversion of money and inclination from settlement problems to 'rewarding' barrow digging' (1970: 156). For others who follow Lanting and van der Waals' critique of these groups these gaps should be much less critical. Nor do all who espouse the Beaker settlement cause necessarily argue for settlement in all phases. Thus Case (1977: 74)—'Nothing gives definite grounds for suggesting an influx of settlers into Britain and Ireland in the Early phase'. But there is general agreement that from a comparatively early stage of Beaker development in this country, mortuary assemblages are paralleled by a series of settlement sites.

Against this model I will argue the radical proposition that there is no clear evidence for distinctive or separate Beaker settlement before the Late phase, in Case's terms for convenience, and that by this phase (if not earlier) there are anyway serious grounds for doubting the meaning of 'separate Beaker settlement'. I will argue that the lacunae in the evidence which Clarke lamented may indeed be real, and that settlement evidence belonging to early groups may be either not separate from that of other contemporary cultural traditions or illusory and largely overplayed. Only in the late phase does the evidence for separate Beaker settlement seem convincing, and these sites may in fact largely belong to the conventional Early Bronze Age period.

Much of the argument involves chronological considerations and so for convenience I set out the summary of chronological schemes presented by Case (1977: 72):

Beaker Pottery Styles	Piggott 1963	Clarke 1970	Lanting & van der Waals 1972
Early	Cord-zoned	AOC	Part of Step 1
Middle	Bell Barrel Short-necked	E,W/MR,N/MR N/NR,BW E.Ang. NI/D,N2	Part of Step 1 Steps 2, 3, 4
Late	Long-necked	N3,N4,S1,S2, S3, S4, SH	Steps 5, 6, 7

It should be emphasised that the three styles of this scheme correspond broadly but not exclusively to Early, Middle and Late phases, so that Middle Style pottery can occur in the Late phase and so on. On this basis the scheme does more justice to the sequence proposed by Clarke, with the East Anglian style for example running parallel to N2-N4 and S1-4 (cf. Case 1977: 72). The scheme also envisages more controversially and on the basis of 'regional distributions' that Late styles such as S1 and N3 emerged in the Middle phase (Case 1977:72). In conventional radiocarbon chronology Case envisages the beginnings of the Early phase at some point in the mid-third millennium bc, on the basis mainly of dates from Skendleby and the possibly analogous, dated assemblage from Goodland, Co. Antrim. It is difficult to accept the evidence of these sites without further support and it remains hard to abandon the comparison provided by the series of radiocarbon dates in the Netherlands in which AOO Beakers have been dated to c. 2300 bc at the very earliest (Lanting & van der Waals 1976: table III). I would therefore prefer to assign a beginning at c. 2200 bc at the earliest to the Beaker presence, represented mainly but not necessarily exclusively by AOC pottery, following the mixture advocated in Lanting and van der Waals' step 1, except in their N.E. England-E. Scotland area. This is in fact compatible with Case's scheme since the upper outer quartile of Middle Style Beakers extends to c. 2100 bc, on the basis of the series of dates from Ballynagilly (1977: fig. 4.1). Though there may be advantages in using dispersion diagrams as outlined by Ottoway (1973b) the method tends to emphasise bunching in the middle at the expense of the extremes of a sample, which in a changing sequence may in fact be the crucial points of transition. For the rest of the Beaker sequence therefore, as far as this can be reliably attempted at all on the basis of a poor number of dates, many of them individual determinations on old samples, considerable overlap can be envisaged. Tentative values of 2100-1700 bc for the Middle Phase and 1800-1500 bc or later for the Late Phase may be envisaged, though the inner quartile values of 2000-1700 bc and 1700-1500 bc respectively may be kept in mind.

Within this tentative framework two further points may be stressed. The first is to re-emphasise the potential unreliability of absolute dating sequences at this stage in Beaker studies, a situation which will last until more systematic dating is undertaken on the Dutch model. This may indeed be held as a reservation on the settlement scheme which will be outlined below. Secondly, however, there seems to be a discernible tendency for certain Beaker groups or styles, such as AOC, Northern/North Northern Rhine, East Anglian and Barbed Wire, to be long-lived and possibly even given an erroneously high starting point in present schemes. The converse seems much less likely, that demonstrably Late styles should have emerged at comparatively early dates. Two recent series of dates support this view, those from Mount Pleasant (Wainwright 1979) and Barnack, Cambridgeshire (Donaldson 1977). At the former, secondary contexts in the ditch of site IV produced a date of 1680 ± 60 bc (BM-668) for an assemblage including AOC, E, W/MR, N/MR and S4 sherds and a higher context still in the same ditch, with AOC, W/MR, S4, Food Vessel and Urn sherds, produced a date of 1324 ± 51 bc (BM-669). These have been seen to establish the contemporaneity of ostensibly diverse Beaker styles (Wainwright 1979:90; Kinnes 1979: 73), with the essential effect of depressing date ranges. There may in fact be a double warning in this series. The dating samples are from ditch contexts, and though the association with the pottery appears good (cf. Waterbolk 1971), there would still be room for older material to end up in the ditch at the same phase as more recent material (cf. Jewell & Dimbleby 1966). Nonetheless the again comparatively late dates from Barnack from a different (mortuary) context do provide some preliminary backing to the implications of the Mount Pleasant series. These were 1620 ± 80 bc (HAR-1645) for the central primary grave with a Wessex-Middle Rhine beaker and associated artefacts, and 1640 ± 80 bc (HAR-1205) for a secondary grave (no. 27). Other stratigraphically later burials had later dates in a coherent series. Only burial 7, unassociated just within the inner ditches had an earlier date, 1850 ± 200 bc (HAR-1158).

Finally as further specific illustrations of this tendency some of the evidence relevant to the four styles cited above (AOC, N/NR, E. Ang., BW) may be rehearsed. AOC sherds have been seen in late dated contexts at Mount Pleasant, and much other evidence suggests that this decorative technique was extremely long-lived. A single date from Antofts Windypit, Helmsley, Yorkshire (1800 ± 150 bc, BM-62), if valid, could be Middle Phase. Stratigraphic associations are varied, and clearly troubled Clarke (1970: 66). Though earlier than S1 beakers at Aldro 116 and Rudston 67, the AOC beakers need not have long preceded them in these contexts. An AOC beaker was contemporary with an anomalous E beaker at Stanton Harcourt, Oxon. AOC sherds preceded BW and N2 sherds at the Sanctuary but at Woodhenge appear to follow W/MR and E sherds (references in Clarke 1970). AOC and E beakers are associated together in many contexts, but the range of E associations is likewise varied and itself cannot be used as an immutable touchstone, perhaps occurring for example at Rusden Low in Derbyshire with one N/NR and shortly followed by two S2 beakers; the record here however is highly uncertain (Bateman 1861: 43-4). In the ditch of the Rudstone cursus cord-impressed sherds are amongst a small assemblage of perhaps Late Beakers (Dymond 1966) and at Willerby 235 (Greenwell 1890) it is uncertain whether a hoard of developed bronze axes precedes or shortly postdates a vessel with AOC affinities (Clarke no. 1409). As for associated objects, the gold earring from Alston Northumberland is the most distinctive. It is of course similar to the E-associated earrings from Radley Oxon. The Orbliston earrings might have come from the same cairn as a lunula (Taylor 1970). Decoration of this kind in European contexts certainly had a long currency, though on different forms, such as the Bennekom ornament probably associated with a Veluwe (2^{If}) Beaker (Butler 1956). Plain earrings of copper or bronze in Britain have late Beaker associations at Tallington Lincolnshire (Simpson 1976: 52) Buxton Derbyshire (Clarke no. 122, FP), and the examples at Cowlam 58 (Greenwell 1877: 222) and in the Migdale hoard are probably of a similar horizon (Coles 1971; Burgess 1974: fig. 26). A wide range is again indicated. The vessels cited from the Rudston cursus and Willerby 235 also suggest that there is a danger in lumping together disparate shapes and styles, united only by a common decorative technique, a criticism very similar to that applied by Lanting and van der Waals to Clarke's Barbed Wire group but rejected—perhaps now questionably—for the AOC group (1972: 32-4).

The continental background can itself be a little further examined, essentially to question our present reliance on the Dutch Veluwe scheme for Beaker development, especially for the duration of the AOO phase or styles. Butler's remark (1956: 67) that associations for this are everywhere rare still holds good though the situation has improved and radiocarbon dates especially are available. On the basis of these a perhaps overly unilinear sequence is being advocated for the Dutch material. However there are now some four types of 2^{II} Beakers advocated (Lanting & van der Waals 1976: 6), with interconnecting associations (*ibid*.: table 1). AOC Beakers (2^{IIb}), themselves of varied shape, therefore represent only one element of this presumed horizon. 2^{Ia} or Maritime Beakers have an element of cord decoration too (*ibid*.: 9; Clarke 1970: 71), and it might be preferable to envisage a situation of considerable overlap of shapes and decorative techniques. Moreover the classic Dutch sequence best fits the Veluwe core area and less well the peripheries, as Lanting and van der Waals themselves point out (1976: 9). In making use of the continental sequence to place AOC Beakers at the head of the British sequence therefore, there is the strong presumption that these were derived from the Veluwe area from a rather unilinear development, and the contribution of other areas is ignored, as well as the non-representation in Britain of 2^{IIa}, 2^{IIc} and 2^{IId} beakers and the absolute rarity of 2^{Ia} Beakers. Rather similar criticisms were in fact made of Clarke's Wessex/Middle Rhine group (Lanting & van der Waals 1972: 30).

Turning to the N/NR group or style, we are faced with a perhaps tighter network of associations, but the possible date range is long. Closed grave finds indicate contemporaneity with N1/D, N2 and N3 Beakers (Clarke 1970: 125, table 6), and the variety of the situation is indicated by further stratigraphic links, of precedence over N2 at Cairnpapple and over N3 at Goodmanham 99, but of secondary position to N2 at Aldro 54 and W/MR at Broxa Moor 4. Precedence is however indicated over Food Vessels, Food Vessel Urns and Collared Urns (Clarke 1970: appendix 4), and Cowie has also stressed the lack of 'contact finds' between N/NR Beakers and Food Vessels Urns (1978: 48). A single unpublished date of 1670 ± 50 bc (Clarke 1970: 128) for two N/NR beakers (no. 647) from Alnwick Northumberland, may also be noted, subject to the usual reservations.

The East Anglian and Barbed Wire groups or styles have equally poor chronologies. Lanting & van der Waals' critique of the latter as a distinct group is hard to resist (1972), but it is equally difficult to see why this decorative technique should be assigned necessarily as early as their steps 3-4, at least as early as the earliest known continental parallels c. 1800 bc, when the bulk of these are of the range 1750-1500 bc (ibid.: 34), a position supported in fact by Clarke's original parallels (1970: 143). In terms of shape and of mutual association, E. Ang. and BW Beakers are closely linked, and these aspects also link both to the preceding discussion of N/NR Beakers. Stratigraphical information is poor, and both styles have scanty associations. The one distinctive association cited by Clarke, of E. Ang. Beakers with a B2/B3 bracer at Brandon Suffolk (Clarke nos. 852-3), is perhaps a little ambiguous. Though bracers of this type have W/MR associations elsewhere (e.g., Clarke 1970: figs. 128, 130, 132, 134), the bracer in question could as well be classified as type C2 and may well have been reworked. Other E. Ang. associations however include N2 (Clarke nos. 551-3) and N4 (nos. 719, 721) Beakers. Likewise Clarke cites hybrids between E. Ang. Beakers and corded Food Vessels (1970: 147-8) and the possible connections between Fengate pottery and Barbed Wire Beakers in terms of shape, decoration and other details of form (ibid.: 132-3) need not be in the direction which Clarke envisaged. Such low dating is not in disagreement with Clarke's estimation of the lower end of the span of E. Ang. Beakers, nor indeed with the view of a contribution to the Ardleigh Urn ceramic tradition (Erith & Longworth 1960: 189). The point to emphasise here is that nothing directly indicates a particularly early date for either the BW or E. Ang. styles. If both are of indigenous origin then the continental situation for BW Beakers would indicate an earlier start, but the question is open whether or not there is a closed BW group as such as the continent. It could also be argued that pot beakers on the continent provide some sort of ancestry for the shapes of both styles, and these were regionally varied and long-lived (Bamford 1970; Louwe Kooijmans 1974; Lanting & van der Waals 1976).

The evidence points to late rather than early dating for the N/NR, E. Ang. and BW Styles and a position perhaps largely outside the mainstream of indigenous development. To qualify this claim however the generally poor and scanty nature of the available evidence should firmly be kept in mind. The chronological preamble to the consideration of settlement has been of some length, but the inherent uncertainties need to be highlighted rather than glossed over.

10.6 'Beaker' settlement: a new pattern

Within this framework, further evidence to doubt the existence of separate Beaker culture groups may be considered. For a supposedly intrusive and identifiable group with specific manifestations in mortuary practice and agricultural improvement, there are remarkably few fresh settlement sites in the Early/Middle phase, or before say 1800 bc. Many of the examples which do recur in discussion of this kind in fact have some close stratigraphical relationship to other assemblages of indigenous affinity, which are likely to be contemporary traditions.

Firstly there are a number of sites with both AOC or E beakers and other indigenous ceramic styles, chiefly Grooved Ware and Peterborough, with no evident chronological separation between the two. Such an explanation is of course relevant, but it is not supported by perhaps the best excavated case available at Downton, Wiltshire (Rahtz 1962) where AOC and E pottery could not be seen to be stratigraphically distinct or later than a range of Peterborough pottery in Ebbsfleet, Mortlake and Fengate styles. Clarke's notion therefore (1970: 57) that distinct foci of different pot styles within certain settlement sites indicate settlement drift through time could therefore be revised to indicate the possibility of distinct but contemporary foci within a given settlement area. This is compatible with the Downton case, and at another well-excavated site, Ballynagilly, Co. Tyrone, it was noted that there were three foci of Beaker ceramics and one of Sandhills pottery within the excavated settlement (ApSimon 1976: 35). Radiocarbon dates from this site have already been mentioned above, there supporting the notion that Sandhills occupation was earlier than the appearance of Beakers. This general hypothesis is compatible with the possibility that Sandhills and Beaker pottery were, at least at first, in contemporary use at Ballynagilly and their distribution centred on separate foci of activity. Another site with a similar situation appears to be Walney Island, Lancashire (Barnes 1956) where AOC pottery was found outside the main concentration of other ceramics notably Grooved ware. Here the site stretches over at least 500 m, and there may well have been several phases of occupation overall. Other sites which might reflect a similar situation but which have been less thoroughly explored include Risby Warren, Lincolnshire (Riley 1957), Archerfield, Gullane E. Lothian (Curle 1908), Hedderwick, E. Lothian (Callander 1929) Luce Sands, Wigtonshire (McInnes 1966), Kirkburn, Lockerbie Dumfriesshire (Cormack 1965, but with only two Grooved Ware sherds recovered) and Carnaby Top 12, Yorkshire (Manby 1974: 26). In these sites the existence of separate foci cannot be demonstrated, except perhaps at Risby Warren, but it is a plausible hypothesis. Whatever the situation the contemporary use of different ceramic styles should be considered in the context of a working hypothesis which seeks to assign Beaker pottery a specialised and probably prestige role. This will further be developed below.

There is a much smaller series of sites in which potentially early Beaker pottery has stratigraphical links with indigenous pottery. This is to except for the present the much larger number of burial and communal monuments in whose ditches such associations or sequences occur (e.g. Clarke 1970: appendix 4). Beacon Hill, Flamborough Head, Yorkshire (Moore 1964) is an apparently small, though damaged, occupation area with AOC pottery associated with a pattern of stakeholes, stratified above Peterborough pottery. A couple of Grooved Ware sherds stratified apparently above the Beaker pottery points to a potentially early date for the latter. Here then is the sort of site which might be taken to support the notion of separate Beaker settlement. At face value however the site is not spectacularly large, though an unknown part has been lost. This point however should serve to qualify the isolation of the Beaker occupation, particularly in view of the heterogeneous nature of the associated flint industry, the Grooved Ware pottery above, and more especially in view of the larger known occupation site nearby at Hartendale, on the other side of Beacon Hill, which has produced quantities of Grooved Ware (Manby 1974: 70) Flamborough may then belong in a sense to the previous category of sites, and reference again to the stratigraphy at Downton will confirm this possibility.

A different category of sites with potentially early Beaker occupation incorporates established or decaying monuments. At both Newgrange and Knowth, Co. Meath in the Boyne passage grave cemetery have been recovered quite sizeable occupations with a useful range of economic evidence, amongst which the most distinctive pottery was Beaker (Eogan 1976; O'Kelly 1973; Wijngaarden-Bakker 1974). These sites have been used for example by Case (1977: 76) to provide examples of the 'exceptionally energetic' farming practices of the Beaker culture in its Middle Phase. The overall context and setting of these sites need however to be kept in mind. These two sites may reasonably be considered amongst the most spectacular communal monuments (the most potent 'signalling devices' in Renfrew's phrase (1979: 222)) of mid-third millennium bc western Europe. The tradition to which they belonged and of which they were the apogee probably continued towards the end of the millennium, on the basis of the radiocarbon dates from the Mound of the Hostages, Tara (O'Riordain 1964), also in Co. Meath. Carrowkeel pottery, the only ceramic style to be deposited inside passage graves, was contemporary with or at least deposited in the same context as Beaker pottery in the enclosure at Monknewtown, also in Co. Meath (Sweetman 1976). We may also note again the lower density of the Beaker presence in Ireland than elsewhere in the British Isles. It is likely then that the cultural tradition to which Newgrange and Kowth belonged, and by extension also the social structure responsible for them, had not completely disappeared by the time of the appearance of Beakers in Ireland. At the two major sites in question however the impressive facades (O'Kelly 1979) had begun to collapse, and it was partly on top of this, hard up against the mounds, that the occupation was centred. It is therefore surely naive to see the occupation, on the flanks of the major monuments of a tradition recent or still continuing, as that of simple everyday Beaker folk. The context suggests much more strongly that this activity was again specialised, and strongly connected with prestige and ritual. The presence of Beaker pottery is again highly significant but it need not be diagnostic of simple occupation by a Beaker group. The same kind of consideration may also apply to the reported occupation including AOC and All-Over-Comb pottery on the surface of the primary mound of the Callis Wold 275 barrow (Coombs 1976). This much smaller mound may however not have been particularly spectacular in its local context.

Clarke contended that there were in general for each Beaker group medium and coarse wares as well as the more distinctive and diagnostic fine wares (1970: 43) and this has served to bolster the reality of early Beaker settlement, as in his AOC and E group (*ibid.*: tables 4 & 5, 57, 73). There is indeed a range of shapes, sizes *and* wares apart from the distinctive and most characteristic fine Beaker pottery. This is however subject to two qualifications. Firstly the *whole* range of non-fine wares in the Later Neolithic is poorly understood, from pottery in contexts such as these to plain or coarse Grooved ware vessels. Indeed there is a general danger in using categories of fine ware to characterise the ceramic evidence as a whole. This is not simply casuistry to avoid evidence contrary to the hypothesis proposed but a plea for much basic research in this field in order that basic categories may be defined rather than merely assumed (cf. Clarke (1976: 472) for essentially the same point). Secondly, a start can be made to this process by not automatically categorising Beaker pottery as domestic when it is not of the finest quality. It is clear, and not only from Clarke's FN and FP groups, that many Beaker pots deposited in graves were not of the finest quality (e.g. Clarke 1970: vol. 2, *passim*). This need not mean that these were essentially domestic pottery or reflect the range of contemporary domestic pottery. This is however in this case to characterise part of the range of poorer wares as Beaker. This seems justifiable on the basis of shape, zonation of decoration and context, but even this in the light of the above remarks may be inconsistent and disputable. A neat illustration of the range of possibilities is provided by the assemblage associated in a grave containing a cremation at Chew Park, Somerset (Rahtz & Greenfield 1977). This includes fine Beaker sherds and others with cruder fabric and decoration as well as plain flat-based pots (ibid.: fig. 86) which are not easily classifiable. In addition a further pit with Grooved Ware and another possible grave with Beaker sherds are known nearby (ibid: 82ff.). The broken condition of these sherds may suggest a reuse from another context, but emphatically they need not be seen as domestic purely on the ground of their quality.

A further series of potentially early Beaker settlement sites cannot be used to support the hypothesis of Beaker settlement as such because of the extremely limited nature of the evidence from them. It was therefore misleading of Clarke to imply that such sites as Shoebury II, Roebury, Methwold and Muirkirk, belonging to his European group, or Shoebury I, Felixstowe Golf Course, Gravesend and Overa Heath, belonging to his Barbed Wire and/or East Anglian groups, were definite settlement sites. Examination of original reports and museum records suggests for the most part that either the quantities of material were very small in these cases or that a settlement context was simply an inference at the time of recording which was not necessarily the same time as the actual recovery. This applies especially to those of the European group. As for the others cited, Shoebury I has small quantities recovered from brickearth diggings, (Colchester Museum records; cf. Clarke 1970: 572, no. 44). Gravesend is unpublished and poorly recorded, while the Felixstowe find might as well be a cremation as a domestic settlement (Smith 1955: 35), and the quantities are again small. Overa Heath deserves more consideration, along with other sites like Hoe in the Wensum basin. These sites include small quantities of fine Beaker pottery as well as rusticated sherds, and a range of flint tools and waste (Apling 1932; Bamford 1970). They are, however essentially burnt mounds consisting of accumulations of fire-shattered flint. They are unusual sites to bear the weight of the Beaker settlement hypothesis, and it is interesting to compare a recent characterisation of burnt mounds and cooking sites in general as connected with prestige activities (Bradley 1978: 84).

The search for fresh early Beaker settlements proves therefore to be much harder than has seemed from a conventional viewpoint. Candidates for this role however do still remain, and they can in no way be ignored. These however in their turn have produced so far mainly AOC pottery, such as Tusculum, N. Berwick (Cree 1908), Ross Links, Northumberland, (Brewis 1928), Newborough, Anglesey (Griffiths 1956), Atherb, Aberdeenshire (Clarke 1970), Peacocks Farm, Cambridgeshire (Clark et al. 1935) or Ardnamurchan, Argyll (Lethbridge 1927); E pottery comes from Plantation Farm, Cambridgeshire (Clark 1933). They are all also largely unexcavated, and the chronological range of AOC pottery has already been highlighted. These sites too therefore are at present insufficient to bear the weight of the Beaker settlement hypothesis.

It is appropriate at this stage to look at contemporary settlement across the Channel, especially in the area of the Lower Rhine, since this is one of the most likely sources at present for British Beakers (whether accompanied by recognisable population movement or not), but also in surrounding regions as well. This is necessary in order to evaluate the likelihood of external colonisation at this stage, in the way attempted for the beginning of the British Neolithic (Whittle 1977). There is a consensus that it is plausible to talk of Beaker people rather than Beaker users in the Netherlands since there are no other contemporary traditions of material culture (Lanting & van der Waals 1976; Clarke 1976). This argument however may be taken in two ways. Superficially it supports the notion of separate and in this case isolated Beaker groups, but on the other hand it ignores the fate of the population of the area in the third millennium bc, represented archaeologically by the TRB and Vlaardingen cultures. Overlap between the latter at least and the Corded Ware culture in the later third millennium bc is supported by radiocarbon dates (Lanting & van der Waals 1976). It is unlikely that the preceding population simply disappeared, especially when there is considerable continuity of settlement (as at Vlaardingen itself, with a Maritime Beaker house in the area of earlier Vlaardingen occupation; Louwe Kooijmans 1974) either at the Corded Ware or Bell Beaker horizon. Logically therefore the 'Beaker people' of the Netherlands can only be the continuation of the previous population, with a changed material culture of which fine beakers are the most distinctive element, unless the previous population can be seen to have disappeared or been displaced, which is unlikely, or to have coexisted, which is generally denied. This must substantially weaken notions of a pan-European Beaker people, but it is still pertinent to consider settlement within the Netherlands and whether developments here might have affected the British Isles.

Settlements are known in some numbers (Lanting & van der Waals 1976) but unfortunately their survival requires covering sediments of estuarine clay, dune sands and so on. In these contexts moreover it can be seen that pits were not a regular component of settlements (e.g. Molenaarsgraaf; (Louwe Kooijmans 1974); compare Myrhøj Jutland (Jensen 1972)) so that traces of such would be extremely difficult to locate or excavate without a covering shield. The question must remain open whether in inland areas there was sufficiently dense or unstable settlement to generate the expansion needed to recolonise the British Isles at this stage, just before 2000 bc. The wider pattern of contemporary colonisation of sandy soils and heathlands as in Jutland or in the Untermaingebeit (Menke 1973) does not decide the issue though it may indeed support the possibility. Only in the lower estuarine and coastal areas are there hints of more complete settlement patterns (Louwe Kooijmans 1974). It could therefore be argued that the well-documented presence of mixed farmers and fishers in these marginal zones, perhaps on a seasonal basis suggests some overall pressure on resources in the Low Countries. On the other hand these zones had been so exploited since the late fourth millennium bc at the latest (ibid.), and moreover settlement in the AOO and Maritime Beaker phases, though represented by traces such as in the coastal area, in the estuarine area and inland (Lanting & van der Waals 1976), seems on the whole transient and scarce (Louwe Kooijmans 1974). This may be a bias produced by the Calais IVb transgression. In the succeeding Veluwe Bell Beaker phase settlements seem more numerous, with several potentially contemporary sites including Molenaarsgraaf spaced out along the Schoonrewoerd stream ridge (ibid.: fig. 19). Here is a fuller landscape, in the 'river clay/wood peat' area, but the units of settlement are small and it does seem unlikely that population from these locations could have seriously competed for land with the sizeable native communities in the British Isles whose settlement was sketched above.

The bulk of what we may call for the present Beaker settlement in the British Isles seems to lie in the Late phase, and the most numerous examples come from East Anglia (Clarke 1970: appendix 6, passim). Other sites can be included from elsewhere, such as the few pits from Cassington and New Wintles nearby in the Upper Thames valley (Leeds 1933; Case et al. 1965; Hawkes & Gray 1969); Gorsey Bigbury and related sites on the Mendips (ApSimon et al. 1976); perhaps Gwithian site XV, level 8 (Megaw 1976); and further afield Northton, Harris (D. Simpson 1976), Udal, North Uist (Crawford & Switsur 1977), and Rosinish, Benbecula (Shepherd 1976) in the north-west of Scotland. In East Anglia sites are relatively numerous and widespread, such as Risby Warren and others on the Jurassic ridge in Lincolnshire (Riley 1957); Fengate, Northamptonshire (Pryor 1978; R.C.H.M. 1969) Fifty Farm Suffolk (Leaf 1935), Hockwold and Methwold, Norfolk (Bamford 1979) around the edge of the Fens; Peacock's Farm and Plantation Farm, Shippea Hill, Cambridgeshire out in the fens on sand ridges (Clark 1933; Clark et al. 1935); Martlesham, Suffolk (Clarke 1970: Martin 1975; 1976) on inland heath, and Lion Point and others on the Essex coast (Hazzledine Warren et al. 1936) It is possible to exaggerate the importance of some of these sites too. For example the collection of sherds and flint from Stoke Ferry, Norfolk is not large, and was collected from dumped dredgings from the R. Wissey in the 1920s and 1930s (Cambridge University Museum records). The status and character of the site must therefore remain uncertain. But the quantities from other sites are impressive, such as Hockwold and Methwold (Norwich Museum) Wattisfield (Ipswich Museum) or Reffley Wood and Lion Point (British Museum). The material deserves the title of the great unwashed of British prehistory and is poorly known in detail (Bamford 1970; Frances Healey, pers. comm.). At Hockwold there is a wide scatter of material from several fields investigated by W. F. Curtis before the war, with concentrations within this. Pits, hearths and stake structures are represented. Pottery includes fine ware Beakers (perhaps up to 50 per cent) and a range of other qualities and shapes including rusticated necked beakers and heavy-walled rusticated coarse ware, not necessarily all based on Beaker forms (fig. 10.1). The flint assemblage is dominated by scrapers but includes a range of other tools. Some animal bone is present and barley cereal impressions are reported (Bamford 1970). The domestic character of the site or sites is clear even though its status (seasonal or permanent, large or small) is obscure. Some Food Vessel and Urn sherds are represented but the Beaker pottery dominates completely. This and other sites like it urgently need to be published in greater detail.

As far as our chronology is reliable at present, I would expect these sites to fall in the range of 1800-1500 bc. One date on old material from Fifty Farm, Suffolk, 1850 ± 150 bc (BM-133) lies outside or overlapping the upper end of this range. Some sites may perhaps fall in the transition between Middle and Late phases in Case's terms, around 1800 bc, and be compatible with the suggested date range for steps 3 and 4 of Lanting and van der Waals (1972; 1900-1800 bc and 1850-1750 bc respectively). Here I would place those few Barbed Wire-East Anglian settlements, in Clarke's terms, which we have at our disposal, if they do not in fact belong to the late phase as argued above. At any rate it is hard to justify an earlier date at present. On a small scale, there is the material from under the Martlesham barrows, Suffolk (Martin 1975; 1976) with accompanying flint assemblage, and other ceramic styles absent. Sites on the Essex coast may belong here, if they do not belong to the category of sites discussed above in which more than one ceramic tradition is represented within a confined area. The major concentration is the Grooved Ware occupation of area 4 at Lion Point (Longworth et al. 1971; Hazzeldine Warren et al. 1936). Sites with Beaker pottery are separate from this but smaller and with much less pottery and flint, of which burnt waste is a major element but amongst which diminutive side scrapers are conspicuous (e.g. Lion Point area 2, site 108; Mill Bay, Dovercourt,

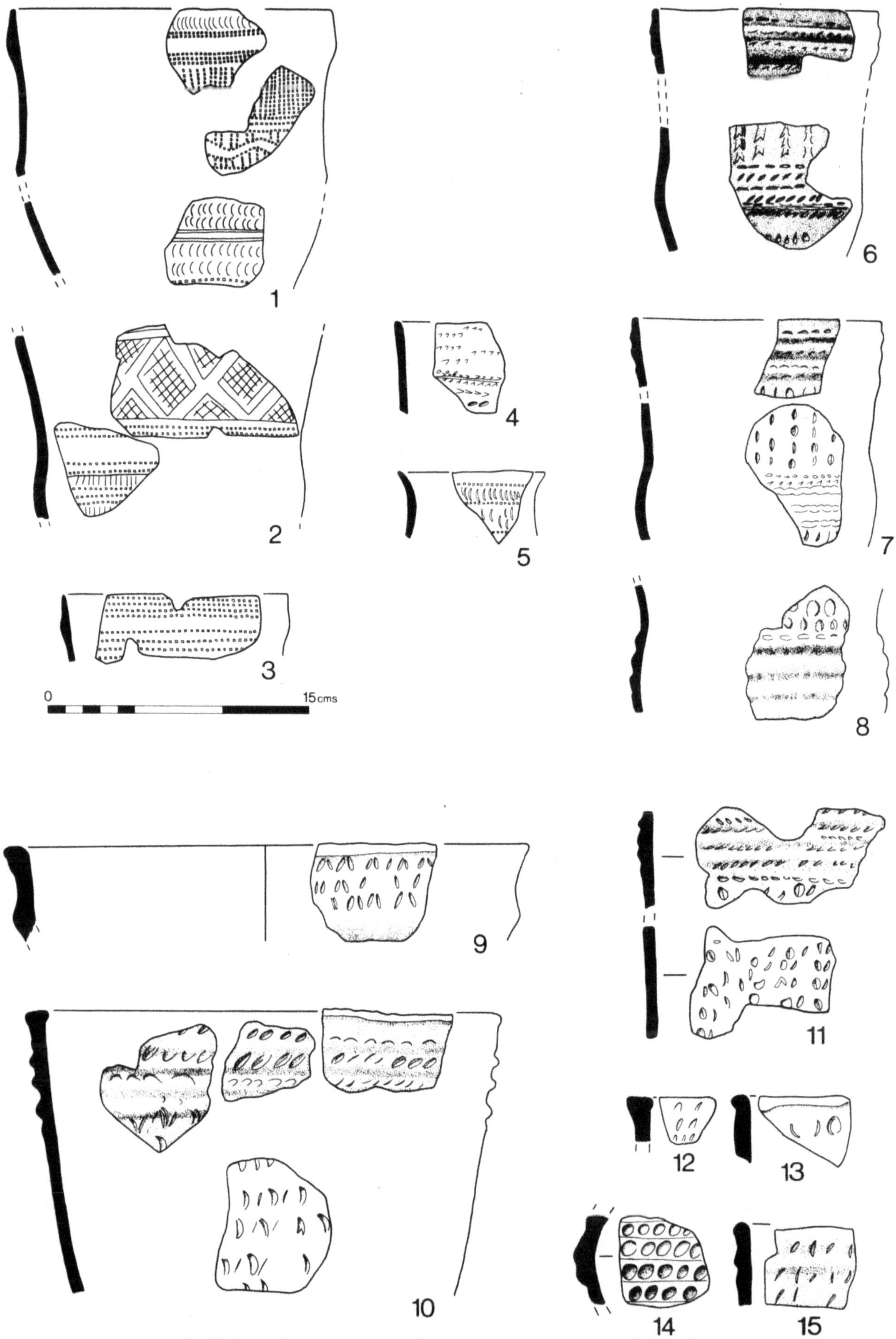

Fig. 10.1 Fine (1-2), medium (4-8) and coarse ware (9-15) pottery from Hockwold, Norfolk (Castle Museum, Norwich).

'camp site' 1; (ibid.); Lion Point, site 114 (Smith 1955). The 'camp site' at area 3, Stone Point, Walton-on-the-Naze (Hazzledine Warren et al. 1936) could be a grave rather than a 'cooking hole'. On the whole the separate sites here with Beaker pottery seem to have had a specialised, distinctive function, which may readily be contrasted with the wider range of material from both Grooved Ware and Earlier Neolithic contexts on the foreshore. Their independent status therefore no less than their date is open to question. On a larger scale, Belle Tout, Sussex (Bradley 1970) presents a big assemblage of material with associated structural and economic evidence which is of uncertain date in the Beaker sequence, though the site has been used to typify the activities of the Beaker people in its Middle Phase (Case 1977: 76). There are a number of problems. Out of an assemblage of over 1000 sherds, some 130 decorated sherds were recorded. The tradition of the plain pottery is unclear as a result of its bad preservation. There is a danger that decorated pottery is again being used to typify the affinities and status of the site as a whole. The decorated pottery has been classed as Beaker on the basis of its shape and decoration, and diagnostic traits suggest that it belongs to the East Anglian style. The debatable chronology of this has been argued above (section 10.5), and the excavator noted anyway that the absence of Barbed Wire decoration and the presence of FP and ribbed decoration could point to a late date (Bradley 1970: 341-2). The motif of fig. 12:10 (ibid.) would also be appropriate to a Late context, and corded decoration on the interior of an everted rim (ibid.: fig. 13.13) could be more appropriate to the Food Vessel tradition rather than 'domestic AOC' pottery. A Middle Phase date is therefore highly debatable, even though the suggested sequence at the site, including two overlapping enclosures does allow for a span of occupation. There is moreover very little distinctive Beaker fine ware at all on the site, and even if the site is Late, as I have argued, its status as a 'Beaker site' has to be given far more critical consideration. Pot shapes can be compared with the East Anglian style, but in an area with poorly known Grooved Ware, Peterborough and Urn traditions, the overall significance of this is perhaps a little unclear. The excavator shares these doubts (Richard Bradley, pers. comm.).

It may be felt that as well as the chronological uncertainties involved the accidents of discovery and excavation cannot accurately be assessed at present, resulting in a very poor sample of settlement over the whole period in question. The point can hardly be denied, for the early activities of amateurs like Curtis cannot easily be matched outside East Anglia. Belle Tout itself is an example of how recent excavation has enlarged on earlier work, here by Toms in 1909 (Toms 1912). Ireland provides perhaps the most dramatic example of all. Though it still has a relatively low Beaker presence, as late as 1938, an editorial in the new series of the Ulster Journal of Archaeology could comment, on the discovery of Beakers from Largantea, that 'the Beaker civilisation ... has hitherto not been recorded in Ireland, save for one or two isolated examples' (Ulster J. Archaeol., third series, vol. 1, 1938). Nonetheless some areas have been within these limits relatively well investigated, such as the Yorkshire Wolds, the fen edges and the Upper Thames valley. In the first, settlements of the 'native' population are abundant (Manby 1974; 1975) with pottery of local Grooved Ware and Peterborough traditions. Potentially early Beaker sites are rare and not certainly separate from these. Late Beaker sites are not represented. Around the Fens, from the end of the transgression c. 2000 bc, there are a series of similar native sites, with Grooved Ware sites mainly along the chalk of west Norfolk (Wainwright

& Longworth 1971; Kinnes 1976) but also represented on the west of the fens with Peterborough material at Fengate, Northamptonshire (R.C.H.M. 1969, Pryor 1978 with earlier references). Separate potentially early Beaker sites are not represented, and probable late sites have been discussed above, such as Hockwold. Beaker material was found at Fengate chiefly in the earlier quarry area; the bulk of it was of late style, in the upper filling of earlier pits. A few sherds came from the recent excavations in the fill of a grave (Pryor 1978: 96). There were also substantial quantities of rusticated pottery, together with in a more limited area, further amounts of Grooved Ware, probably from pits (R.C.H.M. 1969: no. 17, and earlier references).

In the Upper Thames valley there has been sporadic observation of gravel quarrying since its inception on a large scale in the 1920s, and the area can thus claim to be one of the best surveyed in the country. This has produced a considerable density of observed or excavated Beaker burials and stray finds probably from damaged contexts such as burials and/or ring-ditches (for instance at Dorchester: Simpson & Case 1963). Some seventy Beaker pots are known between Lechlade and Goring (Case 1956; 1963; Woodstock County Museum sites and monuments record). Against this distribution Beaker settlement finds are meagre, and seem to be of the Late phase, at Cassington (Leeds 1933; Case et al. 1965), New Wintles (Hawkes & Gray 1969) and perhaps The Hamel, Oxford (Oxford Archaeol. Unit, pers. comm.). The same fieldwork has produced a greater quantity of occupation sites with pottery of other traditions, such as Cassington, Stanton Harcourt and Sutton Courtenay with Grooved Ware (summarised in Wainwright & Longworth 1971); and Cassington, Stanton Harcourt, Foxley Farm, Eynsham (e.g. Leeds 1934; 1938; 1940) and Barton Court Farm, Abingdon (David Miles, pers. comm.), with Peterborough pottery. One collective burial in the Cassington cemetery had sherds of Peterborough pottery (no. 7, Leeds 1934) and pits at Sutton Courtenay in the midst of Grooved Ware occupation held collective inhumations, though they were not associated (Leeds 1923: pits U, V). Several ring-ditches may have primary Peterborough associations (Case 1963) and cremation cemeteries at Dorchester are well known (Atkinson et al. 1951). The different pot types coincide, though in small quantities, at the ritual/sepulchral sites at Dorchester (ibid.); the ditches of the Devil's Quoits, Stanton Harcourt were devoid of material (Gray 1974). Further, the burial and settlement distributions overlap. Each has a spaced pattern, though this may be dependent on the distribution of gravel diggings, with examples spaced at distances of over 500 m (see for example Stanton Harcourt, mapped by Grimes 1960: fig. 57). Within this pattern burials tend to occur in small numbers; at Stanton Harcourt for example 2 graves in Vicarage Field (Case 1956), 3 certain examples on the Aerodrome (Hamlin 1963), and one at Linch Hill Corner (Grimes 1944). Settlement finds coincide with these, and at Eynsham, Cassington, Abingdon-Radley, Sutton Courtenay and Dorchester, though not apparently at Standlake, Yarnton or Summertown where observation has been more restricted. The Late cemeteries at Cassington and Eynsham are larger, and not all burials were accompanied by Beakers. A picture can thus be envisaged in this area as well of indigenous communities widely settled on the best terrace soils with distinctive Beaker-oriented burial assemblages in a range of more or less contemporary mortuary practices (fig. 10.2). Later, putatively after 1800 bc, domestic pottery took on the forms of mortuary/ritual ceramics, as the Grooved Ware and Peterborough pottery traditions faded.

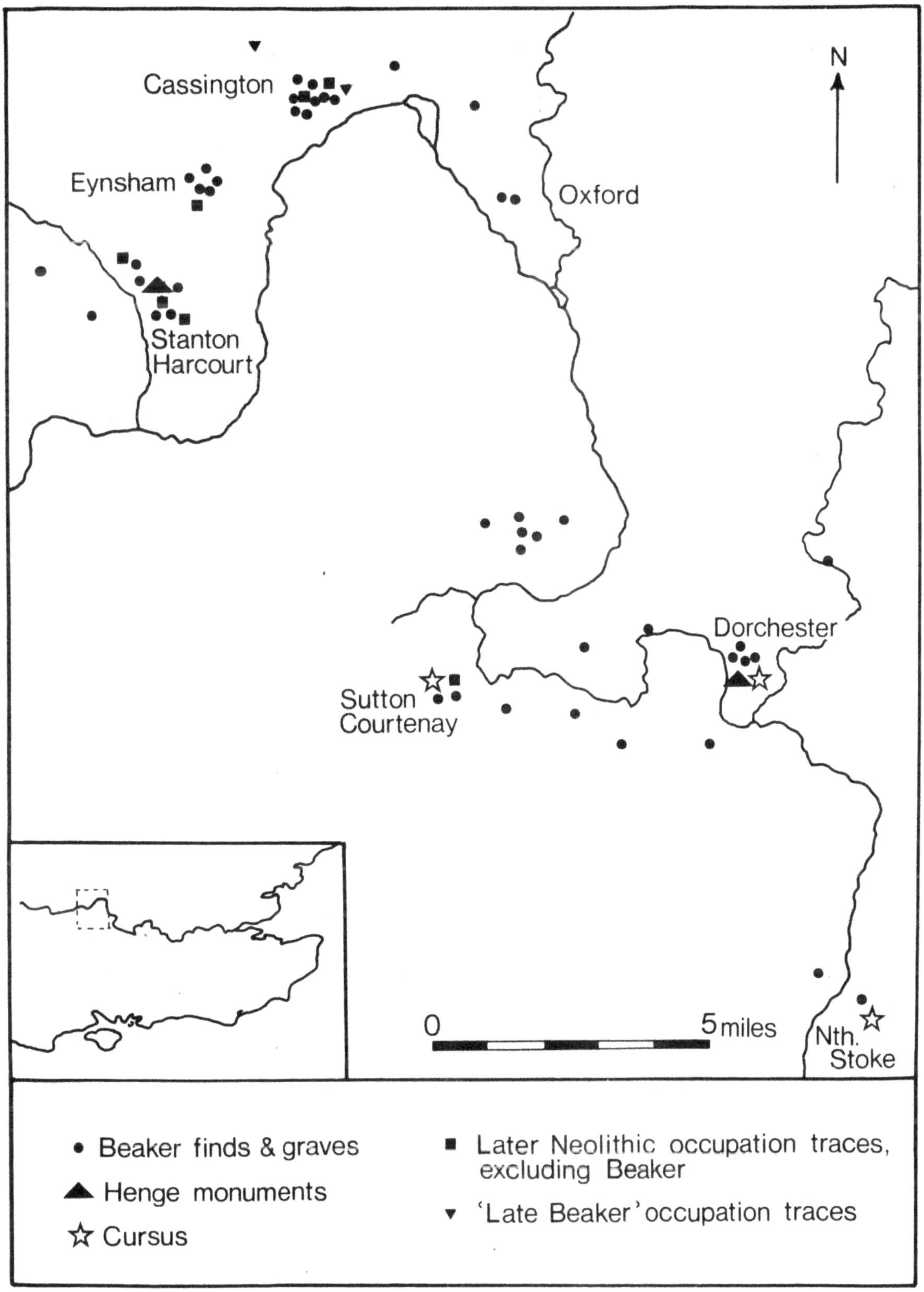

Fig. 10.2 Generalised distribution of Beaker finds and graves, and later Neolithic settlement traces and monuments in part of the Upper Thames Valley (various sources: see text).

The wider context of such a change will be discussed shortly below. Here it is convenient to summarise this interpretation of Later Neolithic settlement. There seem to be two main phases. First, from approximately 2400-1800 bc, settlement evidence is dominated by that of native communities and other activity in the landscape such as the construction of major monuments is similarly oriented. This phase spans both the pre-Beaker and the earlier Beaker horizon. Secondly, from approximately 1800-1500 bc or perhaps a little later, this tradition changed and settlement evidence is represented to a far greater extent after some initial overlap by domestic material of the Beaker Late Phase. This parallels a corresponding change in mortuary assemblages in which Beakers are increasingly left out and new prestige items including Food Vessels, Urns and other pot types are included, along with developed metal products and other artefacts. This indeed is the conventional Early Bronze Age, which many would agree had begun by 1700 bc. In a real sense therefore the convincing Beaker settlements are in fact part of the missing settlement record of the conventional Early Bronze Age, and by the time they can be accepted therefore there is little sense in invoking the ethnic hypothesis. The implications of this view are also discussed further below. There is no need to claim that they represent the full settlement range of the period. This would be a misleading simplification of a complex situation. Apparently domestic contexts have been recorded for Collared Urns at Stockbridge, Hants (Stone and Hill 1938) Rams Hill, Berkshire (Bradley & Ellison 1975) and Stanton Harcourt, Oxon (Case 1951) for Biconical Urns at Mildenhall, Cambridgeshire (Clark 1936b) and for Cordoned Urns at Downpatrick, Co. Down (Pollock & Waterman 1964) amongst others.

The fluidity of the Beaker presence across Europe has been emphasised by Clarke (1976) and Shennan (1978). The aim of this interpretation has been to add the British Isles to those other areas in which the Beaker presence need not imply an intrusive settlement by a numerous exogenous population. The implications of this view must now be considered with other kinds of evidence in discussing aspects of social organisation.

10.7 Aspects of social organization

In what follows we are thus free to consider the role of Beaker assemblages in contemporary communities. As well as seeing them in a different perspective we may also come to pay comparable attention to other more neglected parts of the evidence. In the rest of the paper I shall look at arguably the most striking social aspects of the Later Neolithic period—monument building, exchange networks and status assemblages—through evidence which goes beyond Beakers and their associations. This selection of themes will illustrate that there are indeed, as others have suggested, signs of increased centralisation and ranking in social organisation but that not all the evidence can be used in identical ways to fit a simple model of stratified society. There is no convenient label to denote this state of affairs and indeed this very lack of a modern or recent parallel seems significant in itself.

10.8 Centralisation and ranking: the evidence of monuments

The very size of several monuments in southern Britain, such as Avebury and Silbury Hill, Durrington Walls and its complex including Stonehenge, Marden, Mount Pleasant, and the Dorset cursus with the Knowlton complex, has been taken to support the notion of considerable centralisation and organisation, and hence ranking in the society responsible for their construction (e.g. Renfrew 1973).

It is however open to doubt how confidently such an inference should be made since geographical concentration of effort may not directly entail centralisation of power or rigid social ranking, and in that case this line of evidence can only suggest at best a rather vague model of stratified society. Alternative models for their undertaking can be explored, which also explicitly seek dynamic sources of change.

There are first some important practical qualifications to make. There is room for argument as to the absolute scale of investment of time necessary for the construction of such monuments. It is argued by Startin & Bradley (paper 9 of this volume) that the major henges for example could have been built with an investment in the order of half a million man-hours, rather than of about one million as previously projected. Other vast monuments like Silbury Hill and the Dorset cursus (assuming for the moment for the sake of argument that it is of Later Neolithic date) would still represent major investments of communal effort even if the time factor were to be reduced. The scale of the undertaking can be further reduced however if it can be shown that these the largest monuments of the period were built in more than one phase. It is probable that the cursus has at least two phases (Atkinson 1955) given the apparent terminal on Bottlebush Down, and there is the possibility that lynchets on Gussage Down mask another similar terminal at a point where there is a more marked change of direction than elsewhere along its admittedly irregular course. At Silbury Hill the recognised phases of construction have been set within a sequence of continuous building (Atkinson 1978) but the evident changes of plan together with the frequent dark thin layers in the section suggestive of turf-lines (e.g. Atkinson 1978: 165) might support the possibility of a more discontinuous sequence of events—though the excavator does not support this. Each episode of communal activity at these largest sites could then be considerably reduced in scale.

The time span of construction is another important but unknown variable. There has been a tendency in comparing, for example, major henges with earlier enclosures to assume that both were constructed relatively quickly, on the premiss that the largest monuments entail a larger workforce. This may be a false assumption. An alternative explanation would be of a similarly sized workforce working for a greater number of seasons, which is an important distinction. Excavated evidence provides few means of testing these possibilities. Fresh chalk from the ditch used in the packing of several stone-holes in the main circle of Avebury (Smith 1965: 221) does not support tardiness. On the other hand, radiocarbon dates from different phases of construction at Silbury Hill—(2145 ± 95 bc (I-4136) from unburnt vegetable matter from the primary core (Atkinson 1969), though hardly compatible with the experimental series of determinations on turf from secondary parts of the mound yielding third millennium dates (Stuckenrath & Mielke 1973: 401); 1802 ± 50 bc (BM-841) and 1899 ± 43 bc (BM-842), from antler collagen from the bottom of the second of the two ditches, Burleigh *et al*, 1976: 38)—taken at face value (there are problems of interpretation due to samples of different materials having been used) do suggest a considerable span of time.

Another way in which later Neolithic monument building may have been overestimated is by comparison with earlier monuments. A figure of *c*. 60,000 man-hours has been quoted for the Windmill Hill enclosure (Startin & Bradley, this volume). It is not yet possible, however, to claim that we understand the complexities of such enclosures in the light of continued discoveries by Roger Mercer at Hambledon Hill of subsidiary outworks and enclosures (1980), all of which could serve radically to increase the above figures. There are also some concentrations of such enclosures, as in the Windmill Hill area or in the Upper Thames valley near Lechlade. We do not know if these groups are contemporary or not. If they were, the contrast in terms of man-hours with the Later Neolithic would be further reduced. Earlier Neolithic enclosures may be spread over a considerable time span but each new discovery (e.g. Palmer 1978) also serves to reduce the contrast in overall terms by increasing the level of Earlier Neolithic activity.

A more subtle difficulty has been noted by Fleming (1973: 179) pointing out that man-hour figures may be meaningless without some estimation of the overall 'energy budget' of the communities involved. Figures are available from the ethnographic record which suggest ranges of time invested in subsistence activities by swidden and hoe cultivators, from 220 to 1260 hours per person per year (e.g. Athens 1977), though these tropical and savanna cases need not be directly applicable to this prehistoric case in view of many variables, not least the differing reliance on seed crops. Renewed clearance as described above could have been a time-consuming task, and the scale of work undertaken in even one small area of the Somerset Levels as demonstrated by the Abbot's Way track, which is known over 1100 m and may have covered in all over 2 km between Westhay and Burtle (Coles & Orme 1976), may suggest that the above figures for tropical plot cultivators grossly underestimate the labour requirements of many later Neolithic communities. There may have been regional variation. A broad distinction in these terms might be sought between areas with primary or secondary woodland still subject to clearance and those with established open country, and between areas under continued cultivation and those largely under grass. The last category may have required a lower energy budget to exploit it, and it is interesting that the major monuments of southern England seem to have been in large tracts of open grassland according to soil and molluscan evidence (Evans 1972). The overall point however remains that it is likely that there would have been abundant time available in most regions for the undertaking of communal constructions, or that a relatively small proportion of the population would have sufficed as a workforce without detriment to other subsistence activities.

Both the reduced figures for investment of time in monument building and the corresponding ease with which such labour could have been afforded are therefore important considerations, since they surely serve to reduce the need to see concerted organisation and centralisation of power behind the achievement of construction. With this reduction there are two sorts of interpretation possible. On the one hand one may argue that the scale of organisation required would still necessitate a more ranked society, though this social model is rendered rather vague since there is no clear point at which geographical concentration of effort must entail centralisation of power. Still such a model would find many in its favour. On the other hand, we should be ready to explore alternative models. It is a moot point for example whether stateless societies need centralised direction to undertake large tasks, and other cases would suggest not. Irrigation in the Near East for example preceded rather than followed the establishment of centralised bureaucracy and other state functions (e.g. Redman 1978). It is also important not to let man-hour figures dominate discussion since they distract attention from the function of such monuments, which could be considered separately from the time taken to build them. At the same time some explanation must be sought of the greater size of monuments in southern England. Constructions of several kinds amongst dispersed early agricultural communities in Britain have been seen as foci designed to protect and enhance communal identity and territory (e.g. Fleming 1973; Kinnes 1975; Renfrew 1976). In the Later Neolithic in many areas of the country rather small and modest communal monuments were still built, as in south-west Ireland (O'Nuallain 1975) or in north-east Scotland (Burl 1980) and these monuments can be seen in the same perspective as their predecessors though there are also numerous cases both of increased scale and 'centralising tendencies', as in the Orkneys (Renfrew 1979), and some of more defensive-looking sites, as at Meldon Bridge (Burgess 1976). Since in southern England settlement cannot be seen to have become more nucleated by the Later Neolithic, the major henges too can be ascribed a role of foci of communal cohesion. Their greater size could then be seen not so much as an expression of increased centralisation of power as of an increased need for social cohesion. This size may reflect for example an increased number of individual communities wishing to participate in a common task to promote mutual harmony and reduce the risks of competition (cf. Hodder & Bradley 1979; Pierpoint 1980: 205). It is here that the environmental and economic background is important, for the situation on the chalk after long exploitation since the Earlier Neolithic (Whittle 1978) could have faced communities on or using it with problems of access to arable land, woodland for pannage and timber, and even grazing and with an increasingly undiversified environment. Competition for basic resources would however have been reduced if potential competitors engaged in common undertakings spread over numbers of years. The chalk of southern England in particular seems on present environmental evidence to have been most subject to such pressures, though other areas could have been prone to varying extents.

It is also important to avoid a static view of these major monuments, in view of the changes so evident in the archaeological record, and to take account of their subsequent history. Though artefactual sequences at these sites are prolonged through secondary and tertiary ditch silts, it is striking that on the basis of the excavated sample ditches were so readily left to silt up. Evidence for recutting is rare, and though there were modifications to the internal timber buildings their overall life was limited (Wainwright 1979). These major undertakings, since relatively short-lived in their original state, are evidence of rapid social change, and conditions at their demise may not have been the same at their inception. If the general pattern is one of abandonment or retrenchment relatively soon after construction, there are a number of possibilities. Construction itself might have been sufficient fulfilment of social needs, which seems unlikely. Alternative, less expensive substitutes for producing social cohesion may have been sought, such as barrow cemeteries clustered in many of the same areas; Beaker pottery, absent in primary contexts at major henges but more abundant subsequently; or metal products, such as the developed bronze axe in the secondary silts at Mount Pleasant (Wainwright 1979); but it is unlikely that construction on the reduced scale suggested should have stretched communal capabilities. Social change itself is more likely to have altered the conditions in which the major monuments were created. Two trends would be plausible. First, continued competition for basic resources on the chalk would no longer be wholly satisfied by communal undertakings, leading to their abandonment. It is striking that the one site which was refurbished c. 1700 bc, Mount Pleasant, was now provided with a massive palisade with presumed defensive attributes. Secondly, the organisation of communal labour for geographically concentrated tasks could have encouraged or created some actual centralisation of power, even though this was not a precondition of the undertakings. Both these factors would tend to accelerate social change towards greater differentiation and ranking. Initially however the major monuments themselves need not necessarily have been an expression of such a situation. The 'chiefdom' model derived from labour requirements does little justice to the potential complexity of the situation and does little to account for longer-term change, apart from the observation that chiefdoms tend to be short-lived. Other aspects of the period to be considered now do not firmly support the 'chiefdom' model but rather supply further contradictions as well as further evidence for change.

10.9 Status assemblages, exchange and competition

Many artefacts of the Later Neolithic period may be considered prestige items or status objects on the grounds of the specialised nature of their production, its cost and scarcity. A list would include on these criteria stone axes, battle axes, mace-heads and bracers, flint axes, polished knives and daggers, as well as pots of various styles whose manufacture, as Clarke pointed out with reference to Beakers (1976), may have been lengthy and specialised. In themselves most of the artefacts in question of this period were hardly more valuable than some of the earlier period such as the finest flint, stone and jadeite axes, for which lengthy production procedures must be posited, in the case of the latter for example up to several tens of hours (Coles *et al.* 1974), as well as long-range distribution from restricted sources.

Acquisition of any of these would have been either through local production or from further afield by exchange networks. Exchange networks are likely to have been widespread and important, and there is little evidence to suggest they were subject to redistribution through a controlling agency, as might be suggested by the chiefdom model. Metalwork for example in southern Britain has rather diffuse distribution patterns which seem not to be unduly affected by the presence of major monuments like henges or large barrow groups, unlike the concentrations of Middle Bronze Age metalwork around major enclosures of that period (Bradley 1978: fig. 6:4). Even the more abundant Arreton metalwork of the later Early Bronze Age has a lowland distribution away from other centres inland on the chalk. Specific sources which probably account for much of the distribution of stone axes and some battleaxes of this period are known (Cummins 1979; Roe 1979). Some of this movement, notably of Group 1 axes into south-eastern England, has not produced the characteristic pattern of steady fall-off in numbers with distance but could reflect a more directed distribution. The chronology is awkward since it relies on so few dateable contexts, one of which now includes an Earlier Neolithic context in the south-west (Smith 1979). It is also therefore unclear whether the other similar pattern, of Group VI axes moved into eastern England, extends from the Earlier into the Later Neolithic period. What is striking is how such patterns avoid the area of central-southern England which was the focus of attention when considering monuments. In eastern England much of the production at Grime's Graves dates to the Later Neolithic period (e.g. Mercer 1976) and output was vast. But it is also striking that production was not geared solely to axes to compete with Cornish and other products, but seems to have included a variety of other types including perhaps roughouts for polished flint knives which were extremely common in eastern England as a whole (e.g. Manby 1974). Recent investigations (Sieveking *et al.* 1973) seem not to have recovered substantial associated settlement evidence, and a model here of exploitation by individual or small groups may well be appropriate.

Pottery of all styles may have been acquired by similar networks as Clarke suggested for Beakers (1976). There are several cases of closely similar vessels found a long way apart (cf. Clarke 1970: ch. 8). There are later examples of Food Vessels of similar detail but wide dispersal (Pierpoint 1980: 119). The ceramic situation is however likely to have been varied and complicated, with local production or acquisition playing a part. There are sites in small areas as around Hockwold (Bamford 1970) or Gorsey Bigbury (ApSimon et al. 1976) whose fine pottery is closely similar, suggesting at the most very localised exchange, and 'potter's clay' was recorded at Hockwold and Belle Tout. Indirect evidence from cereal impressions (mainly barley) on Beaker pottery from the southern chalk also suggests some localised production or exchange, since Dennell's observation that soils in this area would have favoured barley rather than wheat from the Earlier Neolithic onwards should provide some measure of geographical control. The size of some of the larger vessels of the period may have precluded easy movement. At one end of the Grooved Ware distribution for example, at Skara Brae, it is likely that some of the largest pots might actually have been built in situ since they are too wide to pass through narrow hut doors—unless they came through the roof (D. V. Clarke, pers. comm.; cf. Childe 1931: 133 and plan). In contrast the broad distributions of finer Grooved Ware bowls, some with quite specific attributes like scalloped rims or appliqué rim knobs (Wainwright & Longworth 1971) are compatible with the exchange model. In some respects this is the best explanation at present of the overall distribution of Grooved Ware which emerges in the later third millennium bc with a far wider pattern of distribution than any Earlier Neolithic pot style or the contemporary, more regionalised Peterborough style. There is a better case here perhaps for relating production to the suggested major centres such as Durrington Walls and MacKie (1977) has sought to explain Grooved Ware as a specific artefact related to groups in control of these centres. But again the patterns are overall more diffuse, and the range of contexts far greater than merely at major centres.

Status objects of all these various kinds could have been a medium for a wide variety of reciprocal and other exchanges (cf. Renfrew 1977) as can be envisaged too in the Earlier Neolithic. But several factors in the Later Neolithic period suggest significant change in the role of such status items towards a use to express status in a competitive fashion, for which again a background of competition for access to basic resources may be argued. One may note in particular from the pre-Beaker phase onwards of the Later Neolithic period the increase in number of types of object in question, the increase in the size of associated groups of such objects, the increase in specialisation of context in which they may appear and finally the increased rate of turnover by which objects fall out of use and are replaced by new substitutes. There are also two aspects of this suggested competition or relative ranking, that within associated groups of objects and contexts and that between them. Because of its very nature the situation was not static and change may be seen to have accelerated throughout the period. An attempt briefly to outline this will be made after discussion of Beaker assemblages as one example of the kind of competition in question.

The Beaker complex provides the currently most accessible example of internal differentiation. It is striking—and often forgotten—that while there are over two thousand finds of Beaker pottery, there have only been a little over two hundred recorded assemblages of associated objects. A loss to the 'accidents of discovery' may be admitted. About

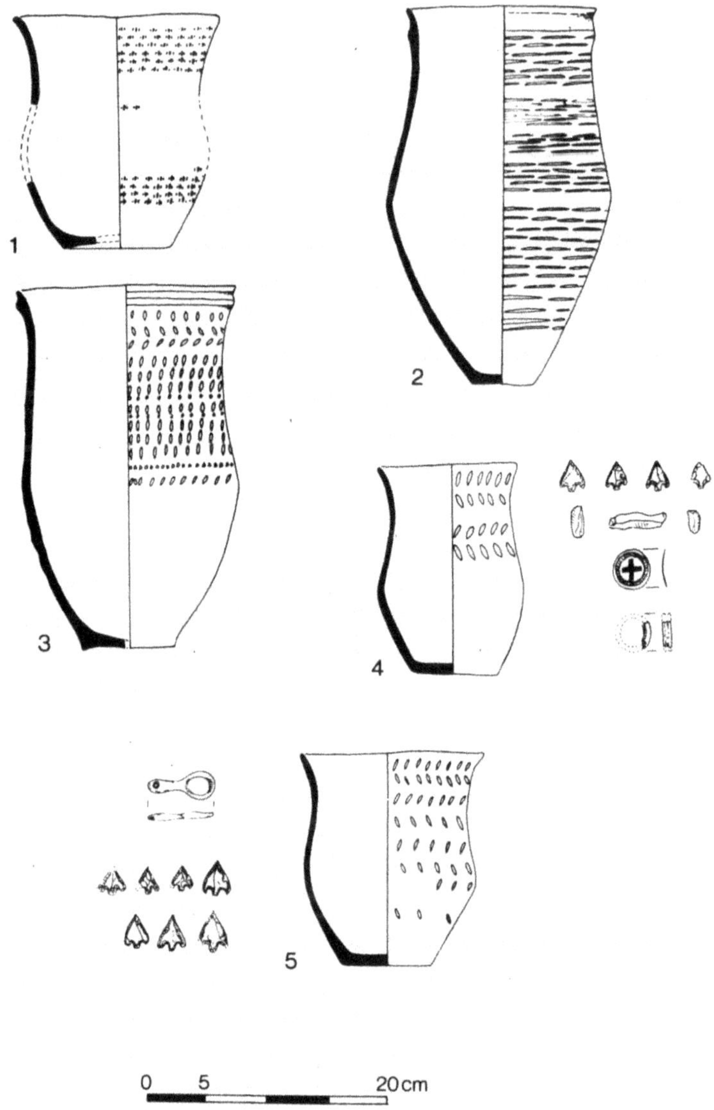

Fig. 10.3 Roughly decorated Beakers classified by Clarke (1970) as 1, European ?; 2-3, Wessex/Middle Rhine; 4-5, Northern/ Middle Rhine. 1: Ashberry, Yorks (Clarke no. 1232). 2: Fengate, Northants (641). 3: Fengate, Northants (640). 4: Monkton Farleigh, Wilts (1111). 5: Stanton Harcourt, Oxon (772). (After Clarke 1970.)

half of this figure however is represented by Clarke's 'Basic' association range (common to all of his groups; appendix 3.2) which includes flint flakes, blades and scrapers, arrowheads, antler picks, additional vessels, shale/jet beads and metal awls. Only the last three categories can have much intrinsic value in the sense defined. Of the total number of pot finds mentioned above, a surprisingly large number have not been of the finest and most distinctive quality. Clarke listed in 1970 (appendix 7.20-22) some 35 occurrences of plain pots, and about 80 and 120 of his FN and FP rusticated wares respectively. Some of these were in settlement contexts, but many were not, these often occurring in graves either alone or with finer ware. For example in Clarke's appendix 7.21 the following FN pots (numbers are given for the sake of brevity rather than the full site names with references) have associated burials: 23 (?) 35, 42 (?), 170, 191, 221, 718, 748, 762, 766 (?) 779 (?), 819, 830, 855, 960. 779, 819 and 855 have other Beaker styles associated as well. The sample is small and the age/sex data minimal but it includes a woman and child at Brantham Hall (855, with an E Beaker), a youth at Wherstead (960) and an adult male at Hasting Hill (221). To this number can be added the cruder pots again often from graves which Clarke assigned to a particular group on the basis of form or some other criterion but which cannot really be categorised as fine ware (e.g. 1970: figs. 31, 229-30, 259, 261 etc.; fig. 10.3).

Single fine ware or cruder Beakers thus constitute much of the Beaker phenomenon in Britain. By themselves it has been argued that they do not represent distinctively higher intrinsic value than many other contemporary artefacts and associations. The British counterparts of suggested central European status kits would be the 'non-basic' Beaker associations and thus represent a very small proportion of the whole complex. Even within this rare group of assemblages, further differentiation can be sought in two other ways. There is variation in the number of different types of artefact represented (between for example the widely endowed assemblages with W/MR pots such as Dorchester, Mere or Roundway and the less fine W/MR pot (Clarke no. 773) from Stanton Harcourt with a single bone spatula). Some small amount of variation is also present in the quality of the pot in such assemblages, since at least two, strikingly at Monkton Farleigh (no. 1111) with a gold cap and a belt ring and less so at Stanton Harcourt (no. 772) with a bone belt ring, have cruder pots associated with distinctive objects. Where aged and sexed, associated burials seem to be predominantly of mature men (Case 1977; Pierpoint 1980) though problems with sample size and quality should make us wary of this generalisation.

The use of 'non-basic' Beaker associated assemblages seems thus likely to have expressed high status, probably linked with important men, and continuing competition is probable. It is more difficult to discern ~~what if anything, controlled access to or use of Beaker pots by themselves~~ what, if anything, controlled access to or use of Beaker pots by themselves and whether this use was exclusive to a restricted group within with barley-beer drinking (Burgess & Shennan 1976) but despite the beaker from Ashgrove Fife with probable lime-meadowsweet beverage (Dickson 1978) other Beaker contents and associations have been more heterogeneous (e.g. sea shells and fishbones at Walney Island (Cross 1939) and Hasting Hill (Clarke 1970: 124); land snails near the head of a skeleton with a beaker at its feet, at Falmer, Sussex (Grinsell 1931: 39); a young person's skull at Winterbourne Monkton G9 (Grinsell 1957: 200); and a ladle at Broomend, Aberdeenshire (Clarke 1970)). Further consideration of the use of Basic aseemblages or pots alone will be given later.

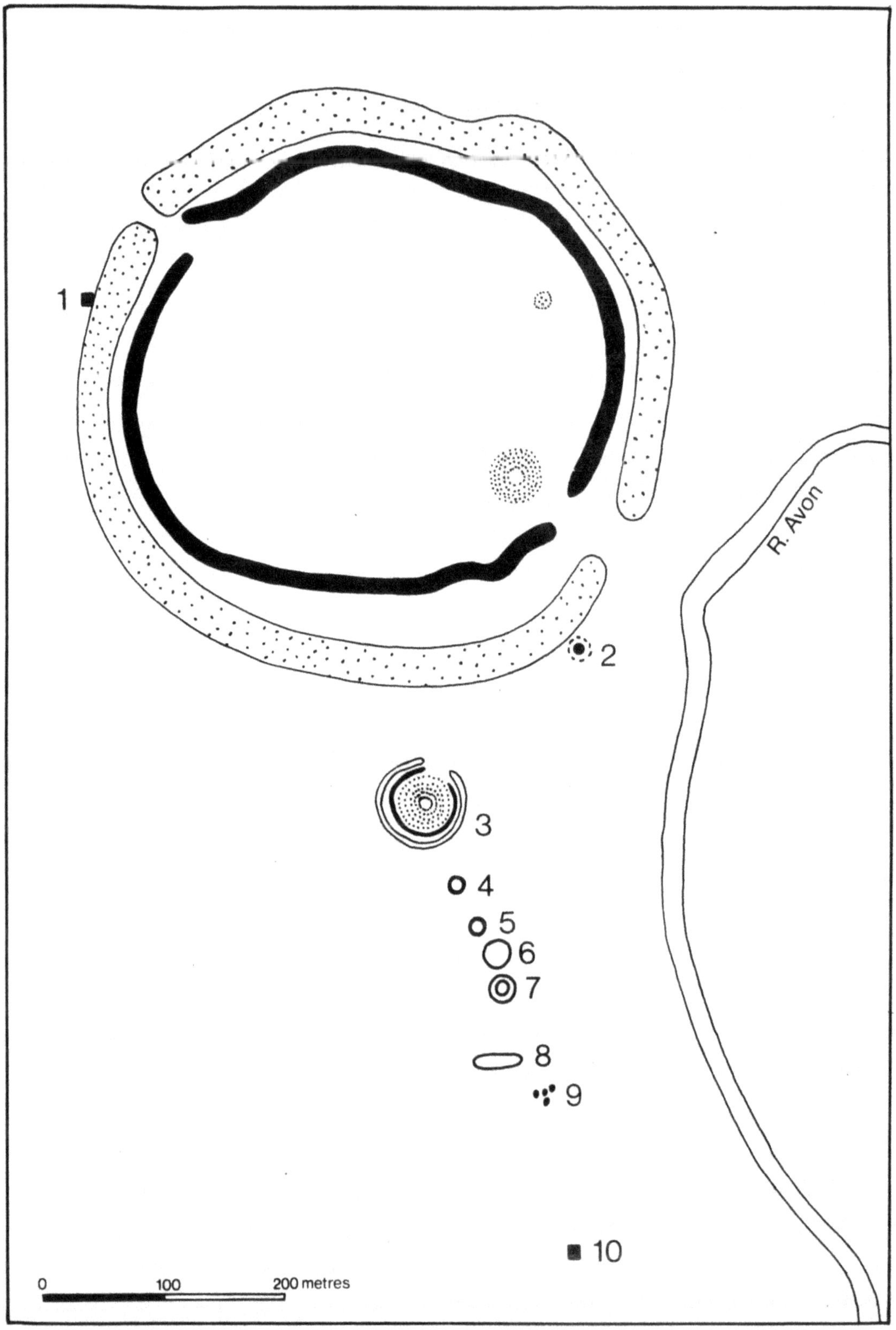

Fig. 10.4 The relationship of surrounding finds to the large henge at Durrington Walls, Wilts. 1: Area of (a) unaccompanied skeleton and (b) Beaker-less non-basic assemblage of late date. 2: Crude plain Beaker with adult woman, possibly in barrow. 3: Woodhenge, with unaccompanied skeleton in ditch base. 4-6: Barrows G70, 69 and 68. 7: Barrow G67: late non-basic Beaker assemblage. 8: Three conjoined barrows; one 'drinking cup' recorded. 9: Woodlands Grooved Ware pits. 10: Inhumation, date uncertain. (Data after Wainwright & Longworth 1971.)

While at this level lack of restricted access would reduce competition, at a higher level some sort of competition between differing networks seems likely. This can be illustrated by the spatial relationship of some Beaker burials to major henges. At Durrington Walls for example Beaker burials were excluded from the henge itself, though Beaker pots are represented within, and moreover there may be more to the relationship than simply the exclusion of all burials from the interior, since the closest burials to the bank are not Beaker associated (fig. 10.4). In the Stonehenge area as a whole (fig. 10.5) it is noticeable that the earlier non-basic associated assemblages have been found at some distance from the complex itself, as at Mere G6a (Clarke, no. 1125), Roundway G8 (1135), Winterslow G3 (1204), and Monkton Farleigh G2 (1111-12). Shrewton G5k (1140) with a N2 beaker is closer. There is a possible exception in the six arrowheads and early beakers in Wilsford G54, but from disturbed contexts (R.C.H.M. 1979: 6). This general spatial relationship may arguably reflect the burials' function of providing alternative foci of status and display to the cohesive focus of the henge itself. It may be countered that all that need be involved is the difference between the spheres of monument use and burials (as with the first example) but the case is supported by the changed distribution of richer Beaker burial assemblages of the later phase. These, such as Winterbourne Stoke G54 (Clarke no. 1195), Durrington G67 (1103) and Amesbury G54 (1039) cluster much closer to the complex itself. A late assemblage at Amesbury G51—with spatulae and skull fragment—has been dated by BM-287, 1788 ± 90 bc (Ashbee 1978). There is also the recently excavated burial with associated bracer and barbed and tanged arrowheads of later Beaker types cut into the upper silting of the ditch at Stonehenge (Atkinson & Evans 1878). These are objects of Beaker type but no Beaker pot was itself found in the sections opened. A radiocarbon date of 1765 ± 70 bc (BM-1582) has been obtained from the skeleton. It is possible therefore to argue that at an early phase richer Beaker burials and major henges were in some sense in competition, or that the users of each were involved in activities contradictory to each other. Only at a later phase, when the use of earthwork henges was lapsing, were the richer Beaker-associated burials present closer to the centre of the complex, and by now their own nature was changing too.

This discussion of Beaker assemblages has served to illustrate the theme of competition for status in one particular case. To illustrate its wider relevance to social change in the period and its relation to other evidence already discussed, a brief outline of the period can be attempted. In the pre-Beaker phase of the Later Neolithic we have seen the emergence in various parts of the country, but particularly in southern England, of a tradition of larger monument building. (Beginning slightly earlier but overlapping with this phase the large passage graves of Ireland provide an analogous case, with carefully built and decorated mounds and selected associated artefacts.) One interpretation of these sites, whether their construction was hierarchically or communally organised, can be that they reflect an increased need for social cohesion in the face of increased competition for basic resources. Another expression of the same social situation can be sought in the ceramic styles which now appear in the record, from the wider ranging Grooved ware pottery to the rather more regionalised Peterborough pottery. A contrasting aspect of the situation however can be seen in the variety of assemblages now apparent, for example those of the Grooved Ware complex (Wainwright & Longworth 1971), the macehead complex (Manby 1974) or the more diffuse Peterborough complex (e.g. Smith 1974). Though neither particularly numerous

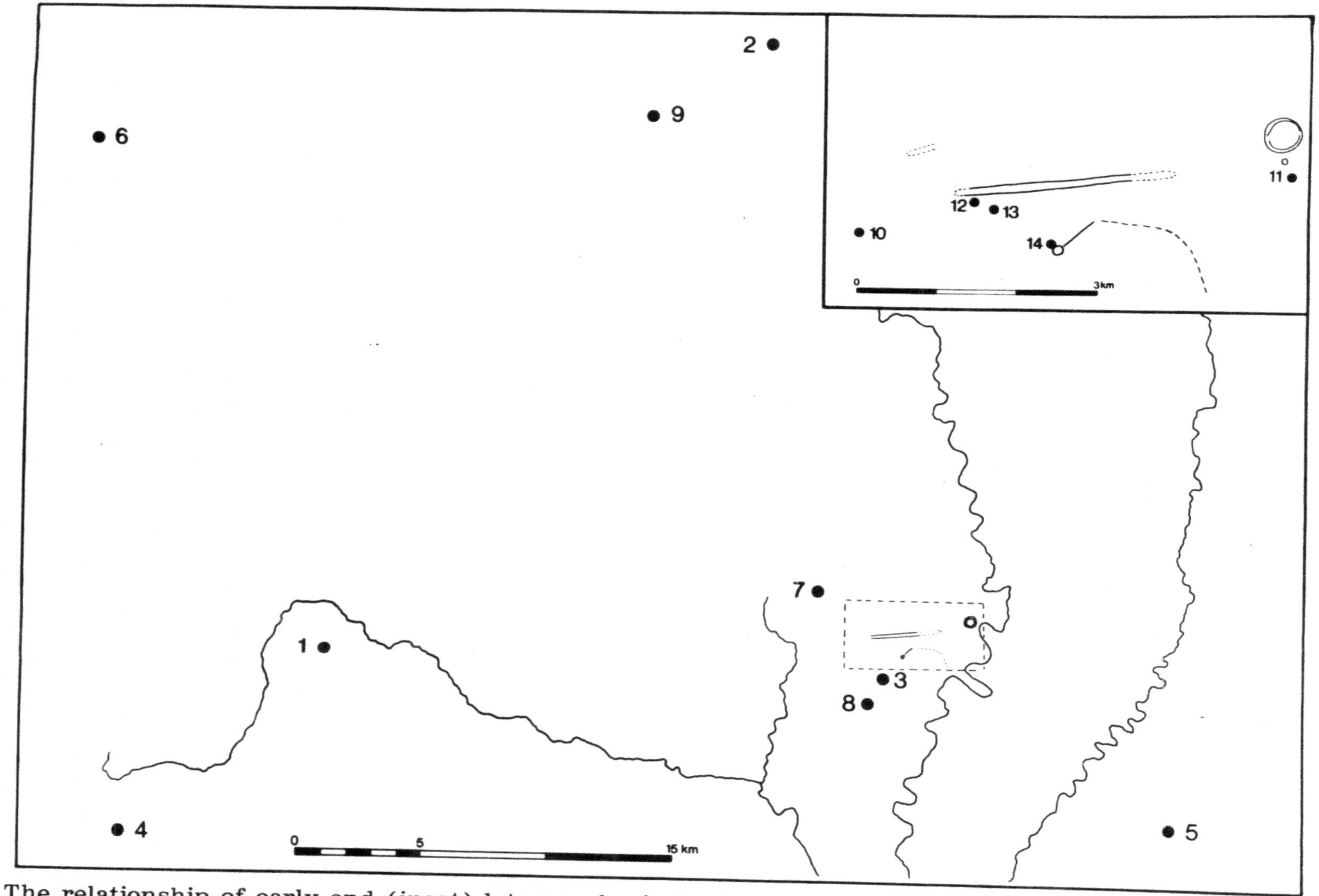

Fig. 10.5 The relationship of early and (inset) late non-basic Beaker grave assemblages to the complex at Stonehenge. 1: Longbridge Deverill G3b. 2: Bishop's Cannings G81. 3: Wilsford (5) G54. 4: Mere G6a. 5: Winterslow G3. 6: Monkton Farleigh. 7: Shrewton G5k. 8: Wilsford cum Lake G1. 9: Roundway G8. 10: Winterbourne Stoke G54. 11: Durrington G67. 12: Amesbury G54. 13: Amesbury G51. 14: Stonehenge ditch. (Data from Clarke 1970; RCHM 1979; Atkinson & Evans 1978.)

nor sharply differentiated, these assemblages could be seen as some reflection of competition for status via the acquisition of more distinctive prestige items, in a reflection of increasing social tensions which monuments and some exchange networks may have been designed to alleviate. If this early, indigenous metallurgy could be put in the same context. In the beginning of the early Beaker phase the largest monuments of southern England were newly built but there is evidence that they were not long maintained. Silbury Hill was unfinished (though its dating is imprecise). Beaker pottery was a new addition to the artefact range. Initially unassociated, its novelty and continental background would have provided suitable exclusivity and high status. Subsequently—and rather rapidly according to the available chronology—high status passed to the users of pots plus associated assemblages containing further novel and distinctive prestige items including more developed metalwork. (In contrast, in the major centre of early metallurgy, in Ireland, Beaker pots were less frequent and not closely associated with metalwork (Harbison 1973)). Pots used by themselves may have reflected a conscious imitation of the possessors of higher status. In an apparently rapidly changing ceramic sequence the lingering of the AOC style could be seen in this light, as a symbolic use of the initially prestigious individual pot. Prestige assemblages may have been used in some sense in competition with the major henges, reflecting a heightening of social tensions. By the late Beaker/Urn phase the major henges were no longer maintained. Social cohesion could perhaps then no longer be maintained by these sorts of sites, however organised, though perhaps to some extent this role was continued by the barrow cemeteries frequently grouped around them. Two which continue in a different form are instructive. Mount Pleasant was refurbished with a massive palisade. Stonehenge continued as a focus but its frequent changes of plan could be seen as a continuing attempt to maintain cohesion and interest in the face of other sorts of competition. In this, there were several changes in Beaker assemblages, from more developed metalwork with imitative flint daggers, to battleaxes, rings and buttons, and ultimately to the shedding of Beaker pots themselves in status assemblages and their relegation to domestic contexts. In metalwork, a succession of dagger types is evident (Gerloff 1975), as well as distinctive and individual gold objects (Coles & Taylor 1971). Other pottery types replace Beakers in status assemblages, such as the various Urn and Food vessel types. It is noticeable that these are now more numerous than before and several of them have more restricted distributions. There is considerable variety in the associated objects, with no distinctive set going with the still widespread Collared Urns. Objects from far and wide including across the Channel were sought. 'Fancy' barrow types were widely used in parts of southern England, some again (in Dorset) with regional characteristics.

By the late Beaker/Urn phase, then, the pace of social change may be said to have accelerated, and the hypothesis of increasing competition for status due to competition for basic resources provides a dynamic explanation of the constant shifts in all aspects of the archaeological record. That so many of these are most evident in southern England right through the period may be related to the environmental and economic background, but the danger of an over-determinist model is obvious and attention should not be distracted from other areas of the country as well. The chiefdom model hardly does justice to the potential complexity of the situation at any stage of its changing development, though the distinction between 'group identity' and 'individualising' chiefdoms (Renfrew 1974) rightly draws attention to some of the major changes discussed here. If the model of chiefdoms is acceptable at all, it might be better applied in the succeeding period, from the Middle Bronze Age onwards at least in southern Britain, with the emergence of defended sites, a metalwork distribution centred on these, and considerable landscape reorganisation. This could be envisaged after the various contradictions explored here, between the need for social cohesion and competition for resources and status, had been resolved. To seek both conflicting social relations and a dynamic source of change has considerable explanatory advantages. It prevents the 'Beaker problem' or the 'Wessex culture' from being seen in isolation. In circumventing a search for the specific content of the associated beliefs and contents of the Beaker complex for example, it allows us to concentrate on the kinds of functions all such phenomena may have served in society and on the changes they may have in turn produced. The resulting diversified model then begins to do justice to the complications of the archaeological record.

No attempt has therefore been made, deliberately, to incorporate awareness of the Beaker phenomenon in Europe into analysis of the period in Britain. This too at this stage of research would be to impart preconceived models and to some extent ignore the dynamics of indigenous social change. It may be more rewarding to examine similarities and divergences in the Beaker phenomenon across Europe, in each regional social setting. Shennan has anyway demonstrated the variability of the phenomenon in central Europe (1978). His characterisation of Beaker assemblages as 'status kits' (1976) obviously provides a point of comparison with the case discussed here. But it is of interest, from the point of view of divergence, to note the much shorter-lived Beaker sequence in central Europe. One hypothesis would be that the fast-developing central European bronze metallurgy more rapidly provided novel, alternative prestige items to replace Beaker 'status kits'.

Finally, astronomical practices can be put in the same social context. If the scientific explanation is to be rejected, the practice of orientation is still of considerable interest. Some monuments of the Earlier Neolithic may have had a general concern with it, while from the mid or the later third millennium bc onwards, starting perhaps with such features as the roof-box at Newgrange, more monuments were built with a specific concern for it. Orientation may have been another attention-focussing device, and since the positions of the heavenly bodies which may have been of interest were so varied from site to site, making uniformity of practice unlikely over wide areas, it is plausible that it could have been used competitively between different communities. There is again no obvious centralisation of the practice on major monuments or concentrations of population. It is notable that the practice appears to end more or less at the end of the period, like so many other features discussed here.

10.10 Conclusions

Many problems remain. Some are empirical, such as the demonstration of the existence of exchange systems in several media. Others are more theoretical, from the definition of status assemblages and the question of whether access to the use of Beakers and similar phenomena was restricted or open or varied from case to case, to the wider need to explain why changes occurred without giving undue prominence to any one particular factor. This paper has however attempted to make the following main points about the Later Neolithic period (in the chronological sense defined):

1. There were important changes in environment and resources by the mid third millennium bc which are likely to have affected subsequent economy, settlement and society, providing a situation in which competition for basic resources is plausible, particularly in southern England.

2. The Beaker phenomenon was not ethnic, nor was it in essence novel. Attention must be paid to its details, particularly the settlement evidence, to provide an empirical demonstration of this view but the main aim should then be to interpret the phenomenon in other ways.

3. Here it was seen as part of wider and longer lasting processes of social change in which communal needs for cohesion were increasingly replaced by more individual acquisition of status in response to competition for scarce resources. Many other features of the period may illustrate these processes too. In this sense the specific content of the Beaker phenomenon is of no concern, only its wider function.

4. Any such interpretation must take into account as much of the evidence as possible, and different aspects of society must be seen in relation to each other. In this way a narrow or static view of the development of society may be avoided, and some appreciation gained of the dynamic sources of change.

5. As predicted in its original formulation, the 'chiefdom' model does little justice to the complexity and change of the period. If acceptable at all, it may be more appropriate to the succeeding period, from the Middle Bronze Age, at least in southern Britain.

6. Astronomy, if 'ritual' rather than 'scientific', does not hold the key to social organisation in the period, any more than any other single aspect of the period, but it is another interesting strand in the situation and a feature which in its social function was perhaps not so dissimilar to many others.

ACKNOWLEDGEMENTS

I should like to thank Richard Atkinson, David V. Clarke, Frances Healey, Martin Jones, Ian Kinnes, and David Miles for discussion or points of information; John Evans for the radiocarbon date from Stonehenge; and Richard Bradley for a critical reading of an earlier draft of this paper. Colin Burgess made useful criticisms at the Newcastle conference; his _Age of Stonehenge_, which appeared as this was being prepared for the press, makes many similar points about the Beaker phenomenon.

REFERENCES

Apling, H., 1932. Bronze age settlements in Norfolk. _Proc. Prehist. Soc. East Anglia_, _6_, 365-70.

ApSimon, A. M., 1976. Ballynagilly and the beginning of the Irish Neolithic. _In_ S. J. de Laet (ed.), _Acculturation and continuity in Atlantic Europe_, 15-30. Brugge.

ApSimon, A. M., Musgrave, J. H., Sheldon, J., Tratman, E. K. & van Wijngaarden-Bakker, L. H., 1976. Gorsey Bigbury, Cheddar, Somerset: Radiocarbon dating, human and animal bones, charcoal and archaeological reassessment. _Proc. Univ. Bristol Spelaeol. Soc._., _14_, 155-83.

Ashbee, P., 1978. Amesbury Barrow 51: Excavation 1960. Wilts Archaeol. Mag., 70/71, 1-60.

Ashbee, P., Smith, I. F., & Evans, J. G., 1979. Excavations of three long barrows near Avebury, Wiltshire. Proc. Prehist. Soc., 45, 207-300.

Athens, J. S., 1977. Theory building and the study of evolutionary process in complex societies. In L. R. Binford (ed.), For theory building in archaeology, 353-84. London.

Atkinson, R. J. C., 1955. The Dorset Cursus. Antiquity, 29, 4-9.

Atkinson, R. J. C., 1969. The date of Silbury Hill. Antiquity, 43, 216.

Atkinson, R. J. C., 1978. Silbury Hill. In R. Sutcliffe (ed.), Chronicle 159-73. London (BBC).

Atkinson, R. J. C. & Evans, J. G., 1978. Recent excavations at Stonehenge. Antiquity, 52, 235-6.

Atkinson, R. J. C., Piggott, C. M. & Sandars, N. K., 1951. Excavations at Dorchester, Oxon. Oxford.

Bailloud, G., 1974. Le Néolithique dans le Bassin Parisien. Paris.

Bamford, H. M., 1970. Some Beaker domestic sites in W. Norfolk and their affinities. Unpub. Ph.D. Thesis, Univ. Edinburgh.

Barnes, F., 1956. Pottery from prehistoric sites, North End, Walney Island. Trans. Cumberland & Westmoreland Antiq. Arch. Soc., 55, 1-16.

Bateman, T., 1848. Vestiges of the antiquities of Derbyshire. London.

Bateman, T., 1861. Ten years' diggings in Celtic and Saxon Grave Hills in the counties of Derby, Stafford and York from 1848 to 1858. London.

Bell, M., 1977. Excavations at Bishopstone. Sussex Archaeol. Coll., 115.

Binford, L. R., 1972. Mortuary practices: their study and potential. In L. R. Binford, An Archaeological Perspective, 208-43. New York & London.

Bradley, R., 1970. The excavation of a Beaker settlement at Belle Tout, East Sussex, England. Proc. Prehist. Soc., 36, 312-79.

Bradley, R., 1978. The prehistoric settlement of Britain. London.

Bradley, R., & Ellison, A., 1976. Rams Hill: A Bronze Age defended enclosure and its landscape. Oxford: B.A.R. 19.

Brewis, P., 1928. Notes on prehistoric pottery and a bronze pin from Ross Links, Northumberland. Archaeologia Aeliana (4th ser.), 5, 13-25.

Brothwell, D. R., & Krzanowski, W., 1974. Evidence of biological differences between British populations from Neolithic to Medieval times, as revealed by eleven commonly available cranial vault measurements. J. Archaeol. Science, 1, 249-60.

Butler, J. J., 1956. The Late Neolithic gold ornament from Bennekom. Palaeohistoria, 5, 59-72.

Burgess, C., 1974. The Bronze Age. In C. Renfrew (ed.), British Prehistory: A new outline, 165-232. London.

Burgess, C., 1976. Meldon Bridge: A Neolithic defended promontory complex near Peebles. In C. Burgess & R. Miket (eds.), Settlement and economy in the third and second millennia BC, 151-80. Oxford: B.A.R. 33.

Burgess, C. & Shennan, S., 1976. The Beaker phenomenon: some suggestions. In C. Burgess & R. Miket (eds.), Settlement and economy in the third and second millennia BC, 309-31. Oxford: B.A.R. 33.

Burl, A., 1980. Science or symbolism: problems of archaeo-astronomy. Antiquity, 54, 191-200.

Burleigh, R., Henson, A. & Meeks, N., 1976. British Museum natural radiocarbon measurements VIII. Radiocarbon, 18, 16-42.

Callander, J. G., 1929. Scottish Neolithic pottery. Proc. Soc. Antiq. Scotland, 63, 29-97.

Case, H., 1951. A cinerary urn from Stanton Harcourt, Oxon. Oxoniensia, 16, 84-5.

Case, H., 1956. Beaker pottery from the Oxford region: 1939-1955. Oxoniensia, 21, 1-21.

Case, H., 1963. Notes on the finds and ring-ditches in the Oxford region. Oxoniensia, 28, 19-52.

Case, H. J., 1966. Were Beaker people the first metallurgists in Ireland? Palaeohistoria, 12, 141-77.

Case, H., 1977. The Beaker culture in Britain and Ireland. In R. Mercer (ed.), Beakers in Britain and Europe: four studies, 71-101. Oxford: B.A.R. S26.

Case, H. J., Bayne, N., Steele, S., Avery, G. & Sutermeister, H., 1965. Excavations at City Farm, Hanborough, Oxon. Oxoniensia, 29/30, 1-98.

Caulfield, S., 1978. Neolithic fields: the Irish evidence. In C. Bowen & P. Fowler (eds.), Early land allotment in Britain, 137-43. Oxford: B.A.R. 48.

Childe, V. G., 1929. The Danube in prehistory. London.

Childe, V. G., 1931. Skara Brae. London.

Christie, P. M., 1967. A Barrow-cemetery of the second millennium BC in Wiltshire England. Proc. Prehist. Soc., 33, 336-66.

Clark, J. G. D., 1933. Report on an Early Bronze Age site in the South-Eastern Fens. Antiq. J., 13, 266-358.

Clark, J. G. D., 1936. Report on a Late Bronze Age site in Mildenhall Fen, West Suffolk. Antiq. J., 16, 29-50.

Clark, J. G. D., 1966. The invasion hypothesis in British archaeology. Antiquity, 40, 172-89.

Clark, J. G. D., Godwin, H., Godwin, M. E., & Clifford, M. H., 1935. Report on recent excavations at Peacock's Farm, Shippea Hill, Cambs. Antiq. J., 15, 284-319.

Clarke, D. L., 1970. Beaker pottery of Great Britain and Ireland. Cambridge.

Clarke, D. L., 1976. The Beaker network—social and economic models. In J. N. Lanting & J. D. van der Waals (eds.), Glockenbecher Symposion, 460-77. Bussum/Haarlem.

Clarke, D. V., 1976. The Neolithic village at Skara Brae, Orkney. 1972-3 excavations: an interim report. Edinburgh.

Coles, J. M., 1971. Scottish Early Bronze Age metalwork. Proc. Soc. Antiq. Scotland, 101, 1-110.

Coles, J. M. & Orme, B. J., 1976. The Abbot's Way. Somerset Levels Papers, 2, 7-20.

Coles, J., Orme, B. & Woolley, A. R., 1974. A jade axe from the Somerset levels. Antiquity, 48, 216-20.

Coles, J. & Taylor, J., 1971. The Wessex culture: a minimal view. Antiquity, 45, 6-14.

Coombs, D., 1976. Beakers from Callis Wold, Barrow 275, Humberside. In C. Burgess & R. Miket (eds.), Settlement & economy in the third and second millennia BC, 143-50. Oxford: B.A.R. 33.

Cormack, W. F., 1965. Burial site at Kirkburn, Lockerbie. Proc. Soc. Antiq. Scotland, 96, 107-35.

Cowie, T. G., 1978. Bronze Age food vessel urns in northern Britain. Oxford: B.A.R. 55.

Crawford, I. & Switsur, R., 1977. Seascaping and C14: the Udal, N. Uist. Antiquity, 51, 124-36.

Cree, J. E., 1908. Notice of a prehistoric kitchen-midden and superimposed medieval stone floor found at Tusculum, North Berwick. Proc. Soc. Antiq. Scotland, 42, 253-94.

Cummins, W. A., 1979. Neolithic stone axes: distribution and trade in England and Wales. In T. H. McK Clough & W. A. Cummins (eds.), Stone axe studies: archaeological, petrological, experimental and ethnographic, (CBA ResReport 23), 5-12. London.

Cunnington, M. E., 1929. Woodhenge. London.

Curle, A. O., 1908. Notice of the examination of prehistoric kitchen middens on the Archerfield Estate, near Gullane, Haddingtonshire, in November 1907. Proc. Soc. Antiq. Scotland, 42, 308-19.

Dalton, G., 1977. Aboriginal economies in stateless societies. In T. K. Earle & J. E. Ericson (eds.), Exchange systems in prehistory, 191-212. London.

Davies, O., 1939. Ancient field-systems and the date of formation of the peat. Ulster J. Archaeol., 2, 61-5.

Dennell, R. W., 1976. Prehistoric crop cultivation in southern England: a reconsideration. Antiq. J., 56, 11-23.

Dickson, J. H., 1978. Bronze Age mead. Antiquity, 52, 108-13.

Donaldson, P., 1977. The excavation of a multiple Round Barrow at Barnack, Cambridgeshire, 1974-1976. Antiq. J., 57, 197-231.

Drew, C. & Piggott, S., 1936. The excavation of barrow 163a on Thickthorn Down, Dorset. Proc. Prehist. Soc., 2, 77-96.

Dymond, D. P., 1966. Ritual monuments at Rudston, E. Yorkshire, England. Proc. Prehist. Soc., 32, 86-95.

Eogan, G., 1974. Report on the excavations of some passage graves, unprotected inhumation burials and a settlement site at Knowth, Co. Meath. Proc. Roy. Irish Academy, 74C, 11-112.

Eogan, E., 1976. Beaker material from Knowth, C. Meath: Preliminary Note. In C. Burgess & R. Miket (eds.), Settlement and economy in the the third and second millenia BC, 251-66. Oxford: B.A.R. 33.

Erith, F. H. & Longworth, I. H., 1960. A Bronze Age urnfield on Vinces Farm, Ardleigh, Essex. Proc. Prehist. Soc, 26, 178-92.

Evans, J. G., 1972. Land snails in archaeology. London.

Fleming, A., 1973. Tombs for the living. Man, (N.S.) 8, 177-93.

Friedman, J. & Rowlands, M. (eds.), 1977. The evolution of social systems. London.

Gerhardt, K., 1979. Anthropotypologie der Glockenbecherleute in ihren Ausschwärmelandschaften. In J. N.Lanting & J. D. van der Waals (eds.), Glockenbecher Symposion, 147-64. Bussum/Haarlem.

Gerloff, S., 1975. The Early Bronze Age daggers in Great Britain. (Prähistorische Bronzefunde, 6, 2). Munich.

Gray, M., 1974. The Devil's Quoits, Stanton Harcourt, Oxon. Oxoniensia, 39, 96-7.

Greenwell, W., 1877. British barrows. Oxford.

Greenwell, W., 1890. Recent reserches in barrows in Yorkshire, Berkshire, Wiltshire, etc. Archaeologia, 52, 1-72.

Griffiths, W. E., 1956. Notes on some of the prehistoric pottery of Anglesey. Trans. Anglesey Antiq. Soc. & Field Club 1956, 1-10.

Grimes, W. F., 1944. Excavations at Stanton Harcourt 1940. Oxoniensia, 8/9, 10-63.

Grimes, W. F., 1960. Excavations on defence sites 1930-40. I. Mainly Neolithic-Bronze Age. London.

Grinsell, L. V., 1931. Sussex in the Bronze Age. Sussex Archaeol. Coll., 72, 30-68.

Grinsell, L. V., 1957. Archaeological gazetteer. In A History of Wiltshire, 1, 1 (Victoria History of the Counties of England), 21-279. London.

Hamlin, A., 1963. Excavations of ring-ditches and other sites at Stanton Harcourt. Oxoniensia, 28, 1-18.

Harbison, P., 1973. The Earlier Bronze Age in Ireland. J. Roy. Soc. Antiq. Ireland, 103, 93-152.

Harrison, R. J., 1974. Origins of the Bell Beaker cultures. Antiquity, 48, 99-109.

Hawkes, S. & Gray, M., 1969. Eynsham, Oxon: Notes & News. Oxoniensia, 34, 104.

Hazzeldine Warren, S., Piggott, S., Clark, J. G. D., Burkitt, M. C. & Godwin, H. & M. E., 1936. Archaeology of the submerged landsurface of the Essex Coast. Proc. Prehist. Soc., 2, 178-210.

Helbaek, H., 1952. Early crops in southern England. Proc. Prehist. Soc., 18, 194-233.

Henneberg, M., Piontek, J. & Strzalko, J., 1978. Natural selection and morphological variability: the case of Europe from Neolithic to modern times. Current Anthropology, 19, 67-82.

Hodder, I. & Bradley, R., 1979. British prehistory: an integrated view. Man, (N.S.) 14, 93-104.

Houlder, C., 1976. Stone axes and henge-monuments. In G. C. Boon & J. M. Lewis, (eds.), Welsh Antiquity: Essays presented to H. N. Savory, 55-62. Cardiff.

Inventaria Archaeologica 1955. Great Britain, 1: Amey's Pit, Site XII, Dorchester-on-Thames (Oxfordshire). (Int. Congress Prehist. & Protohist. Sci.) London.

Jackson, D. A., 1976. The Excavation of Neolithic and Bronze Age Sites at Aldwincle, Northants 1967-71. Northamptonshire Archaeol., 11, 12-70.

Jensen, J. A., 1972. Myrhøj, 3 hustomter med klokkebaegerkeramik. Kuml 1972, 61-122.

Jewell, P. A. & Dimbleby, G. W., 1966. The experimental earthwork on Overton Down, Wiltshire, England: The first four years. Proc. Prehist. Soc., 32, 313-42.

Kehoe, A. B., 1974. Saints of Wessex?. Antiquity, 48, 232-3.

Kerney, M. P., Brown, E. H. & Chandler, T. J., 1964. The late-glacial and post-glacial history of the chalk escarpment near Brook, Kent. Phil. Trans. R. Soc. Lond., B248, 135-204.

Kinnes, I., 1975. Monumental function in British Neolithic burial practices. World Archaeol., 7, 16-29.

Kinnes, I., 1976. Hunstanton. Current Archaeol., 5(51), 124.

Kinnes, I., 1979. Round Barrows and Ring-ditches in the British Neolithic (British Museum Occasional Paper No. 7). London.

Kiszeley, I., 1978. The origins of artificial cranial formation in Eurasia from the sixth millennium BC to the seventh century AD. Oxford: B.A.R. S50.

Lanting, J. N. & van der Waals, J. D., 1972. British beakers as seen from the Continent. Helinium, 12, 20-46.

Lanting, J. N. & van der Waals, J. D., 1976. Beaker culture relations in the Lower Rhine basin. In J. N. Lanting & J. D. van der Waals (eds.), Glockenbecher Symposion, 1-80. Bussum/Haarlem.

Leaf, C. S., 1935. Report on the excavation of two sites in Mildenhall Fen. Proc. Cambridge Antiq. Soc., 35, 106-25.

Leeds, E. T., 1923. A Saxon village near Sutton Courtenay, Berkshire. Archaeologia, 73, 147-92.

Leeds, E. T., 1934. Recent Bronze Age discoveries in Berkshire and Oxfordshire. Antiq. J., 14, 264-76.

Leeds, E. T. , 1938. Beakers of the Upper Thames district. Oxoniensia., 3, 7-30.

Leeds, E. T., 1940. New discoveries of Neolithic pottery in Oxfordshire. Oxoniensia, 5, 1-12.

Lethbridge, T. C., 1927. A settlement site of the Beaker period at Sana Bay, Ardnamurchan, Argyll. Man, 27, 173-4.

Limbrey, S., 1975. Soil Science and archaeology, London.

Longworth, I. H., Wainwright, G. J. & Wilson, K. E., 1971. The Grooved Ware site at Lion Point Clacton. In G. de G. Sieveking (ed.), Prehistoric and Roman Studies, 93-124. London.

Louwe Kooijmans, L. P., 1974. The Rhine/Meuse delta (Analecta Praehistorica Leidensia VII). Leiden.

MacKie, E. W., 1977. Science and society in prehistoric Britain. London.

Manby, T. G., 1974. Grooves Ware sites in Yorkshire. Oxford: B.A.R. 9.

Manby, T. G., 1975. Neolithic occupation sites on the Yorkshire Wolds. Yorkshire Archaeol. J., 47, 23-59.

Martin, E., 1975. The excavation of barrow 1, Martlesham Heath. East Anglian Archaeology, Report no. 1, Suffolk, 5-11.

Martin, E., 1976. The excavation of barrows II, III and IV, Martlesham Heath, 1974. East Anglian Archaeol. Report no. 3, Suffolk, 17-42.

McInnes, I. J., 1964. A Class II henge in the East Riding of Yorkshire. Antiquity, 38, 218-19.

McInnes, I. J., 1966. The Neolithic and Early Bronze Age pottery from Luce Sands, Wigtownshire. Proc. Soc. Antiq. Scotland, 97, 40-81.

Megaw, J. V. S., 1976. Gwithian, Cornwall: some notes on the evidence for Neolithic and Bronze Age settlement. In C. Burgess. and R. Miket (eds.), Settlement and economy in the third and second millennia BC, 51-66. Oxford: B.A.R. 33.

Menke, H., 1973. Zwei Karten zum Neolithikum und zur fruhen Bronzezeit im Rhein-Main-Gebiet. Archäologisches Korrespondenzblatt, 3, 413-16.

Mercer, R. J., 1976. Grime's Graves Norfolk—an interim statement on conclusions drawn from the total excavation of a flint mine shaft and a substantial surface area in 1971-2. In C. Burgess & R. Miket (eds.), Settlement and economy in the third and second millennia BC, 101-11 Oxford: B.A.R. 33.

Mercer, R., 1980. Hambledon Hill. Edinburgh.

Moore, J. W., 1964. Excavations at Beacon Hill, Flamborough Head. Yorkshire Archaeol. J., 41, 191-7.

Morris, D., 1977. Manwatching. London.

Mortimer, J. R., 1905. Forty years' researches in British and Saxon burial mounds of East Yorkshire. London.

O'Kelly, M. J., 1973. Current excavations at Newgrange, Ireland. In G. Daniel & P. Kjaerum (eds.), Megalithic graves & ritual, 137-46. Copenhagen.

O'Kelly, M. J., 1979. The restoration of Newgrange. Antiquity, 53, 205-10.

O'Nuallain, S., 1975. The stone circle complex of Cork and Kerry. J. Roy. Soc. Antiq. Ireland, 105, 83-131.

O'Riordain, S. P., 1974. Tara. Dundalk.

Ottoway, B., 1973. The earliest copper ornaments in northern Europe. Proc. Prehist. Soc., 39, 294-331.

Ottaway, B., 1973b. Dispersion diagrams: a new approach to the display of carbon-14 dates. Archaeometry, 15, 5-12.

Ottaway, B. & Strahm, C., 1975. Swiss Neolithic copper beads: currency, ornament or prestige items. World Archaeol., 6, 307-32.

Palmer, R., 1978. Causewayed enclosure at Crofton (Great Bedwyn). Wilts. Archaeol. Mag., 70/71, 124-5.

Pierpoint, S., 1980. Social patterns in Yorkshire prehistory 3500-750 BC. Oxford: B.A.R. 74.

Piggott, S., 1948. The excavation at Cairnpapple Hill, West Lothian. Proc. Soc. Antiq. Scotland, 82, 68-123.

Piggott, S., 1963. Abercromby and after: the Beaker cultures of Britain re-examined. In I. L. Foster & L. Alcock (eds.), Culture and environment, 53-91. London.

Pitt-Rivers, Lt. Gen., 1898. Excavations in Cranborne Chase, IV.

Pollock, A. J. & Waterman, D. M., 1964. A Bronze Age habitation site at Downpatrick. Ulster J. Archaeol., 27, 31-58.

Pryor, F., 1978. Excavations at Fengate, Peterborough, England. The second report. (Royal Ontario Museum Archaeology Monograph 5). Toronto.

Rahtz, P. A., 1962. Neolithic and Beaker sites at Downton, near Salisbury, Wiltshire. Wilts. Arch. & Nat. Hist. Soc. Mag., 58, 116-42.

Rahtz, P. A. & Greenfield, 1977. Excavations at Chew Valley Lake, Somerset. London: H.M.S.O.

R.C.H.M., 1969. Peterborough New Town: A survey of the antiquities in the areas of development. London: H.M.S.O.

R.C.H.M., 1979. Stonehenge and its environs. Edinburgh.

Redman, C., 1978. The rise of civilisation. San Francisco.

Renfrew, C., 1973. Monuments, mobilisation and social organisation in Neolithic Wessex. In C. Renfrew (ed.), The explanation of culture change, 539-58. London.

Renfrew, C., 1974. Beyond subsistence economy: the evolution of social organisation in prehistoric Europe. In C. Moore (ed.), Reconstructing complex societies (Suppl. Bull. Am. Schools of Oriental Res. No. 20), 69-96.

Renfrew, C., 1976. Megaliths, territories and populations. In S. J. de Laet (ed.), Acculturation and continuity in Atlantic Europe, 198-220. Brugge.

Renfrew, C., 1977. Alternative models for exchange and spatial distribution. In T. K. Earle & J. E. Ericson (eds.), Exchange systems in prehistory, 71-90. London.

Renfrew, C., 1979. Investigations in Orkney. London.

Ricquet, R., 1971. Anthropologie du Néolithique et du Bronze Ancien. Poitiers.

Riley, D. N., 1957. Neolithic and Bronze Age pottery from Risby Warren and other occupation sites in north Lincolnshire. Proc. Prehist. Soc., 23, 40-56.

Roe, F. E. S., 1979. Typology of stone implements with shaftholes. In T. H. McK Clough & W. A. Cummins (eds.), Stone Axe Studies (CBA Res. Report 23), 23-40. London.

Rowlands, M. J. & Gledhill, J., 1977. The relation between archaeology and anthropology. In M. Spriggs (ed.), Archaeology and Anthropology: areas of mutual interest, 43-58. Oxford: B.A.R. S19.

Sangmeister, E., 1972. Sozial-Okonomische Aspeckte der Glockenbecherkultur. Homo, 23, 1/2, 188-203.

Service, E. R., 1962. Primitive social organisation. New York.

Service, E. R., 1971. Cultural evolutionism, theory in practice. New York.

Shennan, S. J., 1978. Bell Beakers and their context in Central Europe. In J. N. Lanting & J. D. van der Waals (eds.), Glockenbecher Symposion, 231-39. Bussum/Haarlem.

Shennan, S. J., 1977. The appearance of the Bell Beaker assemblage in central Europe. In R. Mercer (ed.), Beakers in Britain and Europe: four studies, 51-70. Oxford: B.A.R. S26.

Shennan, S. J., 1978. Archaeological "cultures": an empirical investigation. In I. Hodder (ed.), The spatial organisation of culture, 113-39. London.

Shepherd, I. A. G., 1976. Preliminary results from the Beaker settlement at Rosinish, Benbecula. In C. Burgess & R. Miket (eds.), Settlement and economy in the third and second millennia BC, 209-19. Oxford: B.A.R. 33.

Sieveking, G. de G., Longworth, I. H., Hughes, M. J., Clark, A. J. & Millett, A., 1973. A new survey of Grime's Graves, Norfolk. Proc. Prehist. Soc., 39, 182-218.

Simpson, D. D. A., 1976. The Later Neolithic and Beaker Settlement Site at Northton, Isle of Harris. In C. Burgess & R. Miket (eds.), Settlement and economy in the third and second millennia BC, 221-31. Oxford: B.A.R. 33.

Simpson, W. G., 1976. A barrow cemetery of the second millennium BC at Tallington, Lincolnshire. Proc. Prehist. Soc., 42, 215-40.

Simpson, W. G. & Case, H., 1963. A Beaker-culture burial at Dorchester, Oxford. Oxoniensia, 28, 93-4.

Smith, A. G., 1970. The influence of Mesolithic and Neolithic man on British vegetation: a discussion. In D. Walker & R. G. West, (eds.), Studies in the vegetational history of the British Isles, 81-96. Cambridge.

Smith, A. G., 1975. Neolithic and Bronze Age landscape changes in northern Ireland. In J. G. Evans, S. Limbrey & H. Cleere (eds.), The effect of Man on the landscape: the Highland Zone (CBA Res. Report no. 11), 64-73. London.

Smith, I. F., 1955. Late Beaker pottery from the Lyonesse surface and the the Date of the Transgression. Ann. Rep. London Univ. Inst. of Archaeol., 11, 29-42.

Smith, I. F., 1965. *Windmill Hill and Avebury*. Oxford.

Smith, I. F., 1974. The Neolithic. *In* C. Renfrew (ed.), *British Prehistory: a new outline*, 100-36. London.

Smith, I. F., 1979. The chronology of British stone implements. *In* T. H. McK Clough & W. A. Cummins (eds.), *Stone Axe Studies* (CBA Res. Report 23), 13-22. London.

Stone, J. F. S., 1938. An Early Bronze Age grave in Fargo Plantation near Stonehenge. *Wilts Archaeol. Mag.*, *48*, 357-70.

Stone, J. F. S. & Hill, N. G., 1938. A Middle Bronze Age site at Stockbridge, Hampshire. *Proc. Prehist. Soc.*, *4*, 249-57.

Stuckenrath, R. & Mielke, J. E., 1973. Smithsonian Institution Radiocarbon Measurements VIII. *Radiocarbon*, *15*, 388-424.

Sweetman, D. P., 1976. An earthen enclosure at Monknewtown, Slane, Co. Meath. *Proc. Roy. Ir. Acad.*, *76B*, 25-73.

Tainter, J. A., 1977. Modelling change in prehistoric social systems. *In* L. R. Binford (ed.), *For theory building in archaeology*, 327-52. London.

Taylor, J. J., 1970. Lunulae reconsidered. *Proc. Prehist. Soc.*, *36*, 38-81.

Tinsley, H. M., 1975. The former woodland of the Nidderdale Moors (Yorkshire) and the role of early man in its decline. *J. Ecol.*, *63*, 1-26.

Toms, H. S., 1912. Excavations at the Beltout valley entrenchments. *Sussex Archaeol. Coll.*, *55*, 41-55.

Turner, J., Hewetson, V. P., Hibbert, F. A., Lowry, K. H. & Chambers, C., 1973. The history of the vegetation and flora of Widdybank Fell and the Cow Green Reservoir Basin, Upper Teesdale. *Phil. Trans. Roy. Soc.*, *B265*, 327-408.

Wainwright, G. J., 1969. A review of Henge Monuments in the light of recent research. *Proc. Prehist. Soc.*, *35*, 112-33.

Wainwright, G. J., 1971. The excavation of a Late Neolithic enclosure at Marden, Wiltshire. *Antiq. J.*, *51*, 177-239.

Wainwright, G. J., 1979. *Mount Pleasant, Dorset: Excavations 1970-1971*. London.

Wainwright, G. J., & Longworth, I. H., 1971. *Durrington Walls: Excavations 1966-68*. London.

Waterbolk, H. T., 1971. Working with radiocarbon dates. *Proc. Prehist. Soc.*, *37(2)*, 15-33.

Whittle, A. W. R., 1977. *The Earlier Neolithic of Southern England and its continental background*. Oxford: B.A.R. S35.

Whittle, A. W. R., 1978. Resources and population in the British Neolithic. *Antiquity*, *52*, 34-42.

Whittle, A. W. R., 1980. Two Neolithics? *Current Archaeol.*, *6*, (nos. 70-71), 329-34, 371-73.

Wijngaarden-Bakker, L. H. van, 1974. The animal remains from the Beaker settlement at Newgrange, Co. Meath: first report. *Proc. Roy. Ir. Acad.*, *74C*, 313-83.

Wyss, R., 1969. Wirtschaft und Technik. *In* W. Drack (ed.), *Ur- und Frühgeschichtliche Archäologie der Schweiz. II Die jüngere Steinzeit* 117-38. Basel.

www.ingramcontent.com/pod-product-compliance
Lightning Source LLC
LaVergne TN
LVHW072249300926
849160LV00018B/1620
* 9 7 8 0 8 6 0 5 4 1 3 0 1 *